BIOLOGY *of* AGING

BIOLOGY *of* AGING

JAMES L. CHRISTIANSEN, PhD
Drake University
Des Moines, Iowa

JOHN M. GRZYBOWSKI, MD
Medical Director, Pathology Laboratories
Iowa Lutheran Hospital
Des Moines, Iowa

Chapter 15
contributed by **David Spreadbury**, PhD

With 130 illustrations
including 16 page, 4 color insert

M Mosby

St. Louis Baltimore Boston Chicago London Philadelphia Sydney Toronto

Mosby

Dedicated to Publishing Excellence

Editor in Chief: James M. Smith
Editor: Robert J. Callanan
Developmental Editor: Jean Babrick
Project Manager: Mark Spann
Designer: Gail Morey Hudson
Cover Photos: The Stock Market

Printed in the United States of America

Mosby–Year Book, Inc.
11830 Westline Industrial Drive, St. Louis, Missouri 63146

International Standard Book Number ISBN 0-8016-6363-6

93 94 95 96 97 CL/VH 9 8 7 6 5 4 3 2 1

Preface

HOW THIS TEXT CAME TO BE

Work on this book really began in 1976, when the chairman of the biology department at Drake University asked me to develop a biology of aging course for our new certificate-in-aging program. I suppose he chose me for this project because of my research interest in the aging process in reptiles. When I reviewed the texts then available, I found some that covered theories about the aging process; good geriatric medical texts intended for readers with an MD degree; collections of papers and symposiums; and one text that provided very limited coverage of the problems faced by aging humans. No book then available provided sufficiently desirable coverage of the subject at the appropriate level for my students.

Since then, several texts on the biology of human aging have been published. However, none of them provides the practical medical viewpoint that I believe to be most useful for my students and others enrolled in biology of aging courses.

After some years of developing lectures from the sources then available, my own research experience, and several years of service on the board of a nursing home, I decided to write the book my students and others like them needed. However, I needed the help of a physician to provide the depth of medical experience this book would require. Everyone I contacted in the Iowa medical community recommended Dr. John Grzybowski as one of the most scholarly and knowledgeable physicians I could find. After all, pathologists (who do lots of autopsies) know everything—unfortunately just a little too late to help the patient.

Dr. Grzybowski accepted my challenge, and began a fantastic expansion of my education as he rewrote my chapters, corrected my errors, and supported his changes with fascinating case histories, a good library, and years of medical experience. We are both proud of this collaboration, which can give a social worker, a psychologist, a nursing home administrator, or someone who just wants to know more about a process that will affect most of us, an understanding of what happens both medically and theoretically as humans age.

Using this Text

We have designed this text primarily for students who will work with the elderly as part of their profession. Such students usually lack any background in biology or in medicine. With this in mind, we have designed each system chapter with an introductory section on the anatomy and physiology of that body system.

At Drake University, however, we have found that a number of advanced students also take this course. These include premedical biology majors, pharmacy students, advanced nursing students, and physical therapy students. Because they have had the introductory material, they can skip the introductory anatomy and physiology sections of chapters 3 to 14. At Drake, these students have the option of taking the course for 1 hour less credit, or of undertaking an extensive literature research project. Students who enter the course with extensive backgrounds are graded on a higher standard than that used for students with no background in biology.

MEDICAL TERMINOLOGY AND LITERATURE CITATIONS

Although this text is designed primarily for students with little or no background, we have not shied away from using the basic medical terminology that will be encountered by every social worker, psychologist, or nursing home manager who needs to understand a medical condition or to read a medical history. A glossary (with pronunciations) appears after Appendix C. The definitions found there refer to the word as it is used in the text. Word roots are also provided.

We were reluctant to delete citations of the literature within the narrative. However, sentences broken by frequent literature citations tend to intimidate or distract beginning students. Students interested in further reading or in greater depth of information on a specific topic will profit from the bibliography following each chapter.

ORGANIZATION

The text opens with a chapter on demographics and attitudes toward aging in our society. Chapter 2 discusses cellular aspects, theories of aging, and the evolution of lifespan. It explores the possibility that the aging process may be a genetic "disease" creating a life span that can be altered (within certain limits) by our environment and our actions. The 12 chapters that follow constitute a system-by-system approach to the effect of the aging process on the human body. (Two chapters cover the circulatory system; the sense organs are discussed separately from the rest of the nervous system.) The text ends with an important summary chapter on diet and lifestyles.

CASE HISTORIES

It is important that the student who uses this text has more than an academic knowledge of the processes and diseases affecting the elderly. Understanding—empathy—must exist for the problems faced by both patient and caregiver. For this reason, simplified case histories are included to reinforce the material presented within the systems chapters. These histories illustrate the physical and emotional problems faced by the patient, as well as the difficult decisions that must be made by the caregiver. It becomes apparent that sometimes caregivers, doing their best with the information available, solve one problem but create another.

The case histories are also intended to acquaint students with some of the terminology they will encounter when reading actual case histories. Abbreviations and technical terms are defined or explained, and normal physiologic values are included for comparison with those recorded for the patient.

DIET AND LIFESTYLES

Throughout the text, we point out steps that can be taken to correct or minimize problems that affect later life. The final chapter, "Lifestyles: Environmental Factors and Aging," summarizes what we have said and adds much that is new on the most desirable diet and activities for elderly people, and for younger people who wish to prepare for the longest and most enjoyable later life possible. This chapter also deals with other important aspects of the environment, such as the effects of alcohol and polypharmacy, and it brings together important information about cancer. We are most grateful to Dr. David Spreadbury, a nutritionist at the University of Osteopathic Medicine and Health Sciences, for agreeing to write this chapter and for compiling the extensive list of references that accompanies it.

APPENDIXES

In addition to the glossary, appendixes containing useful reference material appear in this text. They include a brief summary of side effects commonly noted with drugs used to treat conditions often found in elderly patients; a sample case history, showing all the abbreviations and technical terms that are used; and a table of normal physiologic values.

CONCLUSION

Creating this book has been a pleasure and a great deal of work. We believe that it will be a useful addition to the library of anyone who works with and wants to understand our growing elderly population.

James L. Christiansen

Acknowledgments

Many people contributed their time and expertise to make this book possible. We are grateful to all of them, even if we cannot acknowledge everyone here.

At Drake University, Dr. Donald Stratton read and commented on many physiology sections, as did Dr. Harold Swanson. Dr. Dean Hoganson helped with questions in immunology. Drs. Baridon, Graefe, and Heller, pathologists at Iowa Lutheran Hospital, spent much time sharing their experience and helping to resolve medical questions. We thank Dr. Danielson for his many practical observations regarding geriatric questions. And we are expecially grateful to Dr. Albert Mintzer, who revised sections of our chapter on the aging reproductive system and provided several case histories. We must apologize to all these people for the editing of their material, which was necessary to keep this book within its planned limits.

Thanks to the Radiology Department of Iowa Lutheran Hospital for providing many radiologic photographs. Dr. Elmets contributed many photographs of skin conditions. Dr. Sarah Christiansen generated the demographic illustrations. Dr. and Mrs. James B. Christiansen (parents of JLC) spent hundreds of hours reading and commenting on the earliest drafts of the manuscript and working on the glossary.

Jean Babrick, senior developmental editor, and her fantastic crew at Mosby–Year Book built our scholarly ramblings into a readable text.

Our thanks to them all. And a special thank-you to our reviewers. Evelyn Salerno, Phar D, kindly allowed us to adapt her material on drug side effects, and reviewed our adaptations for accuracy. James Stoddard, MD, provided a technical review of the final chapter. And numerous teachers of biology of aging courses contributed their time and their suggestions in reviews as this book moved from prospectus to final draft. They include:

Donelle R. Banks	*California State University/Sacramento*
Paul V. Benko	*Sonoma State University*
Jeffrey A. Chesky	*Sangamon State University*
Douglas M. Dearden	*University of Minnesota/Minneapolis*
Robert A. Koch	*California State University/Fullerton*
Dorothy C. Moses	*The University of Akron*
P. Jamcs Nielsen	*Western Illinois University*
Robert O'Donnell	*State University of New York/Geneseo*
Robert Smith	*Forest Park Community College*
Mary L. Spratt	*University of Missouri/Kansas City*

Contents

1 *Introduction*

A CHANGING POPULATION

Data collected by the World Health Organization (WHO) suggest that the industrialized countries of the world are about to confront a phenomenon that humanity has never seen before. In the immediate future, demographic changes will produce a society dominated in many ways by the concerns and needs of an aging or aged population (Figure 1-1). WHO characterizes this population as "elderly" (ages 60 to 75), "old" (ages 76 to 90), and "very old" (age over 90).

The Census Bureau has projected that, by the year 2030, over 20% of the U.S. population will be over age 65, or "elderly" (Figure 1-2). This compares with 4.1% in 1900 and 11% in 1980. Looking at people, rather than percentages, there were 3 million people over the age of 65 in 1900; there were 27 million in 1980, and the Census Bureau projects that there will be 65 million by the year 2030, when the expansion of this age group should stabilize somewhat (Figure 1-3). These projections assume that there

© MARK KEMPF

Figure 1-1 Inside a nursing home. Demands on the nation's resources for care of the elderly will increase dramatically as the aged population expands.

will be no medical advances that could significantly improve survival beyond age 65 between now and 2030. This assumption is probably untrue. Even as this book is being written, the tremendous medical advances being made, especially in genetic engineering, suggest that the number of people surviving beyond age 65 will be much greater in the future than demographers predict.

In the last century we have seen elimination of diseases such as smallpox, and the prevention of many others such as polio and measles, that had shortened median life span (Figure 1-4). The development of antibiotics alone has nearly eliminated death from bacterial diseases except for those people whose immune systems have become compromised or who have suffered severe damage from some other debilitating disease such as cancer. There is now good evidence that deaths from cardiovascular diseases are declining. This decrease may be the result of more prudent diets or new medical procedures such as coronary bypass surgery, but it is also possible that it is an epidemiologic trend due to unknown causes. That people are living longer is undoubtedly partially responsible for current increases in the number of deaths from cancer, a disease most often associated with an age greater than 50. Some of the cancer increase, however, is clearly due to factors other than increased longevity, such as better reporting.

The net effect of these medical advances will be that the man who might have died at age 57 of heart disease may now live to age 70 or beyond, and the woman who might have died at age 70 of stroke (cerebral vascular accident) may live another 20 years. Cures for all of today's major diseases would still not prolong life expectancy much beyond age 85, seemingly the best current estimate of the human biologic life span. The survivorship curves in Figure 1-5 illustrate how life spans have increased in the United States between 1901 and 1981. As people live longer, increased numbers of them will live to be "very old," and our society will be challenged to provide them with necessary care. Although some elderly persons hold considerable wealth and power, and although only a small percentage of the elderly presently require residential care, the

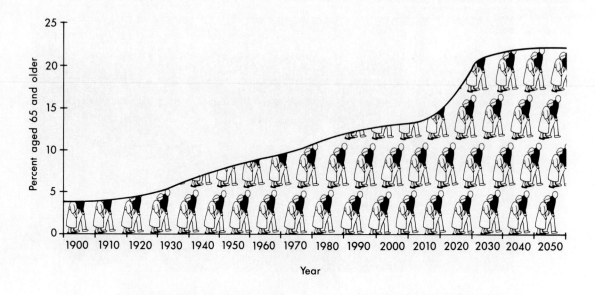

Figure 1-2   Actual and projected percentage of the United States population 65 years old or older. (Created from *Statistical Abstract of the United States; the National Data Book,* ed 110, Washington DC, 1990, U.S. Bureau of the Census).

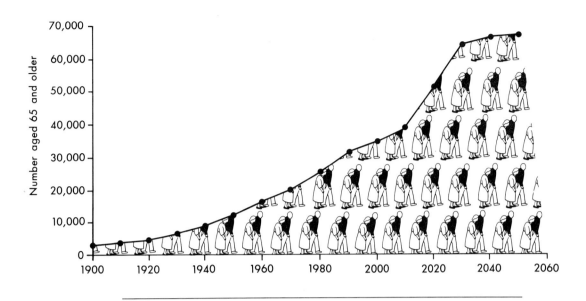

Figure 1-3 Actual and projected number of people age 65 and older through the year 2060.

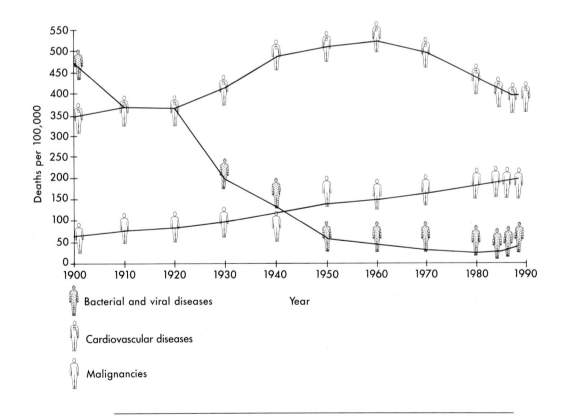

Figure 1-4 Note the decline in infectious disease and the recent decline in cardiovascular diseases while cancer continues to increase.

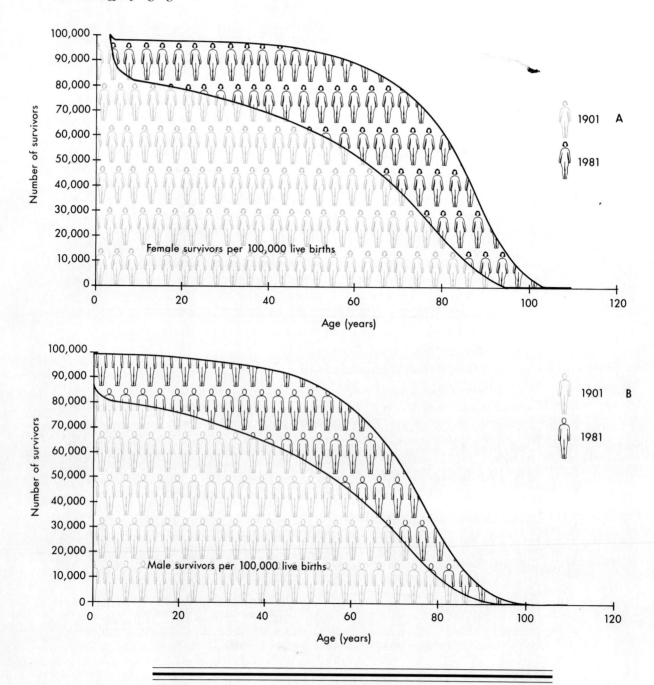

Figure 1-5 These survivorship curves illustrate the number of people surviving to different ages. Note the comparison between the years 1901 and 1981 for females **(A)** and males **(B)**.

majority of them depend to some extent on such resources as Social Security and Medicare. The latter group is where the greatest increase is expected in the next 30 years. In 1920 people over 65 equaled about 8% of the workforce (considered to be those between the ages of 18 through 64). In 1960 the percentage of the elderly had increased to 17%, and by 2020 it could equal 26% of the workforce.

There has been much discussion about how this demographic change—the shift in the dependency ratio—will affect this country's resources, especially the Social Security program. Fewer wage earners will be paying into the fund and more people will be receiving benefits for a longer period. Some observers argue that the federal government may have to cut benefits. However, this is unlikely. The elderly have excellent voting records; it's likely they will soon form one of the largest and most powerful voting blocs ever to exist in the United States. Resources will be diverted from other programs to support the elderly and those who care for them.

Political issues aside, all of this means that the number of elderly persons, including those debilitated by age, is increasing dramatically. Today, the elderly account for over one third of health-care expenditures in the United States. In the near future, physicians, nurses, social workers, counselors, nursing home managers, and numerous others with expertise in working with the elderly will be needed in much greater numbers than are available today. The purpose of this text is to provide these professionals with the biologic basis for understanding some of the problems associated with aging and to look at the limitations these problems place on aging people. The text will use standard medical terminology, all defined and explained, to give the reader a vocabulary sufficient to read and understand most medical reports. We hope, as well, that this text will give the reader a sympathetic understanding of the inevitable consequences of the aging process, a process that will affect us all.

OUR PERCEPTION OF AGING AND DEATH

When we talk about a person's age, it is clear that we are referring to the number of years that person has lived. When we refer to the effects of aging, we imply that with the years there have been a number of debilitating physical events. Properly used, the term *aging* describes the temporal process of growing old. The term *senescence* refers to the accumulation of harmful events that create a noticeable change in appearance or function in the course of aging. The term *senility* refers to the accumulation of the most severe senescent (aging) events.

Because humans have lived with aging and death since the origin of their species, each human culture has developed its own understanding of these processes and formulated its own rules and rituals to deal with them. Humans are one of the few species who exhibit a level of respect for the elderly, and one of very few whose members live substantially beyond their reproductive years. However, human attitudes often change from respect to tolerance when the elderly become very old.

The attitudes that cultures have developed toward the elderly seem to be based largely on the needs of the culture, its resources, and the contributions of the aged. Primitive peoples in harsh environments with limited resources sometimes found it necessary to abandon the elderly who could no longer contribute to the survival of the group. This necessity, over time, became a part of the group's expectations; it was the way the elderly were treated. Other groups, in more favorable environments, did not face such a necessity. They had the luxury of being able to support the elderly and benefit from their knowledge and experience. Respect for the aged became a part of many cultures in such environments; in some cases, this respect extended to ancestor worship.

Because the United States is a blend of varying cultures and attitudes, expectations about the aged differ. Some people may expect that several generations will live together during their entire life spans. Or the aged may expect to live near their relatives, so that help or care is easily available. Other aged people—the majority in the United States—may live independently, and see relatives only infrequently. Many of these are the retirees who move to Sunbelt retirement communities, often thousands of miles from their kin. Both they and their children may see this as the "normal" way of life for the elderly.

In addition to cultural expectations, however, attitudes are shaped by this society's current fascination with youth. The majority of a society tends to set the standards for everyone. A major portion of the population of the United States has most recently been the "baby boomers," who dictated a cultural emphasis on youth. As this group ages, society and the influential advertising industry are both likely to discover that middle age is the "in" group. Eventually, it may even be all right to be old. Right now, however, young is "in" and old is "out." Advertising has convinced most people that appearing aged is undesirable, and the cosmetics industry has made billions of dollars promoting products that mask the effects of aging on skin and hair. Respect for the gray hair and wisdom of age has been replaced with the desire to appear young.

In American society, affluence and the existence of alternatives to family care for the aged have allowed our youth-oriented society to say: "Don't bother us about old people." Families can pay someone else to care for their elderly. They can abandon their elderly as soon as they become a burden. Some people never visit relatives once they are in a nursing home; others only visit once or twice a month. Although many families do all they can for aging parents or grandparents until home care becomes impossible, it is still true that once the relative is in a nursing home and out of sight, he or she may become out of mind as well. This may reflect an underlying refusal to be reminded that we too may face such a future.

Attitudes toward the death of the very old differ greatly from attitudes toward the death of younger persons. Consider the typical response to the sudden death of a child. It is a feeling of crushing sadness. We may say, "What a tragedy. That little 9-year-old never had a chance to grow up, to fall in love, to fulfill the promise of her life." Few can hear of a child's death without an aching feeling of how unfair it is.

Attitudes toward the death of a mature adult are different. We say, "It is sad. She was only 40. She had many good years ahead." Our greatest sympathy goes to the family left behind. A husband has lost his wife; children will grow up without a mother. "We will miss her," we say. The sadness here is for the people left behind as much as for the person who has died.

When an elderly person dies, we say, almost ritually, "How sad. We will miss him." However, we may also say, "It was time; he had a full life," or even, "It was for the best. He hardly seemed to enjoy life any more. He missed his wife, since she died 10 years ago. Walking was painful for him, and he wasn't interested in much anymore." We may even say, "What a relief for his family. It *was* time for this 89-year-old to die."

Three people die. Reaction to their deaths is based on expectations for their futures, had they lived. The child had a full life ahead of her. The middle-aged woman had half a life left. The grandfather had nothing left.

People expect death to result from "old age." They expect the wrinkled skin, the aches and pains, and the forgetfulness that come with age (Figure 1-6). This expectation is so ingrained that even though senescence is a condition that will kill everyone, much less money is spent learning about the basic aging process than is spent on various individual diseases, many of which are only symptoms of age. Consider the people of an earlier America. In the eighteenth and nineteenth centuries, people lost most of their teeth by the time they were 40 years old. Everyone expected to lose his or her teeth by age 40, and no one considered finding out why this happened. Until science developed preventive measures, an abnormal disease process was taken for granted, just as the aging process is taken for granted today. (Although senescence is clearly normal and universal, the concept of senescence as a disease will be considered in Chapter 2.)

Figure 1-6   The effects of aging on various body systems: graying hair, irregular pigmentation, barrel chest.

As this text explores the many physical changes that occur as we age, keep in mind that medical advances are being made that may lessen the effects of aging and make the last years more pleasant and productive. As you will read in Chapter 2, the search for the fountain of youth is not over. There is still hope that the human life span can be significantly lengthened, and that the quality of that life span can be enhanced and made productive and enjoyable to the last day.

SELF-PERCEPTION IN THE ELDERLY

No one really knows what someone else is thinking, but a little introspection can allow us to imagine how others perceive the aging process. Most people do not recognize the fact of their aging, or even acknowledge the effects of age on their performance. We retain an image of ourselves as "normal" rather than old, and any changes are similarly normal and inconsequential. Many elderly people think of themselves in this way and live active lives until they die. They stay involved in life. Almost every community includes individuals who maintain an active, independent lifestyle virtually to the end of their lives.

People who experience physical problems as a result of senescence or chronic disease are more aware of their limitations and of the changes that aging brings. They may be acutely aware of a decline in function, and try to compensate for it. Life can be difficult for these people—even overwhelming. However, many of these elderly have a positive attitude even in the face of a worsening situation. With help, they may have productive and enjoyable last years.

A few people consciously fear growing old. They look for and feel threatened by any indication of aging. Their senescence is a painful, horror-filled time and a major challenge to their mental health. The fear of age-related change combined with the inevitability of it leads these people into despair, grief, and often into depression.

Imagine yourself as an 80-year-old, shopping on a bright winter morning. You look into a store window and suddenly see a reflection. It takes a moment to realize, "That's me!" because what you see is not your image but the reflection of an old man or woman. You see a wrinkled face with irregular pigmentation. Your hair is thinning, or gone. Your clothes sag, and you do not stand as straight as you once did. You still have a clear memory of the youthful body you had most of your life and you want to think of yourself as that youthful person. What you see in the reflection is a contradiction. As you think back over the last 25 years, you realize that conditions have steadily worsened. The thought of the next 25 years is unbearable. To protect the "normal" image of yourself that you must have for a positive outlook on life, you will, at the very least, avoid mirrors. At worst, you will avoid any contact with the reality of the present.

Most elderly people do not make a conscious decision to withdraw from life, but many make a subconscious decision to ignore the impact of age when it begins to interfere with their self-image. They may pay less attention to themselves and neglect such details as personal appearance, clothes, and cleanliness. Family members may become worried about the lack of concern such elderly relatives display toward themselves, their environment, and others.

The director of an Iowa nursing home required all of his staff to read a certain poem before their first day on the job. When orderlies complained about having to clean up a particularly disgusting mess made by an elderly patient, they were asked to discuss with the director or the head of the nursing staff what they believed to be the poem's significance. This poem was reportedly found in the belongings of an elderly woman who died in a nursing home in Ireland. No one knows who wrote it:

Figure 1-7   What do you see?

What Do You See, Nurse?

What do you see, nurse, what do you see?
What are you thinking when you look at me?
A crabbit old woman, not very wise,
Uncertain of habit, with far away eyes,
Who dribbles her food, and makes not reply,
When you say in a loud voice, "I do wish you'd try!"
Who seems not to notice the things that you do,
And forever is losing a stocking or shoe.
Who unresisting or not, lets you do as you will
With bathing and feeding, the long day to fill.
Is that what you're thinking, is that what you see?
Then open your eyes, you're not looking at me.
I'll tell you who I am as I sit here so still,
As I move at your bidding, as I eat at your will.
I am a small child of ten with a father and mother,
Brothers and sisters who love one another.
A young girl of sixteen with wings at her feet
Dreaming that soon now, a lover she'll meet.
A bride soon at twenty, my heart gives a leap,
Remembering the vows that I promised to keep.
At twenty-five now I have young of my own
Who need me to build a secure happy home.
A woman of thirty, my young now grow fast,
Bound to each other with ties that should last.
At forty my young soon will be gone,
But my man stays beside me to see I don't mourn.
At 50 once more babies play around my knee,
Again we know children, my loved one and me.
Dark days are upon me, my husband is dead,
I look to the future, I shudder with dread,
For my young are all busy rearing young of their own
And I think of the years and the love I have known.
I'm an old lady now and nature is cruel,
'Tis her jest to make old age look like a fool.
The body it crumbles, grace and vigor depart,
And now there is a stone where I once had a heart.
But inside this old carcass a young girl still dwells.
And now and again my battered heart swells.
I remember the joys, I remember the pain,
And I am loving and living life over again.
I think of the years all too few, gone so fast,
And accept the stark fact that nothing can last.
So open your eyes, nurse, open and see,
Not a crabbit old woman, look closer, see me.

The writer of this poem was exceptional. She deserves gratitude for sharing the thoughts that may be in the mind of many withdrawn elderly people. We who work with the elderly must be sensitive to this. We must keep in mind the entire spectrum of attitudes and emotions the people we care for may be experiencing.

BIBLIOGRAPHY

Brookbank JW: *The biology of aging,* ed 1, New York, 1990, Harper & Row.

Statistical abstract of the United States; the national data book, ed 110, Washington DC, 1990, Bureau of the Census.

Conner KA: *The aging of America: an introduction to issues facing an aging society,* Hertfordshire, 1991, Prentice-Hall.

Fischer DH: *Growing old in America,* Oxford, Mass, 1978, Oxford University Press.

Glover JW: *United States life tables, 1890, 1901, 1910, and 1901-1910,* Washington, 1921, Department of Commerce, Bureau of the Census.

Olshansky SJ, Carnes BA, Cassel C: In search of methuselah: estimating the upper limits to human longevity, *Science* 250:634-640, 1990.

Thomson D: The ebb and flow of infection, *JAMA* 225(3):269-272, 1976.

2 *Theories of Aging and Senescence*

SOME QUESTIONS ABOUT AGING

Aging and death pose fascinating questions that humanity has always wondered about. Why should a life form begin to deteriorate with age, malfunction, and die? Under ideal conditions, how long could we live? Some simple organisms such as amoebae and some bacteria are able to continue dividing indefinitely, allowing their cell lines to live forever. A few plants live extremely long lives; the bristle cone pine, for example, may live more than 3,000 years (Figure 2-1). Even the mitochondria that exist within all our cells show no signs of aging and are passed on in the cytoplasm following cell division as long as the generations of cells live.

Figure 2-1 Life span varies greatly among organisms. This bristlecone pine is almost 5000 years old.

We speculate about possible causes of cell, tissue, and organism failure. Is it wear and tear, fatal mutation, lack of nutrients, accumulation of waste products, or something else? All of these mechanisms as well as many others have been suggested as explanations for the aging process. When does dysfunction in the body begin? Has it been going on since the gill slits atrophied and disappeared in the early stages of embryologic development? Is senescence genetically programmed or is it an accumulation of environmental factors? This chapter cannot answer these questions but it does provide the authors' views regarding the status of some current theories. These theories are important because the direction of research on ways to prolong and improve the quality of later life depends on the establishment of sound theory.

Studies of aging can be divided into two groups: cellular studies dealing with aging individual cells or tissues, and studies involving changes in the whole organism. Cellular studies often tend to be biochemical and are sometimes referred to as studies of chemical aging. This text involves studies of individual organ systems because they are the largest and most easily studied units in the organism.

About 60 years ago aging and its associated senescence were believed to be simply a matter of the body wearing out. As researchers began to look more closely, however, they discovered that while the *rate* at which a body ages may be influenced by the insults delivered to it in life, the aging process as a whole is not the result of wear and tear. People who live extremely sheltered lives do not significantly outlive those who are active and even break a bone or two. Some researchers suggest human life span would average about 85 years if most diseases were eliminated. The typical human will die before reaching the age of 120. We hold the view that genetic manipulation or manipulation of certain genetic products holds the promise of greatly extending youthful life.

The aging process apparently involves failure of the surveillance, repair, and replacement process typical of a young body (Figure 2-2). The cells that compose most of the body's tissues normally die and are replaced at a rapid rate throughout life. Exceptions are muscle and nerve cells that are formed at birth and may persist as long as the body does. Evidence indicates that cancers begin to grow and are eliminated by our immune system before they are large enough to be seen. Tissues are damaged in day-to-day life and are normally repaired. This maintenance function fails in mice in a year or so, in dogs in 10 to 20 years, and in people in 50 to 100 years (Figure 2-3). When the cells, tissues, and organs are not maintained, senescence appears, causing the organism to deteriorate and eventually die.

THE EVOLUTION OF LIFE SPAN
A Review of Chromosomes and Genes

All genes consist of segments of strands of DNA, deoxyribonucleic acid that make up chromosomes. All body cells (except mature red blood cells) normally contain the same 46 chromosomes and the same kinds of genes (Figure 2-4). Chromosomes contain many thousands of genes most of which are inactive at any one time. Many of the genes that are active when a fertilized egg first divides into two cells are not active when the descendants of those cells have specialized to become brain or skin cells. Other genes, inactive earlier, activate to provide the differentiating cells with their specializations.

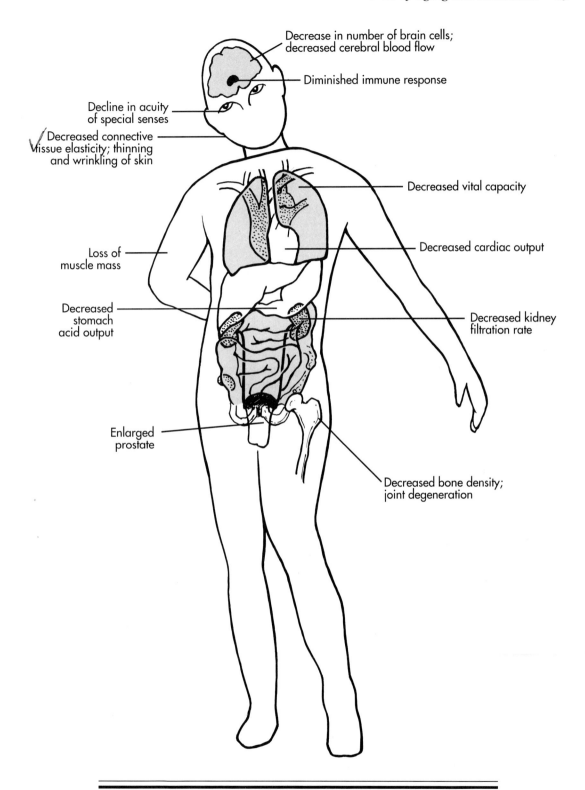

Decrease in number of brain cells; decreased cerebral blood flow

Diminished immune response

Decline in acuity of special senses

Decreased connective tissue elasticity; thinning and wrinkling of skin

Decreased vital capacity

Loss of muscle mass

Decreased cardiac output

Decreased stomach acid output

Decreased kidney filtration rate

Enlarged prostate

Decreased bone density; joint degeneration

Figure 2-2   Although no two individuals age in exactly the same way or at the same rate, all body systems are affected.

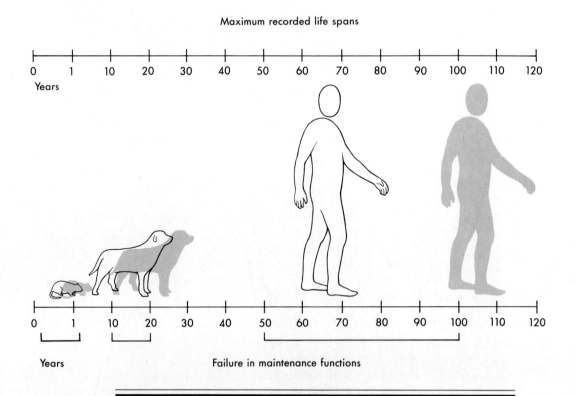

Figure 2-3   Varying life spans in the animal kingdom.

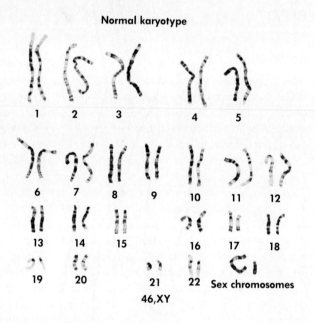

Figure 2-4   This human karyotype shows the 46 chromosomes, cut out of a photomicrograph of the cell nucleus and arranged in pairs. Each chromosome contains thousands of genes.

Our 46 chromosomes are divided into two similar but not identical sets of 23 called genomes. The chromosomes of each pair are similar in that they contain the same kinds of genes. For example, if a chromosome carries a gene for eye color at a certain point (or locus), the other chromosome of the pair will have a gene for eye color at the same locus. The genes may differ, however. One may be a dominant gene for making the brown protein for brown eyes; the other may be a recessive gene that allows the eyes to appear blue. Although genes that control pigment distribution are also involved, for a person to have blue eyes he or she must inherit a gene for blue eyes on each chromosome of the pair.

The concept of evolution states briefly that organisms change with time as a result of random genetic mutations caused by natural radiation and mutagenic compounds that have been on earth since the beginning of the planet. Most of these mutations produce genes that are recessive if they function at all and most are deleterious. Mutations that are beneficial at the time they are produced are rare. The concept of evolution by natural selection suggests that those mutated genes that are common in living things today continue to exist because they presently have or in the past had value to the species. Remember that most mutations are random but environmental factors determine which mutations survive.

Sometimes, mutations develop that have positive value for the species but may yield negative results for some of its members. For example, the inherited behavior that results in an ant colony killing all but one of the potential queen ants is useful in controlling the reproduction of the colony. However, it is rough on the individuals that are not selected to be queen. It would certainly not be out of place for a gene that terminates the life of individuals to be selected if such selection benefits the species, or at least benefits a freely interbreeding, closely related group within the species.

Why We Do Not Live Forever

The hypothesis proposed here suggests that the process of evolution has produced a life span for each species in order to give the species maximum survivability in its niche. For example, an animal with the size and power of an elephant would not be possible if it had the life span of a mouse. In fact, the knowledge required for a herd of elephants to survive (knowing the location of waterholes, salt licks, and available food in a drought) requires years to acquire. Mice, on the other hand, can grow to a size appropriate for their niche, reproduce, and die in a year while still maintaining a successful colony. Even some plant species are genetically determined to be annuals while others live 2 years, and still others may live for thousands of years.

It has been argued that natural selection favoring long-life genes could not have operated in a wild animal population because those animals old enough to show any senescence simply would not survive. We suggest that living forever is the default condition typically found in the most primitive animals (some bacteria and protozoans) on the earth today. Predator destruction of any animal that begins to show weakness is precisely the mechanism for which genes that limit life span by creating senescence have been selected. It is likely that the advanced species on earth today have had genes limiting their life spans throughout most of their evolution.

Why doesn't natural selection favor a genome that allows an individual to live forever, simply producing offspring at a slower rate to avoid overpopulating? The answer lies in two related factors, genetic viability and maximum diversity.

Genetic Viability. If there were only a few immortal living pairs, there would be only a limited number of different genes. When the same small gene pool is used repeatedly, reproductive viability is lost. Recessive genes become concentrated due to inbreeding and produce increased numbers of birth defects. Associated with this inbreeding is a decrease in fertility, often with abnormal sperm produced, resulting in smaller numbers of young. Collectively, these factors may cause the population to die out. Elimination of parents when they have produced enough offspring for the species to be competitive in its niche preserves fertility and reduces the chance of harmful recessive genes being expressed.

Genetic Diversity. A population with a large number of different genes has an advantage over a more inbred population even though some of its genes may be deleterious. When conditions change, a few genes that had been harmful under other conditions may now be valuable when expressed, enabling the individual possessing them to survive when others could not. Therefore, the more genetic diversity present in a species, the greater the chances the species has of surviving when conditions change. An example of this in human medicine is the incompletely dominant gene for sickle cell anemia. When both genes are present, the patient has debilitating disease. But when one normal gene and one sickle cell gene are present, the person has surprising resistance to malaria. When malaria became extremely severe in areas of Africa, the only survivors were those who carried the sickle cell gene. Extinctions have occurred when species lacked the genetic diversity needed for survival under new conditions. The cheetah is an example of the vulnerability of a species that lacks genetic diversity. Members of this species are so genetically similar that they all lack the ability to survive one specific disease. The most crucial diversity is likely located in recessive genes that become visible only when reproduction places them opposite nonfunctional or identical genes.

Evidence that Organisms Measure the Passage of Time

Clocks in Organisms. If the life span of an organism were determined by a gene or genes and was therefore inherited, something in its tissues or cells would have to measure the passage of time. Much has been written about biologic clocks and if one can draw a single conclusion from these works, it is that there is an internal (intrinsic) clock that seems to provide a constant awareness of time passage and that certain external (extrinsic) factors may adjust or "set" this clock and gear certain events to it, and it in turn controls other events.

Biologic clocks have been demonstrated in just about every organism that has been studied. For example, amazingly, many species that navigate great distances fly or swim at ever-changing angles to the sun or stars as they move across the sky (Figure 2-5). To do this, such animals must be guided by an innate knowledge of time passage. Numerous subconscious rhythms have been observed in humans. These range from rhythms of short duration such as heartbeat and breathing, to diurnal rhythms such as the wake-sleep cycle. The human cortisone cycle is correlated with the wake-sleep cycle. Our cortisone levels reach a peak at about 8:00 AM and a low at 8:00 PM. This is slow to change even when the light-dark cycle and activity cycles are altered. It has been suggested that the slow adjustment in this cycle is one of the factors responsible for what so many people have experienced as jet-lag.

Clocks in Cells. Even cells seem conscious of age or time passage. Hayflick's classic experiments showed that fibroblasts (connective tissue cells) produce a fairly

Figure 2-5   Many birds that migrate great distances navigate by the sun by day and the stars by night. Stellar navigation is less clock-dependent than solar navigation because in the northern hemisphere birds use the distant stars, especially those near the North Pole, where star position changes little during the night.

constant number of generations when cultured from individuals of the same age and species. They replicated fewer times when cultured from old individuals of that species. Hayflick also observed that fibroblasts from long-lived species replicated more times than did those removed from short-lived species such as mice. Recent studies indicate that proliferation-promoting genes can be inhibited by antiproliferation genes. These studies provide the first evidence for the genetic mechanism by which a cellular "clock" could turn off cellular reproduction.

If cells somehow retain the knowledge to stop dividing after a certain number of divisions, they may do this by producing something that interferes with the division process. Studies done as early as 1967 determined that fibroblasts would reproduce more generations if grown in a medium made from the plasma from young donors than if raised in plasma from old donors. This indicates that either something in the old plasma inhibits reproduction or that something in the young plasma promotes it. Given the recent discovery of antiproliferation genes, we now suspect that biologic factors can activate these genes, resulting in cellular senescence. Perhaps the product of these antiproliferation genes is present in old plasma.

A Possible Mechanism. A simple mechanism for a cellular clock is suggested by the fact that chemical reactions are taking place in cells all the time. These reactions take time and while the time required for a reaction is dependent on the temperature and many other factors, most of these conditions are quite consistent among the members of a species. The accumulation of enough product from one reaction is the automatic signal for the start of another, and by this mechanism all cells, tissues, organs, and organisms have a time reference. Each completed reaction can be considered the equivalent of the tick of a clock. When we gain understanding of the mechanisms involved, it will become possible to alter them, possibly extending life span and preventing senescence.

Growing old

Senescence as a Disease

Mosby's *Medical Dictionary* defines disease as "a condition of abnormal vital function involving any structure, part, or system of an organism." *The Living Webster Encyclopedic Dictionary of the English Language* defines disease as "an impairment of the functioning of a system of the human body, or of an organ or part thereof; a similar disorder in animals. . . ." Most people define a disease as an *unusual* process that negatively affects the way they feel or function.

The progressive deterioration of senescence, however, *is* usual. In fact, it is universal among animals except for the most primitive forms. There is variation among the details from species to species or even within species, but in all vertebrates it is a relentless process. It eventually kills unless a different disease or accident kills the animal first. As the subsequent chapters of this text look in depth at the impact of senescence upon each of the body's systems, bear in mind that there are things that can be done to alter or slow the effects of this disease (see Chapter 15, Lifestyles) but as yet no one has found a cure. Realizing that a problem exists is the first step toward a solution; our recognition that senescence is a disease holds out hope that a cure or an effective treatment may be found, as has been the case with many other seemingly hopeless conditions.

Some Encouraging Speculation

Each functional gene operates as the construction plan for the production of a specific protein. Each protein is either a part of a structure, cell, or component of body fluid, or it is an enzyme that enables a chemical reaction to take place in the body. It is likely that the gene or genes that limit life span either begin to produce one or more enzymes that are detrimental as a person ages, or fail to produce one or more enzymes that are necessary for life. If this is so, artificial production of the needed enzymes or discovery of the means of destruction of the undesired ones could, in theory, indefinitely prolong life. Even more interestingly, this should not simply extend the life of a worn-out body but should allow the natural repair mechanisms that work so well during our youth to continue operating. We can speculate that it should be possible to live almost indefinitely in a body that retains full function. However, such a discovery would limit the variability and gene exchange that has been the key mechanism in the creation and survival of our species. The social and legal changes in a world without senescence and natural death would be tremendous.

The research that provides hope for slowing or arresting the aging process will probably proceed along the following lines. Genetic research has had tremendous breakthroughs in the last 20 years and many genes that have been found have been discovered only after the protein the gene produces has been isolated. It is likely, therefore, that the key discovery will involve any of a number of proteins that play a part in causing damage, preventing damage, or repairing damage to the body. This protein would probably be produced by a gene that expresses itself only after a certain number of years of life. Such discoveries are likely to contribute to the explanation of one or more of the following theories of aging.

THEORIES OF SENESCENCE

Three inherited conditions hint at the genetic basis for senescence. Each imitates some effects of the "normal" aging process, but at an abnormal point in the individual's lifespan.

Down Syndrome. Instead of having the normal pair of chromosomes, people affected with Down Syndrome have three of the twenty-first chromosome. For that reason, this condition is also called Trisomy 21. Down Syndrome individuals exhibit varying degrees of mental retardation and a number of physical abnormalities including the "Mongolian" fold of the eyes, changes in the hands, tongue, larynx, and some internal organs (Figure 2-6). It is unusual for Down patients to live beyond the age of 45 and death is usually from respiratory or cardiovascular problems.

Down syndrome mimics the aging process in several ways. Affected people have an increased incidence of cancer, above-normal incidence of osteoporosis (loss of cal-

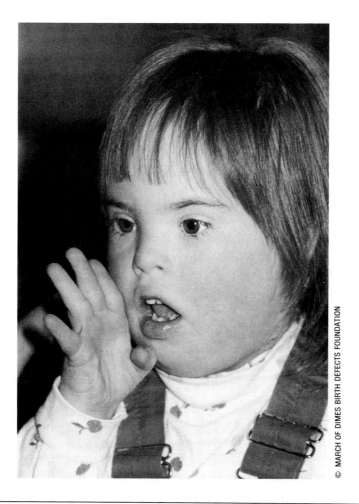

© MARCH OF DIMES BIRTH DEFECTS FOUNDATION

Figure 2-6   A Down syndrome child.

Figure 2-7   Eight- and nine-year-old children suffering from progeria, which mimics some of the effects of aging.

cium from the bones), early formation of cataracts, (opacity of the lenses of the eyes), thinning and increased fragility of the skin, hair loss, blood vessel disease, and increases in the age-related pigment lipofuscin and age-related amyloid deposits. The mental retardation shows many organic similarities to Alzheimer's disease although it does not usually show the same rapid progression.

Progeria. Progeria was first described as a form of premature aging. Signs of the disease usually appear in the first year of life and most children with progeria do not live beyond the age of 15 (Figure 2-7). The fat layer beneath the skin begins to decline and this combined with loss of skin elasticity results in wrinkling. Maturity and tooth development are delayed and the jaw fails to develop to normal proportions. Unlike trisomy 21, there is no increase in lipofuscin deposits and no increase in incidence of cancer and the children have normal to above-normal intelligence. Death is usually from the effects of severe atherosclerosis (deposits of fatty plaque in the blood vessels). There is evidence that the condition is inherited but the mechanism of inheritance is still unknown.

Werner's Syndrome. Werner's syndrome is similar in many ways to progeria but it begins later, at age 15 to 20. It is clearly inherited as a recessive gene and results in hair graying and loss, cataracts, atherosclerosis, diabetes, increased incidence of malignancies, and immunologic problems. Sexual maturity is delayed. This disease clearly involves failure of the collagen-producing mechanism to respond to stimulants that are normally released by the body.

Genetic Theories

Any genetic theory of aging must be based on the concept that the gene responsible for senescence becomes active or the gene responsible for preventing it is deactivated after a certain period of time. Discovery of an antiproliferation gene mentioned earlier supports this concept. When this gene was activated it resulted in decreased cell division. The existence of proliferative and antiproliferative genes would be just another example of the kind of checks and balances that are so common in biologic systems. The situation may be more complex, with many other genes contributing to the strength of the proliferative or antiproliferative genes or their products. Other strengthening or weakening agents could include accumulated waste products, destruction of gene products by protein interactions, deletion mutations eliminating needed elements, cyclic changes associated with internal clocks, hormonal changes, or immune reactions.

Once the age-functioning DNA gene or genes become active, senescence begins. Numerous mechanisms can damage tissues. Direct damage can occur through protein interaction, such as protein cross-linking, in which the peptide components of the proteins in cells link, altering protein structure. Tissues can also be damaged indirectly if possible damage is not prevented. Two examples of this are:
1. Free radicals (potentially damaging chemicals that form in the presence of oxygen) can, if not neutralized, damage cell membranes or the chromosomes within the cells.
2. The immune system may fail to protect against harmful cellular change.

Free Radical Theory

Some products of cellular metabolism are highly reactive parts of molecules called free radicals. They have a strong unbalanced charge and may react with proteins, changing their shape and nature. They also may react with lipids found in cell membranes, affecting their permeability; or they may bind to cell organelles. And they may produce more free radicals in a chain reaction, thus increasing the damage. Other free radicals are taken into the body with food or inhaled air. Because free radicals are so reactive, they have a short life and rapidly combine with enzymes, membranes, or active sites on cell organelles, damaging or destroying them. They may be a major cause of mutation when they react with DNA. The free radical theory suggests that senescence is caused by the accumulation of irreversible damage resulting from these oxidizing compounds. One argument against this theory is that different species and different individuals within species are exposed to approximately the same amount of free radicals in any unit of time, yet they have greatly different life spans.

In recent years there has been considerable interest in using antioxidants such as Vitamins E and C as a means to counter the effects of free radicals and extend life. Research has recently supported the life-prolonging effect of Vitamin C.

Cross-Linking of Collagen and Other Proteins

Collagen, the primary protein of scar tissue, is important in bone, cartilage, tendons and ligaments. It is found scattered in most other tissues and makes up nearly 30%

of the body's protein. When collagen fibers are first deposited in our soft tissues, the molecules are loosely associated or poorly attached to each other and the tissue is flexible. As time passes, active sites on adjacent collagen molecules cause the molecules to become more closely associated, and the tissue becomes less flexible. High blood sugar has been found to promote cross linkage of proteins and is suggested as a mechanism for some of the loss of elastin associated with diabetes. According to the theory, cross linking of collagen and other large proteins in the cells not only causes a stiffening of the tissues but also decreases access of white blood cells that would fight infection or destroy foreign tissue, decreases access to nutrition, inhibits cell growth, and results in failure to dispose of toxins created by metabolism.

Proteins are extremely large molecules that form the framework of cells and most of the enzymes and anchoring points for cellular reactions. Cross-linking of proteins other than collagen can interfere with major cellular reactions in much the same way as free radicals. This is by denaturing the proteins, changing them structurally so that they no longer perform their desired functions. Glycosylation of hemoglobin, a process that occurs in the blood of diabetics as a result of high blood sugar, illustrates how a protein can be permanently altered.

Enzymes are organic catalysts, facilitators of cellular reactions. When an enzyme is absent or nonfunctional, some cellular reactions will progress at a much slower rate while others stop completely. In mitochondria for example, enzymes or components of cellular reactions are held temporarily in a certain order or position so that a desired reaction can take place. If cross-linking were to obscure these "binding points," the required cellular reaction would again be limited.

High blood sugar has been found to promote cross-linking of proteins and is suggested to be a mechanism for the loss of elasticity and some other aspects of blood-vessel disease in diabetics. Apparently some mechanism prevents this excessive cross linking in young people but allows it to occur gradually as we age. Many of the same events occur in short-lived animals but at a much faster rate. Recent research continues to support age-related collagen changes in tissues.

Accumulated Waste Theory

Most waste products are expelled from cells, carried away by the circulatory system, and excreted. Some, such as lipofuscins (Figure 2-8), slowly accumulate in cells, and become noticeable in those of old animals. These yellow-brown pigments are especially common in muscle, nerve and other such cells that are not replaced but may remain throughout life. Although they can be found in young cells, they are so common in old cells that they are called age pigments and are markers for old cells.

Some researchers have speculated that as lipofuscin granules accumulate in the cytoplasm, they may interfere with transport of materials through the cytoplasm or across membranes. They eventually take up so much room that it seems that there is not enough space for other cellular components. We suspect that they are probably an effect rather than a cause of aging. Lipofuscins are chemically inert but other waste products may accumulate in cells as well. These may be chemically active compounds that interfere with cellular enzymes. They include free radicals, aldehydes, and histones. A problem with this theory is that some senescent tissues may lack most of these compounds while others have them in abundance.

Destructive Mutation

The concept of destructive mutation suggests that as people age, natural exposure to the environment results in progressively increasing numbers of mutations being incorporated into their body cells as well as those cells that produce our ova and sperm. Because mutations occur mostly at random, the vast majority are destructive, resulting in poorer performance of the cells carrying the mutation. Studies have shown that liver cells from old mice contain more mutations than do liver cells from young mice. They also have shown that liver cells from short-lived strains of mice (life span of approximately 1 year) have more mutations than do liver cells of long-lived mice (life span of approximately 3 years).

Normally, if a gene were to mutate and produce a new protein that appears on the outside of the cell, the cell containing that gene would likely be destroyed by the immune system due to the presence of a "foreign" protein. Large numbers of mutations could cause large amounts of tissue destruction. Experiments have shown that mice protected from normal causes of mutation (such as radiation) do live longer than unprotected mice, but not much longer.

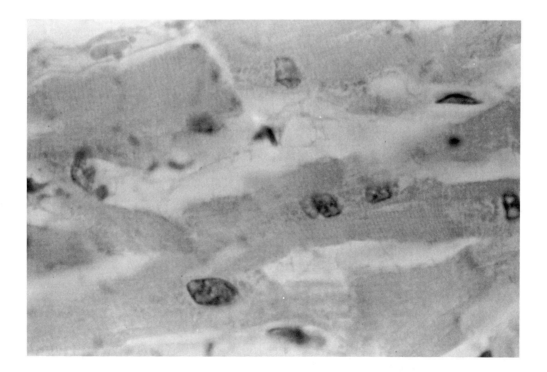

Figure 2-8 Tissue with lipofuscin granules. Lipofuscin granules accumulate in many old cells. They and other waste products are suspected of interfering with some cell activities (See also Color Plate 1, *A*).

Normally, mutation in DNA transcription (duplication of genetic material) is prevented by repair mechanisms present in the cell that make the appropriate correction in the damaged portion of DNA. This mechanism works well in our youth but seems to fail as we age. It could be that here too, failure of the repair mechanism is a cause of the aging process.

Autoimmune Theory

The immune system functions to destroy organisms such as bacteria, viruses, or fungi that are composed of substances foreign to the body. One way it does this is by producing compounds (antibodies) that bind to the foreign substance (antigens) and in some way injure the foreign organism. In the case of viral-infected cells, the immune system may destroy the infected cell along with the virus. Our immune system is programmed to recognize and ignore its own tissue and to be in a state of continual surveillance against foreign materials, mostly proteins, in our body. One autoimmune theory suggests that the body loses its ability to distinguish between its own proteins and foreign ones. The immune system begins to attack and destroy its own tissues at a gradually progressing rate.

Why would the immune system attack a body that it had ignored for many years? The answer may lie in the failure of the body to correct mutations in its cells. As the proteins change and become more different, they are eventually recognized as foreign and destroyed. Immediate destruction of the mutated cell would cause slight damage, but the immune system reacts more slowly with age and may not begin its attack until the mutated cell has reproduced many times. In such cases the destruction may involve large blocks of tissue. Immune system response slows with age, therefore allowing time for production of a large clone of cells before response gets underway. It also has been suggested that immune complexes formed by the antibody and a foreign antigen such as a bacterium may sometimes persist in the tissue instead of being removed. This could stimulate an attack against the tissue by the immune system.

Support for this theory includes the existence of several autoimmune diseases. In the case of myasthenia gravis, antibodies are produced that bind to acetylcholine receptor sites, making it impossible to stimulate a muscle. With rheumatoid arthritis, the attack involves the synovial membrane, possibly as a result of deposition of an immune complex, causing destruction of the joint. Autoimmune antibodies tend to increase with age.

Immune System Failure

Many studies have shown that the immune response declines with age. Not only is it slower in responding to foreign substances but also it seems to manufacture large numbers of antibodies that seem to have no purpose. For this reason the body's ability to fight most major and minor infections declines. Many symptoms of aging result from the accumulated effect of past and present disease that the immune system was or is unable to control. Of course, the aging of the immune system itself has a cause; for the answer to that we must probably invoke one of the other theories. Still, average life span could be lengthened if the responsiveness, power, and accuracy of the immune system could be sustained.

Hormonal Theories

As humans have searched for the key to eternal youth they have often turned to hormones. During the 1920s sexual steroid hormones were believed to be a panacea. More recently, declines in some pituitary and adrenal hormones were suspected of contributing to the aging process. However, repeated experimentation has failed to substantiate a significant impact of most of these hormones on senescence. There are several partial exceptions to this statement. These include the well-studied relationship between post-menopausal appearance of bone loss and estrogen decline in women.

Renewed interest has appeared in a pituitary hormone called somatotropin, or growth hormone. This compound has been very expensive until recently but is now available for experimentation at a much lower price because of its manufacture by genetically engineered bacteria. A number of articles had established that growth hormone levels decline with age and are as much as one-third lower at age 55 than they are at ages 18 to 23. A 1990 article appearing in *The New England Journal of Medicine* reported that people who have suffered such a decline and are treated with growth hormone may respond with increased bone density, increased muscle mass, loss of body fat, and thicker, less wrinkled skin. All of these are clearly factors associated with youth and are among the subjects of greatest concern for many elderly. At this writing no confirmation of this work has appeared yet, but it seems that somatotropin decline may contribute significantly to many of the changes associated with aging.

IN SUMMARY

All of the theories of senescence we have presented result from observation of things that happen as people age. The theories are all correct to some degree; the basic mechanism of aging involves all of them. The genetic theory, based on the assumption that natural selection has favored genes that limit life span, is a basic, unifying concept.

We asked two questions at the beginning of this chapter: why do people die and how long can they live under ideal conditions? The answer lies in the following points:

1. People die because their cells, tissues, and organs fail.
2. Their failure results from genetic programming and environmental factors.
3. People can alter their exposure to the environmental factors. Researchers are starting to work on the genes.
4. In theory, if people could correct the genetic problems and properly select their environment, unlimited life span would be possible with nearly perfect function of both body and mind. Dying of old age would no longer be necessary.

Some observers have suggested that people might not want to live forever. Perhaps this is partially because mythology and literature have often depicted those people who live forever as having lost their humanity. The myth of Dracula is an example. Science does not deal with myths. The deterioration as a result of senescence is well documented and it is certainly feared by many. If this fear were removed because senescence were no longer necessary, we expect most people would look with favor on a future with no programmed end.

BIBLIOGRAPHY

Birkkow G: Time-compensated sun-orientation in animals. *Proc Interna Congr Zool* vol 4: 346-350, 1963.

Breitender-Geleef S, Mallinger R, Bock P: Quantitation of collagen fibril cross-section profiles in human veins, *Human Pathol* 21(10):1031-1035, 1990.

Brookbank JW: *The biology of aging,* ed 1, New York, 1990, Harper & Row.

Brunning E: *The physiological clock,* 1964, Berlin, Springer-Verlag.

Enstrom JE and others: Vitamin C intake and mortality among a sample of the U.S. population, *Epidemiology* 3(3):194-202,

Fischer DH: *Growing old in America,* Oxford, Mass, 1978, Oxford University Press.

Hayflick L: Why do we live so long? *Geriatrics* 43(10):77-87, 1988.

Kohn RR: *Principles of mammalian aging,* Hertforshire, 1978, Prentice-Hall.

Olshansky SJ, Carnes BA, Cassel C: In search of methuselah: estimating the upper limits to human longevity, *Science* (250):634-640, 1990.

Rudman D and others: Effects of human growth hormone in men over 60 years old, *N Engl J Med,* 323(1):1-6, 1990.

Sauer F, and Sauer E: *Star navigation of nocturnal migrating birds.* Cold Springs Harbor Symposium on Quantitative Biology, 1960.

Thornbecke GJ, ed: *Biology of aging and development,* New York, 1975, Faseb & Plenum.

Vance ML: Growth hormone for the elderly? *N Engl J Med* 323(1):52-54, 1990.

Waterman TH: *Animal navigation,* New York, 1989, WH Freeman.

3 The Integumentary System

RELEVANT ANATOMY AND PHYSIOLOGY
Composition and Function

The *integumentary system* is composed of the skin and its appendages. Like all organ systems, it is concerned with homeostasis, or the maintenance of critical factors within the tolerance limits of the body. The skin has the primary homeostatic function of protecting the body from extremes of temperature, moisture, and pressure, and from colonization by harmful organisms. Protection from heat loss is performed by insulating layers of fat and hair; protection from excessive heat is accomplished by the action of sweat glands and evaporation. Many sense organs use the skin to detect changes in temperature, pain, and pressure, alerting the body of the need for protective action. Skin forms a shield that protects us from excessive water loss and acts as a barrier to bacteria and fungi. The integument is also that portion of ourselves that is visible; its condition may have marked influence on how we are perceived both by ourselves and by others.

The skin appendages have specialized functions. *Sebaceous* or oil *glands* lubricate our skin and prevent the hair that emerges from it from becoming excessively brittle. *Nails* protect the ends of fingers and toes. *Hair* and its associated nerves have a sensory function and also serve as an insulator and protector for parts of the body. Our largest skin glands are the *mammary glands*. They will be discussed as part of the reproductive system (Chapter 11) because that is where most people would look for them.

The Embryology of Skin

The skin and its derivatives are formed from two embryologic layers, ectoderm, the outer layer of cells of early embryos, and mesoderm, a highly mobile internal layer of embryologic cells that form such things as blood vessels, muscles, connective tissue, and bones. Research has shown that embryologic tissue has tremendous

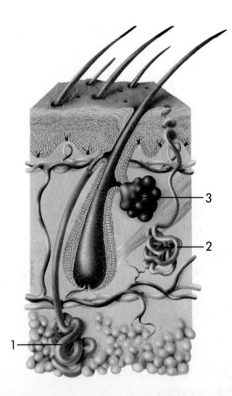

Figure 3-1   The stratum germinativum, the active layer of ectoderm, plunges into the dermis and deeper where the cells specialize to form hair follicles and sebaceous glands.

power of regeneration (the ability to repair or replace damaged tissue). Many tissues lose their regenerative capabilities near the time of birth, but not skin. Significant loss in the skin's regenerative power occurs only with advancing age or in certain disease states.

Most skin derivatives are produced from the ectodermal stratum germinativum. Many structures, such as hair follicles, the sebaceous glands associated with them, mammary glands, and sweat glands, extend deeply into the dermis and even through the dermis into the hypodermic fat (Figure 3-1). Other structures such as fingernails and toenails are limited to the surface.

The Histology of Skin

The *epidermis* is the tough outer surface of skin (Figure 3-2). It provides an almost impenetrable layer protecting the organism from mechanical injury, preventing excessive fluid loss (or gain), and acting as a barrier to pathogenic organisms. The epidermis is produced by a thick layer of dividing cells, the *stratum germinativum,* particularly the basal cell layer. This layer resides immediately above the dermis. Its cells divide, accumulate, and mature toward the outside. The epidermis has no blood vessels of its own and receives its nutrition from a capillary bed located in the superficial dermis below.

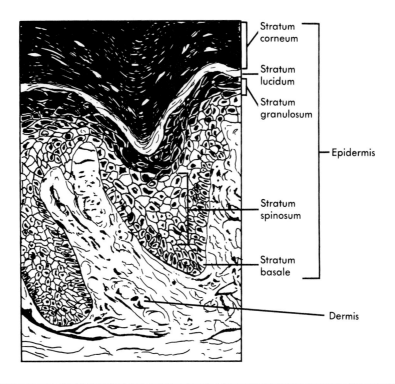

Figure 3-2   The epidermal portion of skin is formed from the stratum germinativum; the much thicker dermis develops from mesoderm. Note the melanocytes, or pigment cells, tucked into the germinativum layer. While ectodermal in origin, they come from a different line of cells than does the epidermis.

The first cells produced from the basal cell layer form the *squamous cell* layer, or *stratum spinosum*. These cells also divide and as they mature they produce keratohyalin granules in their cytoplasm. Keratohyalin is a precursor of keratin, a protein that contributes to the waterproofing of skin. At this point the cells are called the *stratum granulosum*. As these epidermal cells age and are pushed farther from their source of nutrition, they lose their nuclei and cytoplasmic granules and begin to die. In hairless skin they pass through a translucent stage called the *stratum lucidum*. In both hairless and hair-bearing skin the cell remnants finally become keratinized. This outermost layer of the skin is called the *stratum corneum,* or horny layer of skin. In essence this process constitutes cellular aging even in young skin. The skin as a tissue ages as we age; senescence in skin is described later in this chapter.

The brown pigment of the skin is produced by melanocytes located among the cells of the stratum germinativum. Melanocytes have very long, branching processes that weave between the other cells. The amount of melanin, the dark brown pigment produced by melanocytes, varies with the heredity of the person and with other factors such as exposure to the ultraviolet rays of the sun. People whose origins are closer to the equator tend to inherit genes for production of more melanocytes than do people from more northern latitudes. Yellow pigment results from carotenoids, particularly visible in the hypodermic fat of people of the Orient. Other pigments may be deposited in the skin of people with certain liver or blood problems. This is one way that the skin may "mirror" problems deep within the body.

The *dermis* provides support and nutrients for the epidermis. It is much thicker than the epidermis and lies directly beneath it. The dermis is composed of two layers: the upper papillary layer and a lower, thicker, reticular layer. The papillary dermis is about one fifth of the total thickness of the dermis and associates with downward projections of the epidermis to form papillae. Dermal papillae are composed of loose connective tissue, nerve endings, and blood vessels that provide flexibility, sensation, and nutrients for the skin. Deeper in the dermis, the loose connective tissue merges with the denser, more rigid supporting portion, the reticular dermis. (The nerve endings found in the dermis will be discussed in Chapter 14.)

The reticular layer, like the papillary layer of the dermis, is a network of interwoven bundles of elastic and collagenous fibers. This deep layer gives skin its overall strength. The reticular layer contains the nerve endings sensitive to pressure. At the innermost portion of the dermis is a complex of blood vessels that supplies the capillary bed of the epidermis and the entire dermis and subcutaneous tissue with oxygen and nutrients. Constriction and dilation of these blood vessels helps to regulate the loss of body heat. Healthy skin color results from the interplay of vascular factors and pigmentation.

The subcutaneous, or hypodermic, layer lies beneath the skin. It is composed of a few collagenous fibers and large amounts of fat. Women tend to have more subcutaneous fat than men. This fat protects the organs beneath and insulates against heat loss. It also contains a large store of calories and fat-soluble vitamins as a reserve for the famines that still affect much of the world's population. Elastic and collagenous fibers pass through the hypodermic layer and attach the skin to the underlying structures.

THE AGING SKIN
General Skin Changes and Ultraviolet Light

We are all too aware that skin changes as it ages. We usually see increased wrinkling and fragility as time passes, along with decreased rates of healing, irregular pigmentation, dryness, and a variety of skin lesions. The severity of these changes depends on the heredity of the aging person and the rigors of life to which the skin has been exposed. The most severe of these rigors is usually exposure to the sun's ultraviolet rays. The most damaging effect of U.V. rays is probably the mutation of genes that produce cellular proteins. They can also denature proteins already produced. The face, hands, and neck usually show the greatest damage while the skin of the back and buttocks usually remains baby-smooth. Age-related and sun-induced changes ultimately disrupt the homeostatic skin functions to varying degrees and often create problems.

Light-skinned people who have blue or green eyes and/or red hair are particularly affected by ultraviolet light because they have a smaller amount of protective melanin in the skin than do darker-skinned people. Their skin becomes paler and less uniform with age, and experiences more freckling and uneven tanning. The skin shows irregular pigment deposits as well as areas of hypopigmentation resulting from the irregular distribution of melanin in aging skin. These solar-induced changes are best delayed by avoiding excessive ultraviolet light during the early years of life. People should be aware of the potential effects of damage from ultraviolet light from any source and avoid unnecessary exposure. Using sun hats, long sleeved garments, and a protective substance containing the sunscreen paraaminobenzoic acid when one must be out of doors will decrease the amount of ultraviolet radiation absorbed by the skin.

The number of melanocytes in the skin declines with age. This decrease may be as much as 80% between ages 27 and 65. However, those that remain tend to become larger and more active. The pigment they produce becomes unevenly distributed and results in pigment plaques or "age spots." Aggregations of melanocytes result in small, benign, pigmented moles called nevi.

Changes in Skin Thickness

Figure 3-3 shows the thinning of skin that occurs with age. This change is manifested by flattening of the epidermal-dermal interface and possibly by decreased cell division in the stratum germinativum and dermis. Elastic and collagenous fibers in young skin are intact and fine but they become progressively more coarse and less elastic with age as they become cross-linked. This decreases the flexibility and perhaps the thickness of skin and promotes wrinkling.

Many other factors contribute to wrinkling. Changes in blood vessels may impede circulation to the skin, affecting moisture content and overall skin health. Loss of skeletal and muscle mass and the loss of subcutaneous fat also promote wrinkling. Although there appears to be a general decrease in the rate of skin production, this is not necessarily even and skin may become excessively thick in patches and thin and fragile elsewhere. Many elderly people will have a variety of skin tumors resulting from greatly increased rates of division of a few cells. Tindall, commenting on this irregular increase in cell division, stated that "aging skin would appear to be in a race to see if new growths can win before atrophy cancels the entire event."

Temperature Regulation

One of the most important homeostatic mechanisms damaged with age is the skin's ability to prevent heat loss. This results primarily from the loss of subcutaneous fat but other probable causes may include decreased basal metabolic rate or decreased ability to regulate blood flow. Blood flow to the skin may be changed because of inelastic compression of blood vessels resulting from collagen fixation and contraction, atherosclerosis, or even by emotional factors. A person will lose body heat any time the air temperature is lower than 98.6° F (37° C). This typically results in discomfort at normal temperatures. It is not uncommon to enter the home of a elderly man on a warm day and find him wrapped in a heavy sweater with the thermostat turned up to 80°. Intolerance of cold may lead to decreased social or physical activity, much to the detriment of physical and emotional health.

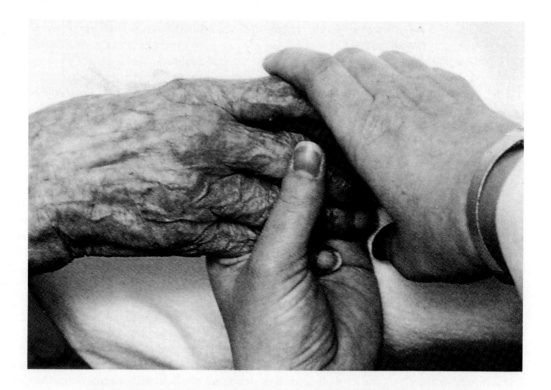

Figure 3-3   Young and old skin differ greatly. Young skin is typically thicker, more elastic, and less fragile than old skin from the same part of the body. This change has been attributed to decreased cell reproduction, collagen cross-linking, and free water loss but the extent of influence of each factor is still in question.

Aging Hair and Hair Production

As skin ages, so do those structures produced by the skin, the skin derivatives. These changes can be observed in hair, sebaceous glands, sweat glands, mammary glands, and fingernails and toenails.

When hair ages, it declines in both quality and quantity. Graying results from the loss of pigment and cellular material. It occurs more rapidly than the aging of the rest of the body in some people and more slowly in others and is obviously under separate genetic controls from other aspects of aging. Gray hairs are usually more fragile than normal (the diameter of the hair shaft decreases) and may become kinky. Graying usually begins at the temples and spreads toward the crown of the head. Unlike scalp hair, graying of the axillary (underarm) hair seems to occur more in line with aging of the rest of the body and may be brown in a 50-year-old who already has white scalp hair. The pattern of hair loss is also genetically determined and may differ from the pattern of graying. In general, scalp hair density declines from about 615 per cm^2 at age 25 to around 485 at age 50. Fine, brittle hairs may remain in the follicles of seemingly bald people but they are of poor quality and break off just above the skin. Often unnoticed is the loss of axillary (underarm), pubic, and other body hair that also occurs. Hair loss from the extremities may be quite complete.

Among women, the increase in facial hair of a bristly quality may be more noticeable than among men. This increase may be due to changes in the androgen/estrogen ratio. In men, as scalp and body hair decrease, there is often an increase in coarse hair in the ears, nose, and eyebrows. These changes may be of concern to aging people of both sexes and raise questions about body image and attractiveness, sexuality, and self-esteem.

Aging Sebaceous and Sweat Glands

Most sebaceous (oil) glands are associated with hair follicles but there are also sebaceous glands on the lips, glans penis, labia minora, and an area of the eyelids. These glands help to keep the hair from becoming dry and brittle and provide a fine coat of oil on the skin that helps prevent excessive water loss. Secretions of most sebaceous glands decline with age but some may increase. Elderly men often suffer from seborrheic dermatitis, an inflammatory skin condition resulting from trapped irritating natural fatty acids that compose the oils. Others suffer from asteatosis, a severe loss of sebaceous secretion common in the elderly of both sexes. The overall effect of sebaceous secretion loss is more brittle hair and more fragile, dry skin.

Functional sweat glands decline in number with age. Studies of axillary (underarm) sweat gland response showed that 60 to 75-year-olds had only a little more than half the sweat production when compared with 15 to 50-year-olds. In spite of this, tolerance of high temperatures does not seem to decline significantly. This may partially be a result of decline in the awareness of high temperatures due to changes in sensory function.

Aging Fingernails and Toenails

Growth of fingernails and toenails declines with age. The quality of the nail produced also declines. The thumbnail of the typical 30-year-old grows at a rate of about 0.83 mm per week but the thumbnail of a typical 90-year-old grows only about 0.52 mm per week. The fingernails and toenails of aging people also typically change in color, consistency, and configuration. Fingernails may become brittle and peel. Toenails may become yellow with age and lose their translucence. They also frequently become thickened and distorted. These changes are often a reflection of diet or gastrointestinal function as much as aging germinal tissue.

Toenails can become ingrown or cause other foot problems if they are not properly trimmed. The skin of the feet becomes dry, thin, and inelastic. Dryness can lead to minute fissure formation and infection. People may be unable to maintain foot care due to other age-related changes. Nail trimming and daily hygiene practices may be neglected, setting the stage for serious mobility problems.

PROBLEMS ASSOCIATED WITH AGING SKIN
Skin and Self-Perception

The elderly population of the United States has become increasingly concerned with a variety of skin conditions. Most of these are not life-threatening but result in discomfort or what is perceived as a less attractive appearance. These conditions along with the normal loss and whitening of hair and the wrinkling of the skin may so affect some persons' opinion of themselves that they may withdraw from society or at least decrease their public activity. Modern cosmetics and even cosmetic surgery have helped many such people to maintain a youthful appearance and a more positive state of mind.

Conditions and Diseases Affecting Aging Skin

Pruritus. The most common complaint associated with skin among elderly people is pruritus, or itching. This may result from dryness resulting from loss of sebaceous secretions as discussed earlier, or from various underlying conditions such as anemia, liver disease, primary dermatologic disease, or infestation by lice, mites, or fungi. The relative lack of moisture and oil in aging skin is a source of both real discomfort and cosmetic embarrassment. Decreased sebaceous gland production may contribute to dry, itching skin but this decrease does not fully explain the occurrence of pruritic skin.

The effects of water loss on aging skin can be minimized in two ways. First, aged skin does not require a daily full soap and water bath. Soap removes skin oils and need only be used in the axillary and anogenital area. The remainder of the skin can be adequately cleansed with plain water. After bathing, the application of an oil or cream to wet skin will assist in minimizing moisture loss. Secondly, adequate fluid intake will assist in cellular hydration. Pruritus leads to scratching, which leads to further itching and subsequent infection. It is usually possible to break the cycle by removing the cause or by treatment with antiinflammatory agents and keeping the fingernails closely trimmed.

Epidermophytosis. This is a technical name for fungal infections. Fungal infections have a variety of common names ranging from ringworm to athlete's foot. Regardless of what they are called, they occur more frequently with advanced age. Pruritus is a common symptom of superficial fungal and yeast infections. Often these infections occur in the axillae of the arms, anogenital regions, and under the breasts. Antifungal agents available today are often effective in treating these problems.

Mites and Lice. When people are put in close proximity to each other as occurs in some nursing homes, infestations of mites or lice can be difficult to control. The bites itch intensely and often become inflamed and secondarily infected from scratching. Removing the offending parasite and treating the remaining symptoms usually works.

Urticaria. Certain allergens entering through the skin, gastrointestinal tract, or lung may cause urticaria, a vascular reaction producing red, elevated patches of skin that itch (hives) (Figure 3-4). For example, consumption of certain foods may produce this reaction in some people. Urticaria is a dermatologic manifestation of the body's immune response. When allergies result from contact with allergens such as soaps, wool, or other clothing components, even the metal of a watch band, the condition is called contact dermatitis. Allergic reactions to drugs are common, particularly in the elderly because many are receiving complex combinations of drugs to treat a variety of different problems. While allergic inflammation may respond to antiinflammatory agents, the most effective lasting treatment is to eliminate the offending allergen. This topic is discussed further in Chapter 9.

Milia. These small white nodules are retention cysts containing the products of sweat or sebaceous glands. They are associated with excessively dry skin in the elderly and with poorly developed sweat glands in the very young. An outbreak of milia in either age group is known as prickly heat. The itching that accompanies this benign condition can generally be managed by lowering the temperature and applying cold, wet cloths.

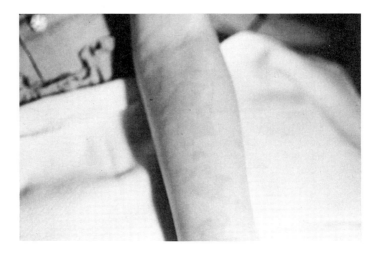

Figure 3-4   Urticaria, or hives, results from a dermal vascular response to an allergen. (See also Color Plate 1, *B.*)

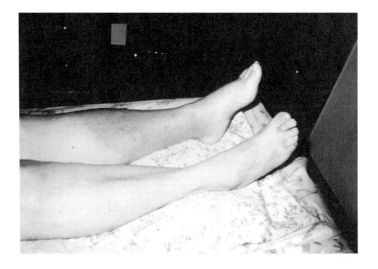

Figure 3-5   Stasis dermatitis often appears on the leading edge of the lower legs as a result of vascular insufficiency. Severe cases may ulcerate. Even when stasis dermatitis has been cured, iron compounds left by extravascular blood may leave the skin discolored. (See also Color Plate 1, *C.*)

Stasis Dermatitis. When a decrease in circulation to the skin occurs, the resulting lack of oxygen and nutrients may cause itching and later ulceration. This condition sometimes occurs with diabetes and other disorders that compromise peripheral vascular competence. It typically develops in the lower legs and may result in large, wet, ulcers infected with bacteria and covered with exudative debris (Figure 3-5). This condition as well as many of those just discussed may cause pruritus. It is important to prevent scratching that could produce further skin damage and infection.

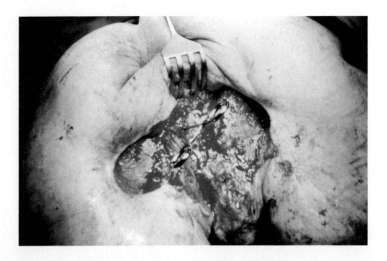

Figure 3-6   Loss of circulation resulting from sustained pressure may produce decubiti, or pressure ulcers. These often result when immobile patients are not moved frequently enough to prevent the circulation loss that occurs when the weight of the body presses on tissue for too long. (See also Color Plate 2, *A*.)

Decubitus Ulcers. Decubitus ulcers, or decubiti (commonly known as pressure ulcers or bedsores), also result from circulation loss (Figure 3-6). They occur in immobile patients when pressure from lying in the same position for a period of time compresses blood vessels, preventing delivery of oxygen to the skin and underlying tissues. The deprived tissue dies and sometimes becomes secondarily infected. For this reason immobile patients must be turned regularly to allow circulation to redevelop in the compressed tissue. It is imperative for care givers to carefully monitor and change the position of immobile patients on a regularly scheduled basis to prevent decubiti. Prevention is by far the best medicine for decubitus ulcers because once decubiti develop, the healing process is very slow. Both decubiti and stasis ulcers are very difficult to eliminate because the underlying causes are chronic and likely to persist.

Herpes Zoster. Also known as shingles, herpes zoster is a disease produced by the virus that causes chicken pox in children. It occurs primarily in older people who have had chicken pox in their youth. One or two percent of the elderly population will have shingles in any one year. The outbreak appears as a series of skin lesions paralleling the pathway of a nerve. These lesions may be excruciatingly painful, wet, and may impair the individual's rest, sleep, eating, elimination, mobility and socialization (Figure 3-7). Shingles is particularly dramatic when it involves the trigeminal nerve and the lesions occur down the forehead to the nose and even onto the cornea of the eye. The lesions begin as vesicles (blisters) containing clear fluid, rupture as they mature, and form craters. They later disappear, usually without permanent disfigurement. A typical attack of shingles lasts 2 to 4 weeks. The pain, however, may persist long after the lesions have disappeared, even for years.

Infection. Response of the immune system to infection declines with age (see Chapter 9). Serious systemic infections may be heralded by mental confusion rather

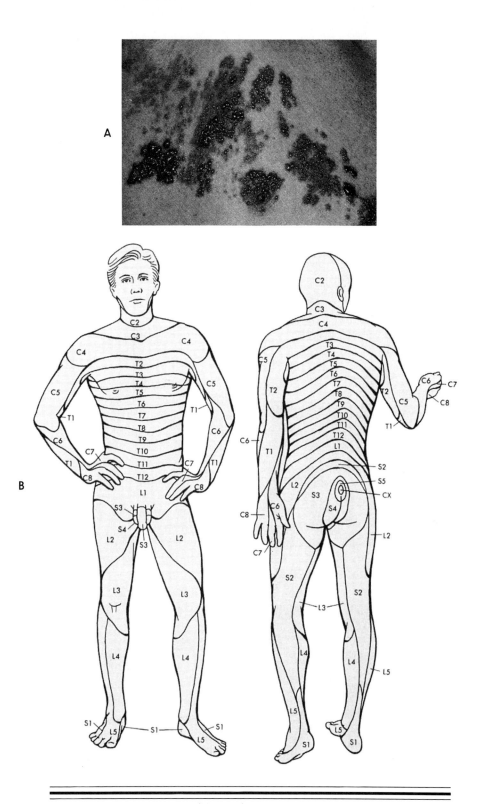

Figure 3-7 Herpes zoster (shingles) results from reactivation of the chicken pox virus, producing lesions along spinal and cranial nerves. The photograph **(A)** shows development of shingles along the path of a thoracic spinal nerve, a dermatome. (See also Color Plate 2, *B*.) A dermatome map of the human body is shown in **(B).**

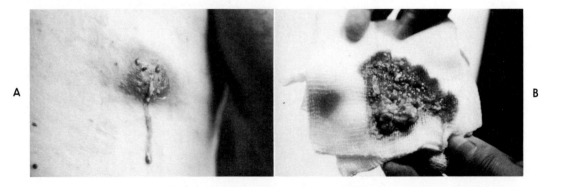

Figure 3-8   The sebaceous cyst shown in **(A)** had become infected. (See also Color Plate 2, *C.*) The exudate it produced when incised is shown in **(B).** Such infections are more common when the immune system weakens with advanced age.

than by a highly elevated temperature. In old age, bacterial and fungal infections occur with greater frequency and severity than in youth. Even the herpes zoster virus probably erupts in response to a weakened immune system. Figure 3-8, *A* shows a sebaceous cyst on the back of a 74-year-old man that had become infected with common skin bacteria. Figure 3-8, *B* shows the large amount of fatty material, pus, and blood that flowed from this cyst when it was excised.

Benign Skin Lesions

The elderly population of today is especially concerned about cancer, partially because of the increased publicity the disease has received and partially because many forms of cancer are more abundant than they were during past generations. It is important for them to realize that the vast majority of lumps and bumps one finds are benign (harmless). All of these can be removed if they cause discomfort or become disfiguring. Recognition that a lesion is benign can prevent an immense amount of worry. Common benign lesions, some of which become more noticeable with age, are identified in Table 3-1 and shown in Figures 3-9 and 3-10.

Malignant Skin Lesions

Malignant lesions may be life-threatening if they spread before being treated or are neglected. They are fortunately less common than benign growths. Most are more likely to occur in older than in younger people. Malignant tumors invade and replace healthy tissue. They can also shed cells, which may be transported throughout the body by the lymphatic and blood vessels. This process, known as metastasis, results in new tumors arising wherever these cells have been deposited. Once extensive invasion or metastasis has occurred, control of the disease (cancer) by surgery becomes much more difficult or impossible and systemic methods such as chemotherapy or radiation must be used. It is therefore important to detect malignant growths in the earliest stages, before extensive invasion or metastasis has occurred. The photomicrograph in Figure 3-11 shows the invading extensions of a squamous cell carcinoma, a moderately common skin cancer. Everyone should be aware of three common types of skin cancer. These are described below and are shown in Figure 3-12.

Table 3-1 Some Common Benign Skin Lesions

LESION	SOURCE	APPEARANCE	LOCATION	COMMENTS
Nevus (pl. nevi) (moles; birthmarks)	produced by melanocytes; sometimes congenital	brown-dark brown; raised; smooth outline	various	Rarely changes to malignant melanoma (see p. 44)
Cherry angioma	Skin blood vessels	bright red, raised, smooth	various	grows throughout life; present in nearly all over 80.
Seborrheic keratosis	excess stratum germinativum activity	raised, sometimes irregular; often become pigmented; thickened skin on top; can be quite large	head, arms, upper body most often	Can become quite large; not life-threatening
Fibroepithelial polyp ("skin tags")	excess stratum germinativum activity	Soft, nonpigmented, usually mushroom-shaped	moist areas; neck, axillae, groin	Slow growing
Solar lentigo ("liver spots") ("age spots")	aggregations of melanocytes	flat, uniformly brown or tan	exposed areas; back of hands	Appear during midlife or later
Senile purpura (ecchymosis)	leakage of blood RBCs into the skin	purplish, under skin surface	various	often seen following minor injury
Telangiectases	loss of elastic support for very small blood vessels	small, abnormally dilated blood vessels; web-like	various	Blanches easily under pressure, unlike other lesions
Actinic keratosis	solar damage to stratum germinativum	hyperkeratotic, often reddened, range from flat to thickened, raised, scaly, warty	sun exposed surfaces	May grow rapidly; in 20% of people with excessive solar exposure, may develop into squamous cell carcinoma
Keratoacanthoma	solar damage to germinativum	raised warty, often ulcerated & hyperkeratotic	face, shoulders	May grow rapidly & looks like a cancer; spontaneously regresses, less common than actinic keratoses

Basal Cell Carcinoma. (Figure 3-12, *A*) This is the most common form of skin cancer and fortunately the least dangerous. Metastasis is rare even with very advanced lesions but invasion and replacement of the surrounding tissues and organs does occur. The lesion is raised, unpigmented, and often has a necrotic, or dying, center with raised margins. The center of an old lesion may therefore be wet or have a scab. Basal cell carcinomas are commonly found on the forehead and other sun-exposed areas and arise from the basal cell layer of the germinativum. In fact, 90% of all basal cell carcinomas occur on the head and neck; the trunk is the next most frequent site. This skin cancer can occur from childhood to old age, but is most frequently seen in male patients greater than 45 years of age who have a history of chronic exposure to the sun. The prognosis for this lesion is excellent if the tumor is removed early. Survival from properly treated basal cell carcinomas approximates 100%.

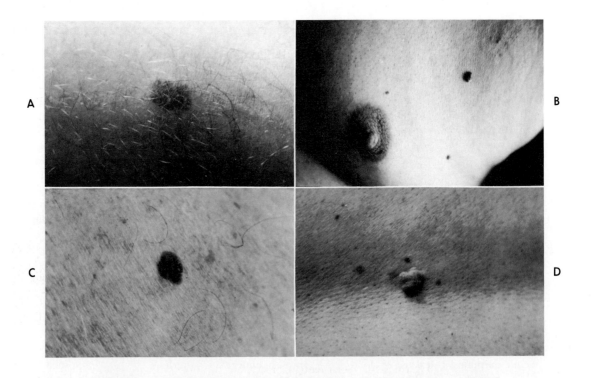

Figure 3-9   There are many types of raised benign lesions: **(A)** nevus or mole, **(B)** cherry angioma, **(C)** seborrheic keratosis, and **(D)** fibroepithelial polyp. (See also Color Plate 3.)

Squamous Cell Carcinoma. The lesion may appear as an irregular, non-pigmented, thickened area of skin and may eventually ulcerate in the center. Although it may occur almost anywhere, it is often found on the face or the back of the hands. If neglected, most of these tumors will invade locally (Figure 3-12, *B*) and some will metastasize, especially those appearing on the lip, tongue, and external genitalia. Metastasis usually occurs fairly late in the life of this malignancy with ample time for the observant person to have the growth removed. Those squamous cell carcinomas arising from actinic keratoses tend to behave differently and are less likely to metastasize. About 95% of squamous cell skin cancers are cured.

Malignant Melanoma. This is clearly the most dangerous and fortunately, the rarest form of skin cancer. Of the deaths in the United States resulting from skin cancer, 75% can be attributed to malignant melanoma. The four distinct forms of malignant melanoma are as follows: (1) The most common form is the superficial spreading melanoma, which can occur anywhere on the body surface. This lesion has a haphazard combination of colors and an irregular shape. The average age of the patient with this lesion is 50 years. (2) Nodular melanomas also occur anywhere on the body and have a wide age distribution. This lesion looks like a pigmented nodule. Invasion of the dermis is often present and there may be lymph node metastasis. (3) Acral (extremity) lentiginous melanoma occurs on palms, soles, nail beds, and mucous membranes. It is usually flat or slightly raised with an irregular pigment pattern and border. (4) Lentigo malignant melanoma is the least common lesion. It is a slowly evolving lesion occurring on sunexposed surfaces of elderly persons. It usually

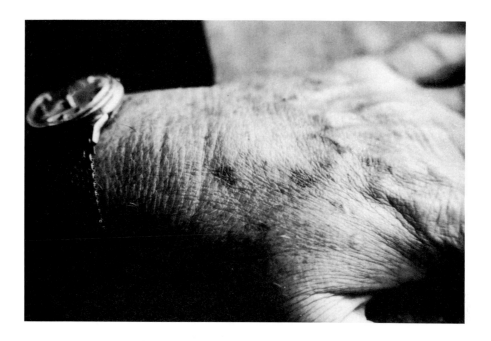

Figure 3-10   Solar lentigo, or "liver spot". (See also Color Plate 4, *A*.)

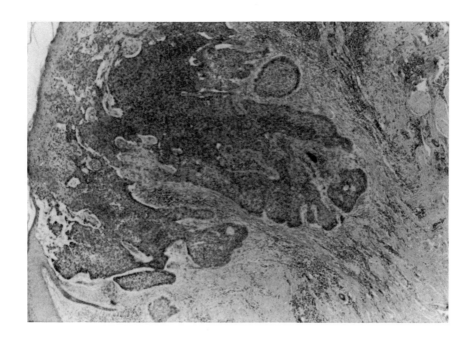

Figure 3-11   Basal cell carcinoma is invasive and but only rarely metastasizes (spreading malignant cells). Note the extensions of this tumor originating in the basal cell layer of the epidermis and projecting deeply into the dermis. (See also Color Plate 4, *B*.)

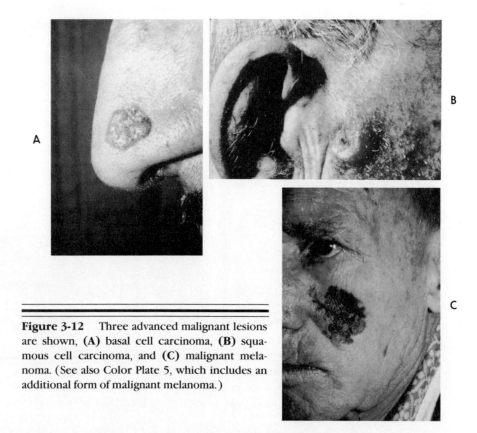

Figure 3-12   Three advanced malignant lesions are shown, (**A**) basal cell carcinoma, (**B**) squamous cell carcinoma, and (**C**) malignant melanoma. (See also Color Plate 5, which includes an additional form of malignant melanoma.)

undergoes many color changes but its slow growth makes it somewhat less dangerous than the others.

Although malignant melanoma (Figure 13-12, *C*) is a rare disorder, it is becoming more common, seemingly because people today tend to expose themselves to the sun to a greater extent than did their ancestors. Decline in the protective ozone layer of the atmosphere may also be a factor. Malignant melanomas are usually darkly pigmented (rarely unpigmented) and often metastasize early. The classic pattern involves a birthmark that darkens suddenly (it often turns blue or black) and begins to grow. Black pigment may spread away from the original growth under the skin as malignant melanocytes rapidly reproduce and generate excesses of melanin. This can result in a "patriotic" appearance to the tumor: red, white and blue. Metastasis occurs both through the lymphatics and the blood vessels. Like other forms of skin cancer, the center of the growth may become necrotic and the lesion may eventually become wet or scabbed. Once metastasis has occurred, the prognosis for malignant melanoma is poor. These cancers are much more common among light-skinned than dark-skinned people and become more common with age. It is important to understand that malignant melanoma can occur in moles. In fact, 50% of these lesions develop from previously existing birthmarks. They may appear as any of the forms just listed, with the possible exception of lentigo melanoma, which appears to develop from a solar lentigo. Development from moles and lentigos occurs not because these lesions are premalignant, but because of the predominance of melanocytes in them.

CASE HISTORIES

❖ Herpes Zoster (Shingles)

G.L. is a 71-year-old woman who noticed a burning and tingling pain that had persisted over her left chest for 1 week. She was admitted to the hospital after developing blisters in the same region. There were no fever, chills, sweating episodes or other evidence of bacterial infection. Her medical history was unremarkable with the exception of mild hypertension treated with a diuretic (thiazide) to reduce blood volume and lower blood pressure.

At the time of admission to the hospital, her vital signs (blood pressure, pulse, respiratory rate, and temperature) were normal. The results of her physical examination were normal except for the skin findings. There were numerous erythematous, or red-based, vesicles covering a narrow strip of the left chest wall. The blisters started under her left arm and ended about 5 inches above her navel. This location corresponds with the distribution of the spinal nerve originating just under the 7th thoracic vertebra. A sensory area supplied by the branches of a single spinal nerve is called a dermatome; in this case it is known as the T7 dermatome. In addition, a few scattered blisters were noted on her arms, chest, back, and abdomen. Some of the vesicles were small; others were large and turgid containing a clear fluid. Other blisters had broken and crusted over. A diagnosis was made of herpes zoster with dissemination.

Herpes zoster is a blistering disease resulting from reactivation of the chicken pox virus and it usually follows a single dermatome. G.L.'s therapy included antiviral medication (acyclovir) administered locally through a cream and systemically through an I.V. (intravenously). Prednisone, an oral antiinflammatory steroid, was given to reduce inflammation. Because the lesions and surrounding areas are very painful, a narcotic (Demerol) was used to control the pain.

During her 8-day hospital stay, new lesions developed while older ones crusted and healed. Most lesions continued to be in the left T7 dermatome but a few scattered new lesions continued to appear. Lack of internal organ involvement was confirmed by routine blood tests and the absence of clinical evidence. Her CBC (complete blood count) showed a mild anemia, which was not investigated further. At the time of her discharge she was still experiencing pain in the left T7 dermatome area.

Comments. Shingles usually remains limited to a single nerve root distribution, or dermatome. In this case the few lesions occurring outside of the dermatome might have suggested a more severe and generalized disease course. Disseminated infections can develop into pneumonia, encephalitis, or hepatitis, and can result in death. Acyclovir was used to prevent these serious infectious complications.

❖ Basal Cell Carcinoma

R.R., a 73-year-old white government worker, was observed to have an ulcerated area 6 mm in diameter on the tip of his nose. The lesion was first noted during hospitalization for major abdominal surgery. During that time, R.R. required a nasogastric tube, which was fixed to the tip of his nose with adhesive tape. Upon removal of the tape, he noticed a small ulcer that did not heal as expected. Although R.R. works mainly indoors, his hobbies include many outdoor activities such as wood chopping and horseback riding. His face is well tanned. Because of the appearance of the lesion and the habits of R.R., the lesion was diagnosed as a basal cell carcinoma. Removal was easily achieved with no apparent complications and the diagnosis was confirmed by histologic examination.

Approximately 2 months later, during a routine checkup, a second lesion was discovered next to the location of the first. This too was a basal cell cancer and was removed with no problems.

Comments. Basal cell carcinoma is the most curable of all the skin malignancies. Unless it is allowed to grow untreated for a long period and to invade the underlying muscles, bones, and nerves, surgical removal is adequate and complete therapy. If undetected fingers of the tumor extend beyond the surgical margins and escape removal, the remaining cells will continue to grow and the lesion will probably recur, as may have been the situation in this case. However, since these tumors are associated with sun exposure, multiple lesions can occur in adjacent similarly exposed areas.

❖ Squamous Cell Carcinoma

H.C., a 63-year-old white farmer, went to his physician with a non-healing ulcer on his lower lip. The ulcer had been present for 6 months, but had recently begun to ooze. Examination confirmed the presence of a firm, ulcerated, raised, hyperkeratotic lesion 1.1 cm in diameter and involving the vermillion (pink) border of his lower left lip. It was also noted that his face was deeply tanned and wrinkled, giving his skin a "weather beaten" appearance. Examination of the mouth, neck, and associated lymph nodes revealed no other abnormalities. The entire tumor was removed and found to be a well-differentiated squamous cell carcinoma extending into the reticular dermis. Tumor cells were not present in the surgical borders. No further therapy was given.

Two years later a routine examination revealed a mass beneath the skin of the neck. It was located in the left anterior portion of the neck next to the larynx, was freely moveable, and measured 2.1 cm. It was surgically removed and found to contain well-differentiated squamous cell cancer. The patient refused further therapy and has remained symptom free for the last 5 years.

Comments. Although not all skin cancers are the result of solar (ultraviolet) damage, many can be traced to excessive sun exposure. This is true for basal cell, squamous cell, and some melanotic tumors. The lower lip receives more direct sun than the upper and has a thin epidermal layer that does not offer as much protection from damaging ultraviolet radiation as does other skin. In general, the better differentiated tumors behave in a less aggressive manner than the poorly differentiated ones. This tumor seems to have confined itself to local invasion and metastasis of a single lymph node. There is no evidence of spread to other organs.

Squamous cell carcinoma is a rather common skin lesion arising from the squamous cell layer of skin. Because it is known that it can arise from an actinic keratosis, it is prudent to have any suspicious lesion biopsied and sent for pathologic examination. These tumors are more aggressive than basal cell carcinomas and may metastasize early, spreading throughout the body and giving rise to more tumors. Those arising on the lip, tongue, or external genitalia are often especially aggressive.

❖ Malignant Melanoma

J.B., a 48-year-old man, went to his doctor to have his hypertension (high blood pressure) evaluated. His physician noticed a slightly pigmented area just below the right scapula; questioning revealed that a pigmented lesion had been removed from that area 4 months previously. That lesion had not been submitted for pathologic examination. Upon discovery of what appeared to be recurrent growth, the entire involved area was surgically removed. Pathologic examination revealed abnormal pigmented cells in the dermis typical of either malignant melanoma or atypical dermal nevus. The remainder of the physical examination and laboratory evaluations were normal with the exception of mild hypertension.

At J.B.'s 4-month follow-up visit, two freely moveable, 1.5 cm subcutaneous masses thought to be lymph nodes were felt in the right axilla, or arm pit. When these nodes were removed they were found to be 2 to 3 times normal size and to contain grossly black areas. The pathology department diagnosed them as metastatic malignant melanoma. Chemotherapy was started. Repeat laboratory testing showed elevated LDH (lactate dehydrogenase, an enzyme that is associated with cell metabolism) and abnormal liver function. One month later the LDH continued to rise while the liver function deteriorated further. Nuclear scans and x-rays revealed developing masses in the lung, brain, liver, and bone. There was no apparent response to the chemotherapy, and the patient died 14 months after his initial visit. The autopsy revealed multiple black and white tumor nodes in the liver, lungs, kidneys, brain, heart, thyroid, adrenals, and lymph nodes. They ranged in size from a few mm to 7 cm.

Comments. Malignant melanoma cells arise from melanocytes and produce some of the most aggressive and difficult to treat malignancies. Isolated malignant cells can be difficult to distinguish from normal cells. It is easier to diagnose lesions when their relationship to normal structures can be seen. Consequently the first biopsy was difficult to diagnose clearly. At autopsy the tumor was found in its pigmented (melanotic) and nonpigmented (amelanotic) forms. LDH is an enzyme that is often elevated in malignancies, especially those associated with rapid growth and necrosis of tumor cells.

BIBLIOGRAPHY

Ackerman AB: *Histologic diagnosis of inflammatory skin diseases,* Philadelphia, 1978, Lea & Febiger.

Agate JM: *Pressure sores.* In Pathy MJS, ed: *Principles and practice of geriatric medicine,* New York, 1985, John Wiley & Sons.

Hanna MJD and MacMillan AL: *Aging in the skin.* In Brocklehurst JC, ed: *Textbook of geriatric medicine and gerontology,* London, 1978, Churchill Livingstone.

Marks R: *Skin disorders.* In Pathy MJS, ed: *Principles and practice of geriatric medicine,* New York, 1985, John Wiley & Sons.

Okun MR, Leon ME: *Gross and microscopic pathology of the skin,* Boston, 1976, Dermatology Foundation Press.

Orlando JC: *Pressure ulcers: principles of management.* In Reichel W, ed: *Clinical aspects of aging,* Baltimore, 1983, Williams & Wilkins.

Slemanowitz VJ, Rizer RL, Orentreich N: *Aging of the skin and its appendages.* In Finch, Hayflick eds: *Handbook of the biology of aging,* New York, 1977, Van Nostrand Reinhold.

Tindall JP: *Geriatric dermatology.* In *Clinical aspects of aging,* Baltimore, 1983, Williams & Wilkins.

4 *The Skeletal System*

RELEVANT ANATOMY AND PHYSIOLOGY
Functions and Structure

A typical human skeletal system (Figure 4-1) is composed of 206 bones, their associated cartilages, and the system of ligaments that holds the bones together, forming joints. The entire skeletal system is derived from mesoderm, one of the three basic embryonic tissues. The skeleton forms the framework that supports the weight of the body. It enables precise movements by providing strategically placed anchoring points for muscles and by providing movable articulations, or joints. Other functions include storage of minerals and fat and providing protection for the brain, spinal cord, and other fragile organs. A function usually ascribed to bone is the manufacture of blood; this will be discussed in Chapter 9.

Human bones are classified by shape (long, short, or irregular) and function. Long bones are located in areas requiring larger movements for strength or speed, such as in the arms or legs. Short bones with many articulations, like those in the hands, are located where fine, delicate movements are needed. The cranium protects the brain with an egg-shaped set of bones. The spinal cord is protected by a series of hollow, irregular bones with many surfaces for muscle and bone attachment.

Bone tissue can be categorized as either dense or spongy. The outer portion (cortex) of a typical long bone is composed of dense, compact bone and the inner portion is spongy, trabecular bone. The densest bone is present in high stress, weight-bearing, or protective areas such as the diaphysis, or shaft, of long bones. This hard, cortical tissue contains approximately 99% of the calcium of the body. Spongy or trabecular bone is relatively soft and contains cavities filled with the red marrow that is responsible for the hemopoietic, or blood manufacturing, function of bone. In adults red marrow is found primarily in the sternum, each end, or epiphysis, of the long bones, and in the ridges and crests of the ilium and tibia. The trabecular areas of other bones are filled with fat in adults. The end of a femur (thigh bone) showing both dense and spongy bone is shown in Figure 4-2.

The skeleton is separated into axial and appendicular portions. The *axial skeleton* comprises the skull, vertebrae, ribs, and sternum. This central bony axis of the body is flexible while affording stability and support to the head, trunk, and abdomen. The vertebrae and their separating disks and spinal nerves often develop problems with age; therefore it is useful to know the kinds of vertebrae and their numbering system. There

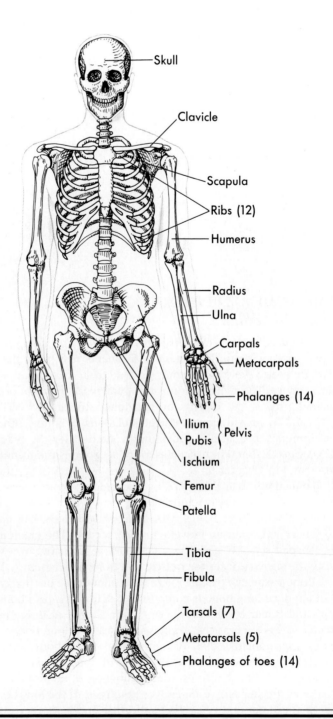

Figure 4-1   The human skeleton provides efficient support and attachment points for muscles but many of its bones are subject to the ravages of age.

are seven cervical, or neck, vertebrae and they are designated C1 to C7. The ribs are attached to the 12 thoracic vertebrae designated T1 to T12. The lumbar or lower back vertebrae are numbered L1 to L5. The sacrum consists of five fused vertebrae associated with the pelvis. Attached to the end of the sacrum are four vestigial vertebrae that make up the coccyx. The vertebral column also provides an anchor for the movements of the

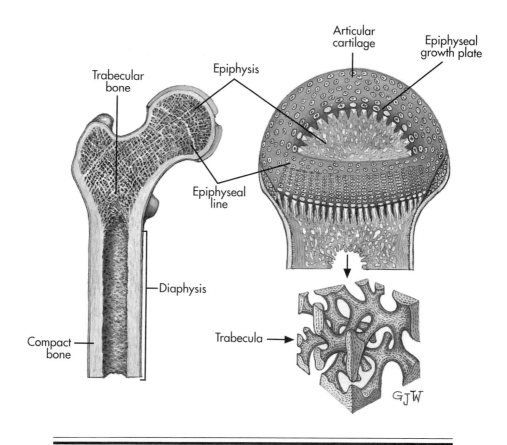

Figure 4-2   The head and neck of a femur showing dense and trabecular bone. Broken hips in the elderly commonly involve fracture of the femoral neck.

appendicular skeleton.

The *appendicular skeleton* comprises the bones in the upper and the lower extremities. The pectoral, or shoulder, girdle consists of the clavicles and scapulae. The long bones of the arms attach to the pectoral girdle. The pelvic, or hip, girdle consists of three fused, flat bones connected to each side of the sacrum. The uppermost of the leg bones, the femur, fits into the acetabulum, or hip socket. The hip is one of the most troublesome joints as the body ages, because it is subject to both fracture and a wear-and-tear form of arthritis.

Bone Growth and Composition

Long bones such as the femur, tibia, or humerus lengthen from birth through the adolescent years. This growth originates in both directions from cartilage plates, contributing some growth to the epiphyses on the ends and more rapid growth to the diaphysis in the middle. This center of growth is called the metaphysis or the epiphyseal cartilage plate. (See Figure 4-2.) Influenced by increased levels of sex hormones after puberty, the cartilage becomes quiescent. At this point the diaphysis and epiphyses fuse, ending growth in bone length. Only a thin epiphyseal line remains where the metaphysis once was. This process of bone maturation is considered a result of aging even though it is usually complete by age 20.

Bone, or osseous tissue, has both inorganic and organic components. Most of the inorganic material consists of calcium salts (about 70% of bone weight) deposited on the organic (protein) matrix (about 30% of bone weight). Magnesium, potassium, sodium, and carbonate ions are also present. As organisms live and are exposed to different environmental factors, their bones may accumulate undesirable compounds such as radioactive strontium. A lengthy accumulation of radioactive minerals in bone may result in bone cancer or in damage to the blood-forming tissue of the marrow.

About 95% of the organic matrix of bone is composed of collagen fibers arranged primarily along the axis of stress. These fibers form the framework for the deposition of calcium salts in the form of hydroxyapitite crystals and are responsible for the tensile strength of bone. The rest of the organic matrix is a mucoprotein mixture generally referred to as "ground substance." Cartilage and connective tissues such as ligaments are composed primarily of the protein collagen.

The Histology of Bone

Because bone is a dense tissue, its fine structure is dictated by its blood supply. Blood enters bone through a canal system containing arteries and veins. The largest blood vessels, entering from the marrow cavity or from the outside, pass through Volkman's canals. Branches from these vessels pass through smaller Haversian canals. Each Haversian canal is surrounded by rings of inactive bone cells, or osteocytes. The osteocytes are surrounded by bone they laid down when they were active osteoblasts (discussed below). Each set of osteocyte rings and its concentric lamellae of surrounding bone is called an osteon (Figure 4-3).

The only communication between osteocytes and the blood supply of the Haversian canals occurs in tiny extensions of the osteocytes, through small channels called canaliculi. This poor circulation leaves the osteocytes with limited levels of nutrient and is at least partly responsible for their low activity rate. Poor circulation makes it difficult

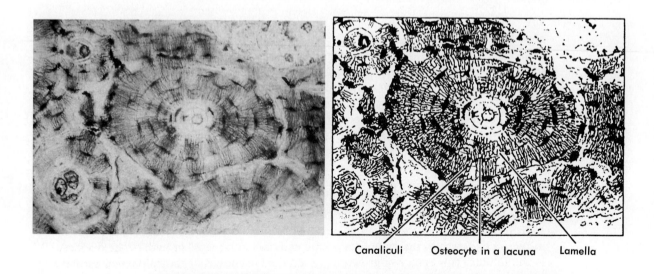

Canaliculi   Osteocyte in a lacuna   Lamella

Figure 4-3   The greatest concentration of osteoblasts and osteoclasts is found inside the endosteum lining the marrow cavity. In the solution channel osteoclasts are dissolving bone. The osteons are old solution channels filled with new bone.

for antibiotics to penetrate and control bacterial infection when it occurs in bone. Some of these infections last for years and result in progressive bone destruction despite appropriate treatment. Such infections are called osteomyelitis.

The outside of bone is covered with a membrane called the periosteum. A corresponding membrane called the endosteum lines the marrow cavity. The periosteum is associated with many pain nerve endings. A relatively minor bruise can cause a painful swelling. A concentration of bone-producing osteoblasts and larger, bone reabsorbing cells called osteoclasts resides just inside the endosteum. Osteoblasts arise from mesenchymal stem cells and have a single nucleus but osteoclasts have more than one nucleus and arise from monoblasts, a type of embryonic white blood cell. A smaller number of osteoblasts and osteoclasts reside under the periosteum.

Bone Remodeling and Blood Calcium Regulation

Bone gives the impression of being a solid, permanent substance. This is misleading, however, because bone is continually being absorbed and replaced. The skeleton is in a constant process of remodeling due to this reabsorption and new bone formation. This process is a normal bone activity and continues throughout life. It usually produces efficient bone structure as it responds to the mechanical stresses applied by daily activity. In addition, calcium can be quickly mobilized from bone by activation of bone absorption, or it can be stored by new bone formation.

When blood calcium is low, secretory activity of the parathyroid glands increases. Under the influence of parathyroid hormone, osteoclasts dissolve bone and release the calcium and other components into the blood. Figure 4-3 shows the formation of a solution channel as a group of osteoclasts dissolve their way into dense bone. As the concentration of calcium in the blood becomes elevated, the parathyroid glands slow or terminate their secretion of parathyroid hormone.

Under the influence of the thyrocalcitonin, osteoblasts remove calcium from the blood and deposit it in bone. Thyrocalcitonin, produced by the "C" cells in the thyroid gland, increases the rate of osteoblast formation and slows osteoclast activity. This alters the balance of the system toward bone deposition. It is released by high blood calcium and results in a lowering of the blood calcium level. Therefore, when osteoblastic activity exceeds osteoclastic activity, new bone deposition exceeds bone destruction. The result of this balance is a fluctuating turnover of bone and more importantly, a mechanism for regulating blood calcium.

Excessive amounts of calcium in the blood produce lethargy and muscle weakness. The interaction between the parathyroid hormone and thyrocalcitonin mechanisms maintains a blood calcium level that fluctuates within the physiologic tolerance of the body. Throughout this process calcium continues to be lost through the urine. It is therefore necessary to at least match the urinary loss with dietary absorption in maintaining the normal blood calcium of 8.5 to 10 mg/dl. Calcium absorption from the intestine depends on the availability of active vitamin D. In adults, a deficiency of vitamin D, or sunshine that is required to activate vitamin D precursors in the skin, will result in decreased calcium absorption from the intestine and ultimate bone loss. In healthy people these processes occur unnoticed. The serum calcium level remains reasonably constant while bones remain strong through the complex bone formation and reabsorption processes just described. Should the level of blood calcium decline below critical concentrations, the patient may go into tetany, a condition of severe, sustained muscle contraction. This condition ultimately involves breathing muscles and causes death if not corrected.

Cartilage

Cartilage is a specialized connective tissue in the skeletal system. It is smooth, white or yellow, and is a resilient supporting tissue. The distinctive properties of cartilage include its limited vascularity and low metabolic rate, its capacity for continued growth, its rigidity, and its resistance to compressive and shearing forces. There are three types of cartilage: hyaline, fibrous, and elastic. *Hyaline cartilage* is the bluish white cartilage found covering the ends of bones making up synovial joints, the nasal septum, the ends of ribs, and the rings of the trachea. *Fibrous cartilage* is a white cartilage that is particularly resistant to tension and is found in the symphysis pubis, knee ligaments, and the intervertebral discs. *Elastic* (yellow) *cartilage* is found in the epiglottis and in the outer ear. Because of the part they play in joints, hyaline and fibrous cartilage are most important in discussion of the skeletal system.

Hyaline cartilage serves as a smooth surface for articulating bones. It covers the ends of bones as they meet in a joint, providing a smooth, lubricated, wear-resistant surface. The thickness of articular cartilages varies from 1 mm to 7 mm from the smaller to the larger joints. Unlike cartilage involved with the formation of bone, it does not ossify with maturity.

Hyaline cartilage is an avascular (without blood vessels) tissue and it depends on the synovial fluid and the vascular synovial membrane, the perichondrium, for nutrition.

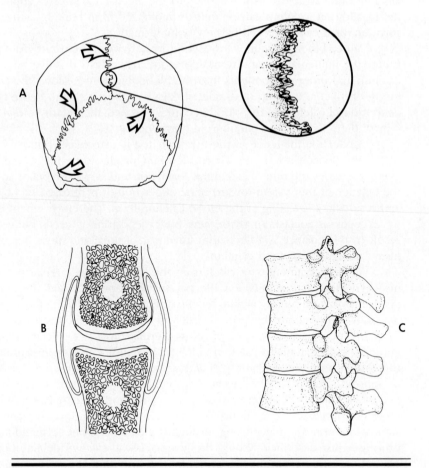

Figure 4-4   The three groups of joints: **(A)** immovable (suture); **(B)** freely movable synovial joint; **(C)** partially movable joint (vertebrae).

The matrix of hyaline cartilage is composed primarily of bound water with collagen, and hydrophilic proteoglycans in combination with hyaluronic acid. Proteoglycans have the unique ability to release water when compressed and to rebind it when the pressure is released, giving the cartilage its resiliency. In weight-bearing joints, this property allows for a "pumping" action that provides the cartilage with nutrients. Therefore, if a major weight-bearing joint is immobilized, the cartilage becomes poorly nourished and begins to atrophy. Atrophy will continue until joint motion and weight bearing are resumed. Maintaining mobility for the health of joints becomes particularly important in the day-to-day life of elderly people with skeletal problems. Destruction of the articular cartilages and the arthritis and joint failure it produces affects the lives of hundreds of millions of people.

Joints

Joints are the contact points between bones. However, not all joints permit free movement. They are divided into three groups: *fibrous* (fixed or immovable) *joints, cartilaginous* (slightly movable) *joints,* and *synovial* (freely movable) *joints.* Figure 4-4 shows examples of the three groups.

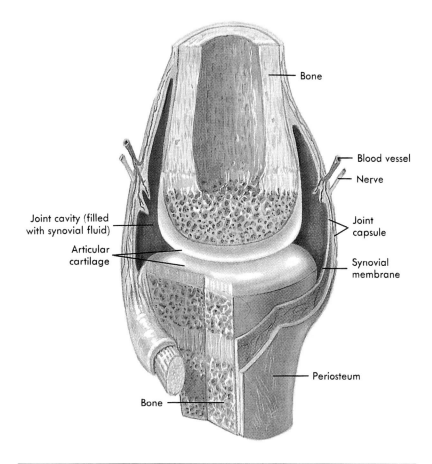

Figure 4-5   Note that synovial fluid fills the space between the articular cartilages. This fluid is secreted and absorbed by the synovial membrane and both nourishes and lubricates the cartilages.

Synovial joints occur where two opposing bones are covered by hyaline articular cartilage and are joined by a fibrous tissue capsule enclosing a joint cavity containing synovial fluid (Figure 4-5). The bones are linked by the fibrous capsule but are separated by the articular cartilages. A thin layer of cells, the synovial membrane, lines the entire joint cavity except for the contact surfaces of the articular cartilages. It secretes and absorbs synovial fluid. As mentioned earlier, the synovial fluid provides both lubrication for the joint and nourishment for the articular cartilage. The fibrous joint capsule surrounding synovial joints is composed of ligaments that extend from one bone to the other and encase the entire joint.

Ligaments and Tendons

Ligaments connect one bone to another and are connective tissue structures composed of collagen. Tendons connect muscle to bone or to other structures. Ligaments and tendons, like the majority of cartilages, are supplied with oxygen and nutrients passively rather than directly through a network of blood vessels. They are more closely associated with blood vessels than is cartilage, however. Because of the reduced circulation, infection can be as difficult to control in these structures as it may be in bone.

THE AGING SKELETAL SYSTEM
General Skeletal Changes and their Impact on Stature

Changes occur in the skeletal systems of older adults that can affect their ability to meet the requirements of daily living. Changes in the shape and contour of joints occur frequently. The joints tend to appear larger than the surrounding tissue, and joint contours may become irregular. This may result from reactive bony growth secondary to wear and tear, or to the atrophy and decline in muscle mass that occur with aging. The articular hyaline cartilage becomes thinner with age, allowing bones to slide over one another with an audible click or grating sound. Because of collagen cross linking, joints are also less flexible in the elderly. Overall aging changes may cause slower, more deliberate movements; balance may be harder to maintain. The range of motion of joints may decrease, and movements may become jerky and even painful.

As individuals age, bone density changes. Although growth in length of bone ceases after fusion of the diaphysis and epiphysis, slow growth in width normally continues. This growth declines as an individual ages, and starting at about age 40 there is progressive bone loss as osteoclastic activity exceeds osteoblastic activity. This decrease is characterized by gradual reabsorption of the inner surface of the long and flat bones, coupled with a decrease in the deposition of new bone on the outside surface. The long bones appear to be externally enlarged but in fact are internally hollow and the vertebrae are thinned. When the loss is so great that it can be seen radiologically, the patient is said to have osteopenia, or clinical bone loss. Some of this decline in bone mass is probably attributable to diet. Until age 50 maintenance of the adult skeleton requires about 100 grams of calcium per year to replace bone that has been reabsorbed and lost in the urine. This replacement calcium is often not available in the diets of elderly people. Vitamin D deficiency may occur as well, interfering with calcium absorption by the intestinal tract. Vitamin D is added to milk to help compensate for such deficiencies. This is of little help however, if the person losing calcium does not drink milk. This is often the case with elderly people.

Posture and structural changes occur primarily because of calcium loss from

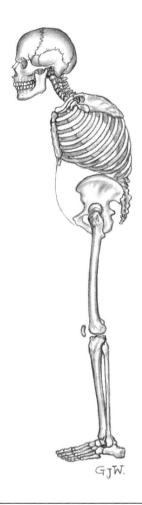

GJW.

Figure 4-6 Skeletal changes with age.

bone and because of atrophic changes in muscles and cartilage. The stature of aging persons also changes. A loss in height of about 1.5 cm occurs for every 20 years of age after age 40. As stature decreases, long bones, especially in the upper extremities, take on a disproportionate length. The trunk shortens as the intervertebral disks become thin; the shoulder width decreases and the chest and pelvis widen. Changes in the vertebrae, such as vertebral collapse, or their disks can produce increased lower spinal curvature, lordosis, kyphosis (often described as hump-back), or scoliosis, a sideways or lateral curvature of spine. In the very old, often many of the limbs cannot be completely extended because of decreased ligamental flexibility. This lack of extension results in a permanent "stooped" appearance and accentuates short stature (Figure 4-6). Regular physical activity may inhibit development of these conditions.

Slippage or herniation of one or more intervertebral discs is often associated with both unusual stress and age-related changes (Figure 4-7). These discs are composed of fibroid cartilage with a soft center. They absorb shock and add flexibility to the spinal column. Sustained or extreme uneven pressure can cause the semisolid center to shift to one side, or even herniate through the outer connective tissue. If this shift causes pres-

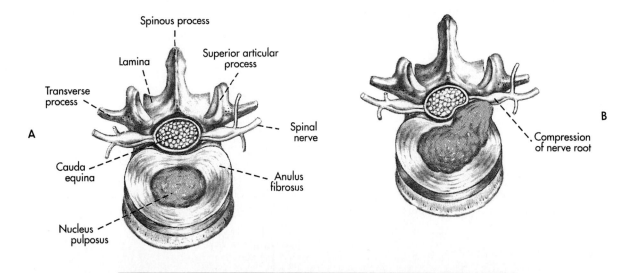

Figure 4-7   **(A)** A normal intervertebral disc and, **(B)** a herniated disc. A combination of unusual stress and age-related degeneration of the fibroid disc capsule contributes to slipped and herniated discs. These may cause considerable back pain and may contribute to or result from poor posture.

sure on the spinal cord or spinal nerves, severe pain and loss of function may result. Most people past 40 show evidence of calcification, increased pigmentation, or other age-related changes in the discs that probably contribute to disc fragility.

Activity and Bone Loss

Many tissues of the body degenerate simply because they are not used. Because new bone is formed in response to the stress placed upon it, bone loss occurs during aging. This happens because elderly people are typically less active than they were when they were younger. There are often good biologic reasons for this decreased physical activity. Because of the risk of fracture or accelerating the deterioration of severely arthritic joints, one must be careful in recommending new strenuous physical activities to those who may have already suffered substantial bone loss.

Normal and Pathologic Fractures

Decrease in bone strength creates an increased likelihood of fractures. These fractures are termed either *normal* or *pathologic.* Normal fractures are those resulting from abnormal stress such as an unusually severe fall. Pathologic fractures result from abnormalities within the bone, which cause local areas of weakness. Sometimes the weak areas can fracture during normal activities of daily life, such as walking with a bag of groceries. Many fractures experienced by the elderly are pathologic fractures. These frequently result from advanced bone disease such as osteoporosis or metastasis of cancer to the bone. Sometimes they are caused by cancers originating in the bone.

The Healing of Fractures

A fracture heals in stages. This healing process may be slow or incomplete in an elderly person. The events following a fracture or reduction of a fracture (the realigning of a broken bone) are described in Table 4-1.

Aging Joints

While some level of osteopenia (bone loss) will be experienced by every 90-year-old, most people will suffer no ill effect from it. Such is not the case with the effect of aging on joints. Senescence and disease affect the freely movable, synovial joints, causing considerable pain, loss of movement, and even deformity in some cases. A typical synovial joint is shown in Figure 4-5. We will discuss several joint conditions including osteoarthritis, and rheumatoid arthritis in the following section.

Diseases and Conditions Associated with Age

Fractures. In elderly people fractures may result as much from decreased bone strength as from trauma. The majority of fractures occur in people between ages 55 and 75, a clear indication that the fractures are more related to weakened bone than increased trauma. Interestingly, the incidence of fracture declines after age 75, possibly because those people genetically or dietarily predisposed to weakened bone will already have shown the effect by this age. Certainly other factors are also involved, including the fact that most people become both lighter and less active as they pass age 75. These factors will decrease some of the stress that could contribute to fractures. It is often through an unexpected fracture that a specific bone disease is discovered.

Healing of fractures is usually slow in the elderly even when there is no demonstrable underlying pathologic process. All of the steps in the healing process may be slower, especially new bone formation. For this reason the bone may require pinning or even total replacement, as is required with many hip fractures (Figure 4-8). Such surgery is needed to prevent a long period of bed rest. Most of the frail elderly cannot tolerate prolonged immobilization and its associated acceleration of muscle atrophy, loss of

Table 4-1 Stages of Fracture Healing

1. Hemorrhage begins immediately with blood flowing into the surrounding tissue from broken vessels in the Haversian system and areas of the endosteum and periosteum. This forms a fracture hematoma and the blood soon clots.
2. Cells adjacent to the fracture die as a result of the loss of circulation.
3. New blood vessels begin to invade the fracture area, restoring circulation. With them new osteoblasts developed from stem cells near the endosteum and periosteum invade between the broken fragments of bone.
4. Osteoclasts and other phagocytes invade and begin dissolving the dead bone and removing other debris.
5. Osteoblasts begin laying down new trabecular bone near the marrow cavity (forming the internal callus) and later near the outside of the bone (forming the external callus).
6. During the next several weeks, remodeling of the bone takes place as new bone is deposited along the fibrous network throughout the break area and most of the calluses and remaining dead bone are reabsorbed.

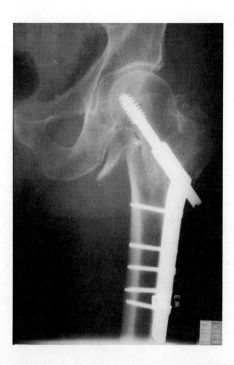

Figure 4-8   Pinning a hip fracture may reduce the immobilization period of an elderly person.

bone density, respiratory problems (often leading to pneumonia), and blood clot formation in the deep veins of the legs. With surgery, hip fracture patients can often be up and walking in a matter of a few days.

Paget's Disease of Bone. This disease is sometimes called osteitis deformans and is described as an excessively rapid and uncoordinated turnover of bone. Osteoclasts and osteoblasts both show marked increases in activity in localized areas of the skeleton. The disease begins with increased osteoclastic resorption of bone. Osteoblasts seemingly go to extremes to replace this lamellar bone and lay down woven strands of new, immature, weak bone. The resulting bone will have soft, porous areas and within these areas, the osteons and their Haversian canals will have disappeared. A network of blood vessels will be present, however, and the bone at these points will have a higher temperature because of the increased circulation. Figure 4-9 shows a section of bone with Paget's disease.

This disorder often lacks significant harmful symptoms. It affects men slightly more often than women. The disease is rare in people under 40 but the incidence increases from 1% in the fifth decade to almost 10% by the tenth decade. Ninety-five percent of the patients are diagnosed without symptoms as a result of x-ray findings and an elevated serum alkaline phosphatase. Alkaline phosphatase is an enzyme found in osteoblasts and other cells. Its level in the blood will rise whenever there is excessive bone formation. Five percent of Paget's disease patients complain of pain. In time, swellings and deformities of the bones occur, especially along the femur and tibia. There may be skull enlargement and scoliosis or lateral curvature of the spine. Severe systemic cases may produce a variety of problems, including cardiovascular disease and enlargement of the heart with increased cardiac output. Pressure on the brain can result in dementia.

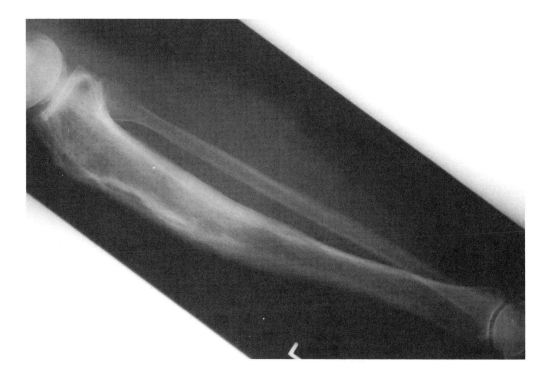

Figure 4-9 The areas of dense and translucent bone are typical of Paget's disease of bone. This results from the high rate of uneven bone turnover.

Damage to the cranial and spinal nerves can cause many problems, including deafness and blindness. Patients with Paget's disease may also become deaf because of calcification of the stapes ligament in the middle ear cavity.

Paget's disease may be caused by a virus. The osteoclasts of Paget's disease patients have been found to contain measles virus-like structures, and measles antibodies bind to these osteoclasts. This may be a slow virus disease having the same relationship to measles as shingles (herpes zoster) has to chicken pox (see p. 47).

Osteoporosis. This condition manifests itself as a loss of both the mineral and organic matter of bone. The bone produced appears normal but is weakened. About 10% of people over 50 years old show bone loss so severe that spontaneous fractures of the vertebrae of the lower thorax and lumbar regions occur. In many of these cases the supporting parts of the vertebrae may have collapsed—the vertebral disk material herniates and prolapses into the adjacent vertebra. A fractured osteoporotic vertebra is illustrated in Figure 4-10.

One-fourth of all white women in the United States develop osteoporotic compression fractures of the vertebrae; 50% develop these fractures by the age of 75. Osteoporosis produces enlargement of the medullary portion of the long bones and an increase in the porosity of dense bone. This is because more osteoclasts are located near the endosteum and marrow cavity than elsewhere. All of the bones of the body are affected, although the greatest effect is on the vertebrae, long bones, and pelvis. The most severe result of this disease is fracture, as the bone becomes less able to support the body's weight and the stresses of normal daily activity.

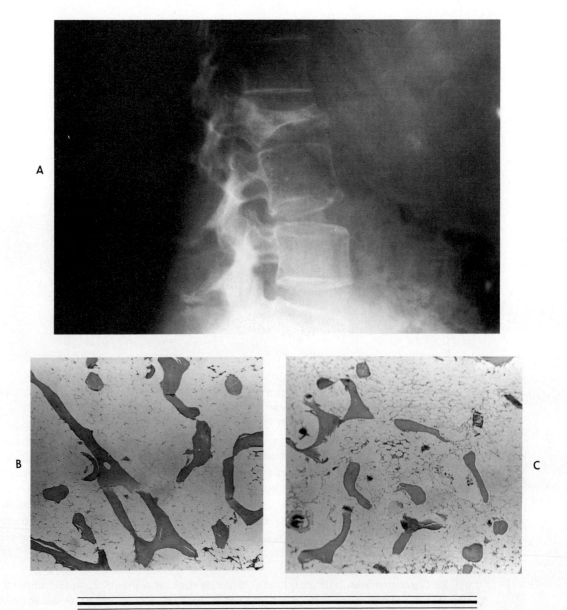

Figure 4-10   Vertebrae weakened by osteoporosis may collapse. This nuclear magnetic resonance photo (**A**) shows a fractured vertebra with a herniated vertebral disk (See also Color Plate 7, **B** and **C** for photomicrographs of normal (**B**) and osteoporotic (**C**) bone.)

Osteoporosis results from a shift toward greater osteoclastic activity without a corresponding increase in deposition of new bone by osteoblasts. It is clearly more severe in women than in men; women lose as much as 8% of their remaining bone each year for a variable period of time after menopause. The rate of bone loss is about 0.75 to 1% per year for women starting at age 30 to 35 and beginning for men at age 50 to 55. Ultimately women will lose 40% to 50% of their bone. Maximum rate of loss in men is usually only about 3% per year. Also, men usually start with more bone mass in relation to body weight and begin losing it later in life. Estrogens apparently inhibit the action of parathyroid hormone on osteoclasts, allowing normal bone growth in young women. The loss of most of the estrogens after menopause seems to result in the excessive bone loss associated with the early stages of osteoporosis. This can usually be controlled or sometimes reversed with such therapy. However, such therapy may have unwanted results and is used only with caution. Higher levels of estrogens may increase the rate of growth of estrogen-sensitive cancers such as cancer of the breast or uterus.

The clinical features of osteoporosis are fairly homogeneous. Bone loss occurs in aging men and women of all races. This decline is associated with menopause in women and declining testosterone production in men. During this time nearly normal bone remodeling occurs and urinary secretion of calcium remains within normal limits. Men develop more dense bone tissue with skeletons that are usually heavier and more massive than those of women. One reason for this may be the greater stress men place on their bones as a result of their genetically larger size and usually more strenuous physical activity. In American society until recently, at least, it has usually been the man of the house who changes the tire, mows the lawn, builds the new addition, and does most of the other heavy jobs. Within normal limits bone becomes stronger because of compression stress (regardless of age) as long as adequate amounts of calcium and vitamin D are present. The male hormone testosterone may have a direct beneficial effect as well, but no experimental evidence of this exists.

Diet also contributes to osteoporosis. Modern diets often do not have enough calcium to maintain proper calcium balance. To some extent this results from a decreased efficiency in calcium absorption found in the elderly. Some of the osteoporotic elderly show a marked decrease in the percentage of dietary calcium that can be absorbed. This means that these people need a much higher than normal calcium intake. At the same time, consumption of milk, a major source of calcium, tends to decline with age. Many people lose the ability to digest lactose (milk sugar). The undigested lactose provides food for bacteria that cause intestinal discomfort including bloating, gas, and diarrhea. Milk is often avoided for this reason. Primitive peoples probably gnawed on the bones of the animals they ate, providing an ideal balance of calcium and phosphorus. This practice is frowned upon in most restaurants today. Alcoholics often develop osteoporosis but this effect may be dietary because the diet of alcoholics is often deficient.

Osteomalacia. Pathologic fracture can result from osteomalacia just as it can be caused by osteoporosis. Osteomalacia is the result of a deficiency of vitamin D, essential for the absorption of calcium in the intestines. The disease, sometimes called adult rickets, is corrected with administration of vitamin D along with adequate calcium. In osteomalacia there is a decrease in the ratio of calcium to organic matter. This decrease happens because as bone remodeling occurs, organic bone is manufactured but the lack of calcium prevents calcium deposition on the organic matrix, or osteoid. The resultant bone is more fragile than normal.

There may be a decrease in vitamin D activity in most cases of osteoporosis and classic osteomalacia, especially in postmenopausal women. Large doses of vitamin D may be required to produce normal calcium absorption. It has been suggested that failure of conversion of vitamin D to the physiologically active metabolite, 1,25 dihydroxy vitamin D, may be responsible for decreased calcium absorption. Even though treatment with vitamin D or 1,25 dihydroxy vitamin D increases calcium absorption in osteoporotic patients, it does not arrest bone loss as it does in osteomalacia.

Steroid Induced Osteopenia. Many elderly people are receiving corticosteroids, synthetic hormones related to the adrenal gland hormone cortisol. These hor-

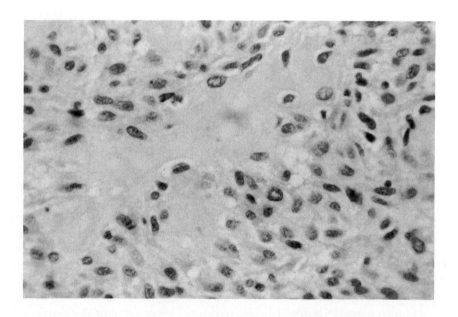

Figure 4-11   Osteogenic sarcoma is a rare but dangerous cancer of bone. Note the numerous malignant cells with an area of new bone being deposited (center left).

mones may have a variety of side effects, including bone loss. Steroid-induced osteopenia resembles osteoporosis histologically. It occurs most frequently in women over 50 and is more severe in trabecular bones (vertebrae and ribs) than in the long bones. Bone reduction occurs through at least two mechanisms. Corticosteroids decrease the synthesis of bone collagen by osteoblasts and apparently decrease osteoblast formation. Perhaps osteoclastic activity is also stimulated either directly or by increased parathyroid hormone secretion resulting from impaired intestinal absorption of calcium and the resulting low blood calcium. It has been suggested that corticosteroids may either interfere with vitamin D or interfere with calcium transport in some other way. When the steroids are withdrawn, a rebound in osteoblastic activity usually occurs, with rapid production of new bone.

Osteogenic Sarcoma . This rare cancer is an extremely dangerous neoplasm. It is most frequently seen in the rapidly growing long bones of children. It is uncommon in middle age but appears again in the elderly. Metastasis is usually rapid, occurring mainly through the bloodstream. The tumor and its metastatic growths usually maintain their osteogenic characteristics, producing spicules of bone and a few cells recognizable as osteoblasts and osteoclasts (Figure 4-11). The most effective treatment for early osteogenic sarcoma is amputation of the affected limb.

Most cancers referred to by the public as "bone cancers" are actually metastatic lesions to bone of other kinds of cancer whose primary tumors did not arise in bone.

Osteoarthritis. A noninflammatory degenerative bone disease, in its mildest forms osteoarthritis affects nearly everyone over 50 years of age. Men and women are

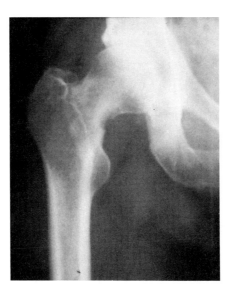

Figure 4-12   Spurs of bone projecting into a joint damaged by osteoarthritis.

affected equally although men may experience the effects earlier. It is often described as the result of the wear and tear that our joints experience. However, the variation in intensity with which the disease strikes indicates the cause is much more than that. Sometimes the people with the most severe osteoarthritis are those with the least strenuous lifestyles.

The disease first appears as irregularities on the articular cartilages of the weight-bearing joints, knee, hip, and spine. Biochemical changes take place in the cartilage itself, decreasing the hydrophilic proteoglycans associated with the collagen network. It may be that some of the proteoglycans and even some of the collagen fibers are destroyed by enzymes in the synovial fluid that are normally prevented from penetrating the articular cartilage. Others suggest that it is an autoimmune problem.

Microfractures develop as the roughened surfaces of the damaged cartilages grind against each other. The cartilages erode, pieces may break off, and bone eventually becomes exposed. The bone may respond to the trauma by growing more dense and forming spurs that project into the synovial cavity (see Figure 4-12). These spurs, along with pieces of broken cartilage, form "joint mice" that slide around the synovial space and can temporarily block movement. They may later slide out of the way again, permitting normal use of the joint. This damage causes varying amounts of pain and stiffness. Inflammation of the joint may eventually occur but usually does not revolve around the synovial membrane. The stiffness increases over the years until, in some instances, the joint becomes completely nonfunctional. Osteoarthritis usually progresses slowly. Its symptoms are chiefly pain and stiffness, with deformity of joints coming later. The pain is treated with analgesics, such as aspirin. Severe cases require surgical replacement of the affected joint.

Rheumatoid Arthritis. Unlike osteoarthritis, rheumatoid arthritis is an inflammatory process capable of affecting any joint. Usually the more mobile and frequently used joints such as fingers and wrists are affected. Large weight-bearing joints are less

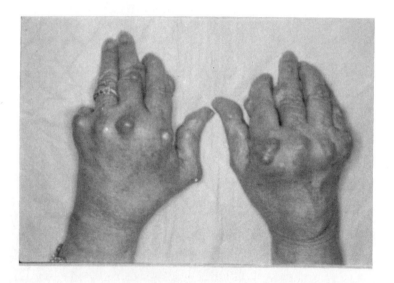

Figure 4-13 Rheumatoid arthritis may produce severe ankylosis of the fingers and other joints.

uniformly involved. This painful disease produces a more rapid and severe deterioration of the involved joints. Women are affected three times more frequently than men. Approximately 2% of the adult population of the United States has this disease. Unlike osteoarthritis, rheumatoid arthritis usually begins with young adults. It can, however, begin at any age. The severe crippling consequences may, in time, immobilize many of the affected people.

Rheumatoid arthritis begins with inflammation of the synovial membranes. These membranes secrete the synovial fluid that lubricates joints. Contained in connective tissue capsules called bursae, they surround each joint space and are found wherever one body part must slip past another with minimum friction. Rheumatic diseases therefore may not be limited to joints, although joint inflammation is a major symptom of the disease.

The joints swell as the synovial membranes become inflamed and thickened. The synovium is transformed into a vascular mass of inflammatory tissue called pannus. The pannus is associated with destructive enzymes that erode the articular cartilage and create a swollen, red, inflamed joint that is very painful. The inflammation can completely destroy the cartilage. Eventually bone is exposed; connective tissue forms between the articulating bones, and later ossifies, fusing the bones. When the bones are fused the condition called ankylosis of the joint results (Figure 4-13). Evidence indicates that rheumatoid arthritis may be the result of an autoimmune mechanism by which deposition of an immune complex results in damage to the synovial membrane.

The course of untreated rheumatoid arthritis can be progressively downhill. There are no cures for this disease. Therapy is primarily palliative, using antiinflammatory agents or aspirin in the early stages and immunosuppressive agents, corticosteroids, or gold when the disease is more advanced. In most cases the disease can be arrested. In a few, in spite of all possible therapy, the disease remains progressive. People who develop rheumatoid arthritis early in life are often crippled when they become old. Therefore, while this is not particularly a disease of the elderly, its consequences may be most severe with that group.

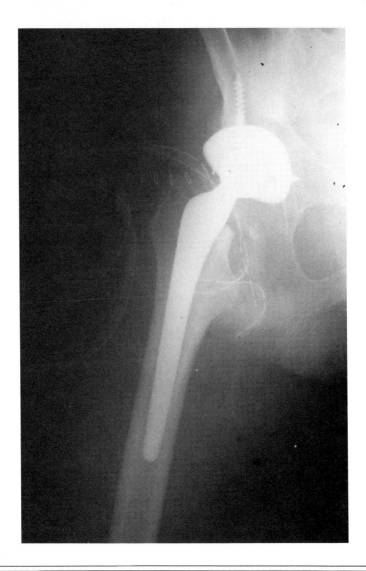

Figure 4-14 X-ray showing total hip replacement. Note the metal components cemented in place. The articulating surfaces are polyethylene.

The only hope for regaining the use of a joint destroyed by arthritis is presently substitution of an artificial joint, or prosthesis. During this surgical procedure, the damaged articulation portions of the joint are removed and replaced with a metal alloy and polyethylene prosthesis. Recovery from hip replacement usually takes about 8 weeks. Some early prosthetic joint replacements failed because of bone deterioration around the cement used to hold the prosthetic joint in place. This failure is less common today. About 300,000 joints are replaced with prostheses each year. A prosthetic joint in place is shown in Figure 4-14.

Other Arthropathies. Arthritis can be caused by bacterial infection (septic arthritis) or by deposition of sodium urate crystals in the joint (gouty arthritis). In the latter instance the disease gout results from the accumulation of uric acid in the blood as a result of either failure of the kidneys to excrete the compound or excessive production. The uric acid forms sodium urate that can crystallize in the soft tissues of the body, especially the joints. This results in inflammation, swelling, pain, and ultimately, destruction of the joint.

Bursitis. This condition results from injury or excessive friction in a bursa. The symptoms are pain on use and swelling caused by accumulation of fluid within the enclosed space. A specific type of bursitis occurs over the great toe and produces a *bunion*. A bunion forms when the base of the great toe is pushed into the other toes. With the displacement, the bone of the toe is moved to a point of prominence that is easily traumatized. The repetitive trauma produces the bursitis. In other places the swellings may be called "water on the knee (or elbow)."

CASE HISTORIES

❖ Nonunion Fracture

A.G. is a 79-year-old bedridden man with chronic hip pain. His illness began a year ago when he fell and broke the neck of his right femur (right hip). A steel pin was surgically implanted to join the broken pieces. Followup x-rays showed reabsorption of the fractured femoral head and no evidence of healing. The avascular necrosis (death of bone due to lack of blood supply) that resulted in partial reabsorption of the head resulted in weakening and collapse of much of the remaining bone. The hip could not support weight and was painful with any motion. Many other health problems discouraged further therapy, but the hip pain increased and A.G. was hospitalized for a hip remodeling procedure.

A.G.'s medical history revealed that he was diabetic, requiring daily injections of insulin, and that he had had a stroke that left him partially paralyzed. His mental state was impaired, possibly because of cerebral vascular narrowing caused by the diabetes. Prior surgeries included an old appendectomy and bilateral cataract operations.

A.G.'s preoperative examination revealed difficulty with memory (the history was provided by A.G.'s wife). His left arm was paralyzed and left leg was weak and clumsy. Any movement of his right hip caused obvious pain associated with a grating sensation that could be felt by the examining physician. X-rays revealed the right femoral head to be markedly irregular and deformed; portions were absent, consistent with avascular necrosis.

A local anesthetic was used because of A.G.'s debilitated state. The necrotic head was removed along with the old plate and screws. The post-operative recovery was uneventful until 3 days after surgery, when A.G. became confused and febrile. His white blood cell count (WBC) (important in fighting infection) fell from 14,800 to 4,300 while the bands (immature white blood cells indicating inflammation) rose from 4% to 54%. The patient died on the fourth postoperative day. An autopsy documented bilateral lobar pneumonia.

Comments. Many factors can cause a fracture to fail to heal. In this case the fracture so severely compromised the blood supply to the femoral head that the bone became necrotic and partially dissolved. The diabetes was a possible contributing factor, along with its attendant vascular narrowing and tendency toward slow healing. The resulting pain restricted the patient's quality of life to such a degree that his wife insisted the operation be performed. The extended shallow breathing associated with a general anesthetic creates an environment where pneumonia can develop. The postoperative pneumonia demonstrates the inability of some elderly diabetic patients to mount an effective response to a bacterial challenge. In this case the decreasing white blood cell count and the large number of circulating immature WBCs indicated a severe infection.

❖ Osteoporosis

M.G. is a 71-year-old grandmother whose first symptom was lower back pain. Her medical history included normal menopause at 48 years, peptic ulcer disease diagnosed at age 42, glaucoma, and a hiatal hernia. Her physical examination revealed mild scoliosis (sideways curvature of the spine), and her neurologic exam was normal. Laboratory testing revealed no significant abnormality. However, radiographs of the back showed increased lucency (decreased density to x-ray) of all the bones, mild scoliosis, and lordosis (increased lower spinal curvature)—but no evidence of fracture. A diagnosis of osteoporosis was made and the patient was treated with pain medication, physical therapy, and vitamin D and calcium supplements.

During the next 5 years she had multiple hospitalizations for recurrent back pain. Repeat spine x-rays demonstrated a compression fracture of the 11th thoracic vertebra, which was probably the source of her initial pain. The bone continued to lose density, the scoliosis and lordosis worsened, and she developed an almost constant right leg pain believed to be caused by nerve compression from the collapsing vertebrae. She required the use of a cane for ambulation. Eventually she could not care for herself and was placed in a nursing home. On her last hospitalization for pain, her radiographic examination revealed fractured 9th and 10th ribs sustained in a fall during her bath, the old T11 compression fracture, compression of two lumbar segments, and narrowing of all her vertebral disc spaces.

Comments. Primary osteoporosis is a generalized decrease in bone density without obvious cause. It is more common after menopause in females and after 60 years in males. The only truly diagnostic tests involve determination of bone density by biopsy, dual beam photon emission, or CAT scanning. Routine x-rays can only suggest the diagnosis; laboratory testing is noncontributory. The disorder is often subclinical and may manifest with a fracture after minimal trauma. Some cases, however, are severe enough to result in numerous fractures, disabling pain and significant disability.

❖ Paget's Disease of Bone

E.H. came to his physician for an insurance examination at the age of 47. A routine blood test showed a markedly elevated level of alkaline phosphatase, an enzyme. His prior medical history revealed an appendectomy at the age of 14 and mild adult onset diabetes controlled with weight loss. He smoked cigarettes for 36 pack years (2 packages of cigarettes each day for the past 18 years). His physical examination showed no evidence of significant disease but a few minor changes were noted. These included a grade II/IV aortic ejection murmur, and a small reducible left inguinal hernia. Laboratory studies revealed a fasting blood sugar of 121 mg/dl (normal is below 110), an alkaline phosphatase of 53 Bodansky units (normal is up to 4.5) and normal serum calcium, phosphorus, and acid phosphatase. Alkaline phosphatase is an enzyme that is important in bone and liver metabolism. The chest x-ray was within normal limits but a bone scan showed increased activity in the right femoral head and acetabulum (the femoral socket), in the upper left femur and in multiple portions of the skull and lower vertebrae. A bone survey by x-ray revealed mottled dense bone in these areas, characteristic of Paget's disease.

During the next three years E.H. remained asymptomatic but then noticed the gradual onset of pain in his right leg. The pain was present in his hip whenever the leg was used. When it became severe enough to require the use of a cane, he was tried on a weekly course of mithramycin, 700 μg, I.V., and dramatically improved. This compound is a drug that interferes with calcium metabolism and has been found effective in treatment of Paget's disease. The pain essentially disappeared but within 5 months he developed thrombocytopenia (low platelet count and therefore low blood clotting ability) and the medication had to be discontinued.

Comments. As in this case, Paget's disease of the bone often is diagnosed without symptoms. In 20% of the cases there are no symptoms throughout the patient's life. In most cases therapy is limited to the relief of symptoms. In Paget's disease, the source of the enzyme alkaline phosphatase, was the abnormally high deposition of bone (osteoblastic activity). Such activity can also be seen in certain tumors, such as prostatic cancer, that have metastasized to the bone. Because this malignancy is also associated with a high acid phosphatase, the finding of a normal acid phosphatase helps rule out this diagnosis.

❖ Osteoarthritis

D.C. is a 55-year-old farmer with progressively worsening hip pain. His illness began in childhood when a congenital defect in his right leg was found. This resulted in rotation of his leg when walking and subsequent degenerative joint changes. In an attempt to reduce the arthritic pain and correct the deformity, a surgical repair with corrective rotation was done when he was 28 years old. The right leg is now 2 inches shorter than the left, and D.C. has developed groin pain. In addition, his side-to-side gait has caused degenerative joint disease in the fifth lumbar and first sacral vertebrae with possible spinal nerve compression producing tingling feelings that radiate down his right thigh and into his right calf. Because the symptoms are not completely disabling, he has been told to wait until a new, cement-free, total hip joint becomes available. However, because of the pain, D.C. has decided to have his hip replaced with the current total hip procedure.

Prior to the operation, a computed tomography (CT) scan of the lower spine was done that did not show any disc herniation or nerve compression but did reveal significant bony degeneration of the L5 and S1 joints. An electromyelogram (EMG) for evaluation of nerve conduction was normal, and his pain was believed to be entirely due to the degenerative hip joint. Because of the prior surgery and the abnormal anatomy, a custom-made prosthesis was obtained.

His preoperative physical examination indicated only the orthopedic abnormality. The right leg measured 2 inches shorter than the left. The left hip and leg were normal on examination but the right showed marked limitation of movement with 10 to 15 degrees of flexion contracture (he could not completely straighten his leg), only 10 degrees of external rotation, and no internal rotation. His gait was side-to-side with a significant limp favoring the right side.

During the operation a custom-produced right total hip joint was substituted for his damaged one and his leg length was increased 1 inch. D.C. is currently recovering with no neurologic complications. He expects continuing to need a shoe lift on the right foot, but his range of motion has greatly improved and most of his pain is gone.

Comments. Osteoarthritis can be caused by many processes. This patient's problem began as the result of a birth defect and resultant damage to the femoral head. The abnormal bony anatomy resulted in a wearing down of the cartilage, pain, and further deformity. Most cases of osteoarthritis can be considered to result from multiple, silent injuries to the cartilage and bone that eventually cause pain and limitation of motion. Because failures have occurred when cement is used, the ideal replacement for D.C. would have been one without cement. However, cement-free replacements had not been perfected at the time D.C. needed relief.

❖ Rheumatoid Arthritis

At the age of 22 years S.E. first began to notice pain and stiffness in her hands. The symptoms were worse in the morning and disappeared as the day progressed. They were limited to the joints of her fingers and hands. Physical examination by her family physician revealed only mild swelling of the proximal interphalangeal (finger) joints, and her x-rays revealed nothing abnormal. Laboratory findings were limited to an erythrocyte sedimentation rate (ESR) of 80 mm (a reading above 20 is a sign of inflammation) and a positive test for rheumatoid factor. She was suspected of having rheumatoid arthritis and was placed on a program of rest, gentle exercise, and high-dose aspirin.

During the next 2 years she experienced 5 flareups of the disease with worsening pain, visible swelling of the affected joints, mild fever, and weakness. Symmetrical symptoms developed in her feet, hips, wrists, and elbows. Repeat x-rays showed displacement of the metacarpophalangeal (hand and finger) joints with deviation of the hand relative to the arm, early osteoporosis, and soft tissue swellings. Various antiinflammatory drugs were tried without success.

She was started on corticosteroids to control the inflammation and experienced enough symptomatic relief to resume her duties as a mother and homemaker. In spite of the symptomatic improvement, the disease progressed. Over the next 6 years her physical deformities worsened, her functional capacity declined, and the x-ray appearance of her joints deteriorated. Each time the corticosteroids were reduced, she experienced a severe symptomatic episode sometimes requiring hospitalization.

In time she could no longer get around and became bedridden. After an episode of illness in her children, she developed pneumococcal pneumonia and had a protracted hospital course during which she could not be weaned from the corticosteroids. One month following her hospital stay for pneumonia, she was readmitted with pneumococcal septicemia (body-wide infection with the pneumococcal bacteria), disseminated intravascular coagulation (diffuse blood clotting in vessels), and shock. After a 10-day hospital course she died.

Comments. Although the vast majority of people crippled by rheumatoid arthritis are more than 50 years old, this case history illustrates that this disease process can be devastating for the young as well. Rheumatoid arthritis (RA) is thought to be an autoimmune disorder affecting primarily the smaller peripheral joints in a symmetrical pattern. As such it often manifests with antibodies to human immunoglobins (the rheumatoid factor) and to nuclear proteins (antinuclear antibodies, ANA). Such autoimmune problems are more likely as age advances.

Because it is an inflammatory condition, rheumatoid arthritis is associated with nonspecific markers of inflammation such as elevation of the erythrocyte sedimentation rate (ESR). Often the disease subsides or is controlled with antiinflammatory agents, but occasionally it progresses to a severe, debilitating disease as in this case. Current therapy recognizes the dangers associated with long-term corticosteroid use and limits this drug to short courses for the control of very symptomatic flare-ups of the condition. S.E.'s death is directly related to the continued use of steroids which, although necessary to control her severe symptoms, interfered with her ability to fight off infections.

❖ Gouty Arthritis

D.N., an 80-year-old retired physician, was first diagnosed as having gout at the age of 40. His symptom was a painful, hot toe. Aspiration of the joint produced synovial fluid with 13,000 WBCs and intracellular uric acid crystals. His serum uric acid was 10.4 mg/dl (normal is less than 8). He was treated with episodic symptomatic therapy using colchicine, an antimitotic drug. Over the years he developed tophi (granulomatous deposits of uric acid appearing as subcutaneous nodules) in the hands, feet, ears and elbows, and his pain periodically reappeared. In 1975 he began taking allopurinol, an inhibitor of uric acid. His foot pain now is much reduced and the tophi have dramatically decreased in size but are still present.

Comments. In cases of gout, uric acid crystals precipitate in synovial fluids, causing an inflammatory arthritis. In some patients the crystals form tumorous growths under the skin (tophi). Untreated, the arthritis may become crippling and irreversible.

BIBLIOGRAPHY

Braxel US: *Common metabolic disorders of the skeleton in aging.* In Reichel W, ed: *Clinical aspects of aging,* Baltimore, 1983, Williams & Wilkins.

Courpron P, Drugnet M: *Bone disorders.* In Pathy MJS, ed: *Principles and practice of geriatric medicine,* New York, 1985, John Wiley & Sons.

Grob D: *Prevalent joint diseases in older persons.* In Reichel W, ed: *Clinical aspects of aging,* Baltimore, 1983, Williams & Wilkins.

Guyton AC: *Textbook of medical physiology,* ed 7, Philadelphia, 1986, WB Saunders.

Krane SM, Holick MF: *Metabolic bone disease.* In *Harrison's principles of internal medicine,* ed 12, New York, 1991, McGraw-Hill.

McCartney DJ: *Arthritis and allied conditions, a textbook of rheumatology,* ed 11, Philadelphia, 1989, Lea & Febiger.

Need AG and others: Osteoporosis, new insights from densitometry, *J Am Geriatrics Soc* 38:1153-1158, 1990.

Tonna EA: *Aging of skeletal-dental systems.* In Finch CE, Hayflick L, eds: *The biology of aging,* New York, 1977, Van Nostrand Reinhold.

5 *The Muscular System*

RELEVANT ANATOMY AND PHYSIOLOGY
Types and Functions of Muscle Tissue

The muscular system moves the body, parts of the body, or substances within the body. We tend to think of this system as consisting only of the large skeletal muscles we see in our arms, legs, and torso. However, the entire system of muscles is divided into three types: *skeletal, smooth,* and *cardiac* (Figure 5-1). These muscle types all arise from the embryologic layer known as mesoderm. Each type of muscle tissue has a particular structure and function.

Skeletal Muscles. These are the large muscles associated with the skeleton. Their fibers are striated (see Figure 5-1, *A*), and are often arranged in groups called fascicles. The fibers are subdivided into subunits called myofibrils. The striations result from layers of muscle protein, or myosin, stacked so densely within the myofibrils that they hinder the passage of light when viewed with a microscope (Figure 5-2). The framework of muscle cells is composed of the protein actin. Skeletal muscles contract more rapidly and with greater force than smooth muscles. They are *voluntary,* meaning that they respond to conscious thought.

Skeletal muscle fibers are not properly called cells because they are formed early in life from many nuclear divisions without the usual associated cytoplasmic divisions. Each fiber therefore has many nuclei (is multinucleated) and is larger than most of the body's other cells. New muscle fibers are not normally produced after birth, and growth of a muscle usually occurs only from enlargement of existing muscle fibers. However, normal muscle contains a scattering of embryonic cells (satellite cells) that can produce new muscle tissue following an injury. Each muscle is composed of many muscle fibers, which are innervated by *motor nerves* either individually or in groups. Skeletal muscles are surrounded and interwoven with connective tissue called *fascia,* which is gathered at the ends of each muscle to form tendons that attach the muscles to bones.

Smooth Muscle. These muscle cells lack branches, do not form large multinucleated fibers, and lack the microscopically visible cross striations present in the other two muscle types (Figure 5-1, *B*). Smooth muscle is also called *visceral muscle* because it is the contractile tissue of the soft organs of the body including the skin, arteries, esophagus, stomach, and intestines (the viscera). Smooth muscles contract slowly

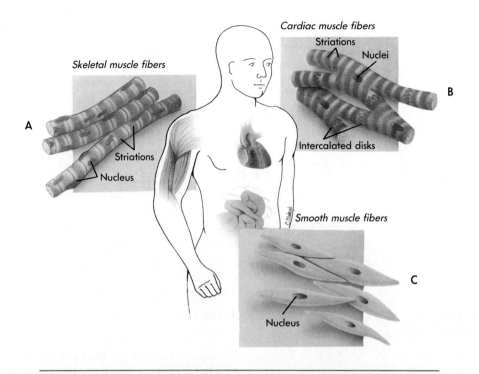

Figure 5-1 Three types of muscle fibers: **A,** Skeletal; **B,** cardiac; and **C,** smooth (visceral). Visceral muscle lacks striations; the numerous nuclei around each muscle fiber and the distinct striations distinguish skeletal muscle from other muscle types. The single nucleus in a branching striated fiber and intercalated disks distinguish cardiac muscle.

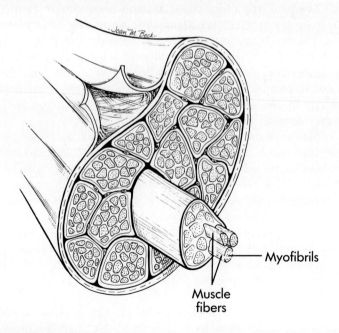

Figure 5-2 The relationship between muscle, groups of muscle fibers (fascicles), and their subunits, myofibrils, is shown here. The dense protein, myosin, forms the striations of skeletal and cardiac muscle.

but they do not tire easily and can exert continuous, even force over a long period. They are also *involuntary,* meaning that they are not under conscious control.

   Cardiac Muscle. This might be viewed as a tissue that combines the advantages of visceral and skeletal muscle. Cardiac muscle is found only in the heart and is a little slower in its contraction than skeletal muscle. It is involuntary, as is smooth muscle, but its fibers are striated. The striations, however, are not as dense as those of skeletal muscle. Each cardiac muscle fiber is a distinct branching cell with a single nucleus. The branches are connected to the branches of adjacent cells by *intercalated disks,* structures that are not present in smooth and skeletal muscles. The intercalated disks are formed from the cell membranes at the ends of the cell fiber branches. Cardiac muscle cells create an interwoven network which forms the basket-shaped heart. This arrangement of muscle cells decreases the size of the space it encloses (the atrial or ventricular cavity) when the muscle is contracted, enabling the blood to squeeze out into another part of the heart or into the arterial system.

Stimulation of Muscle and the "All or None Principle"

   According to the "all or none" principle, muscle fibers contract completely or not at all. Contraction occurs when skeletal muscle is stimulated by a *neurotransmitter* called acetylcholine, a compound released by *motor neurons,* the cells of motor nerves, at neuromuscular junctions (Figure 5-3). Acetylcholine is manufactured by the neuron and accumulated in junctional vesicles. When the neuron "fires" the neurotransmitter is released and diffuses across the junction between the nerve and the muscle. There it binds to acetylcholine receptors, most of which are located in junctional clefts in the muscle. When enough acetylcholine has been bound to cause the muscle fiber to contract, it is said that the stimulus has reached *threshold.* There is no mechanism for varying the intensity of contraction once threshold has been attained. Muscle fibers must therefore contract all the way or not at all—the "all or none principle." The acetylcholine is destroyed by the enzyme acetylcholine esterase preventing continued stimulation of the fibers and allowing them to relax. The acetate and choline products of this de-

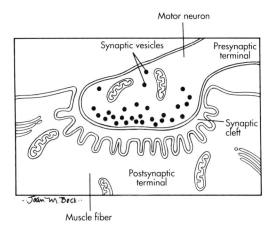

Figure 5-3   After acetylcholine is released from a motor nerve, it diffuses across the neuromuscular junction. It stimulates the muscle fiber to contract when enough has been absorbed by the muscles' acetylcholine receptors.

struction are reabsorbed by the motor nerve to make more acetylcholine. (Some smooth muscles of the body are stimulated by another neurotransmitter, norepineph-rine.) This neurotransmitter is used only by the sympathetic nervous system, which will be discussed in a later chapter. Other neurotransmitters function in the brain and other organs.

Because any muscle fiber (or group of muscle fibers stimulated by a single motor neuron) either contracts or it does not, the amount of stimulus a muscle fiber is given does not change the amount of force exerted by the muscle. The force produced by a muscle can only be varied by changing the number of fibers stimulated in that muscle. A person might stimulate only 15% of the fibers in a particular muscle to lift a light object but three times that many to lift a heavier one. In a healthy muscle some fibers are always being stimulated and are always in a state of contraction, even when the muscle is resting. This condition is referred to as muscle tone and is one of the factors that helps healthy muscle retain its strength.

The Environment of Muscles

Muscles are very sensitive to changes in their chemical environment and to hor-mones. Muscle weakness, for example, may result from either an excess or deficiency in thyroxin, a hormone produced by the thyroid gland. The same is true for a variety of hormones from the adrenal glands. Both potassium and calcium levels are critical for normal muscle function. Deficiency in calcium can result in tetany, a condition of con-tinued spasmodic contraction of opposing muscles resulting sometimes in death. An ex-cess of calcium, hypercalcemia, depresses skeletal muscle action and produces weak-ness. A deficiency in potassium results in weakness and sometimes paralysis of skeletal muscles. An excess of potassium can affect the heart, where it may produce arrhythmia and weakness of the cardiac muscle.

Oxidation of Sugar in Muscles and Oxygen Debt

In the presence of oxygen, blood sugar, or glucose, can be metabolized to carbon dioxide, water, and energy by the process of *oxidative respiration*. Therefore oxygen is as essential in liberating energy from glucose for the normal functioning of muscle cells as it is for respiration in other body cells (Figure 5-4). Sometimes muscles require more oxygen than they receive. When this occurs, the normal oxidative respiration is re-placed by another energy-releasing mechanism, one that does not require oxygen. Un-der these *anaerobic* (without oxygen) conditions, the muscle cells release the energy in glucose through the manufacture of lactic acid. By this means the muscle may obtain enough energy for continued life and activity in the absence of oxygen. This is a short-term solution, however, because lactic acid is toxic and excessive amounts of it can cause a muscle to spasmodically contract (cramp). To rid the muscle of excess lactic acid, energy will be needed beyond the usual energy requirements of the muscle. This build up of lactic acid creates an *oxygen debt,* because more oxygen will be required to release that extra energy by cellular respiration. As oxygen becomes available, some will be used to convert the lactic acid back into energy stores. It should be noted that glu-cose is stored in the body as glycogen.

The Principle of Opposition

A second principle of muscle action is the *principle of opposition*. This states that all muscles work in opposition to other muscles. There are virtually no exceptions

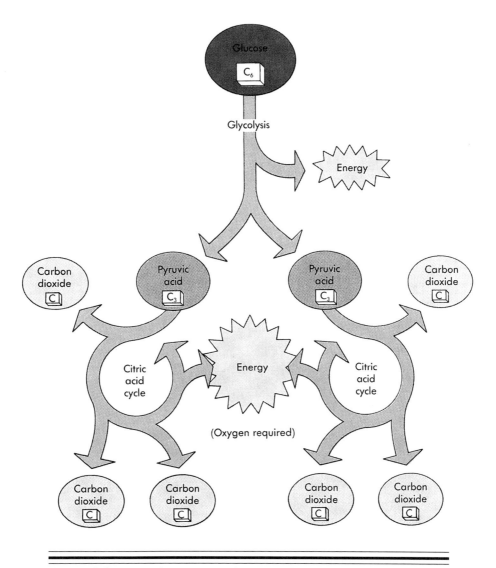

Figure 5-4   Anaerobic energy releasing path to lactic acid. Most energy comes from aerobic respiration (at the bottom of the diagram).

to this, even among smooth muscles. Muscles contract, but no muscle is capable of extending itself. To be extended, it must be stretched by other muscles or, in a few instances, by elastic connective tissue or another opposing force such as gravity. Table 5-1 lists skeletal muscles by the kind of action they produce when working in opposition to another muscle or force. Cardiac and smooth muscles also work with opposing forces. The muscles of the atrium of the heart work in opposition to the muscles of the ventricle by forcing blood into the ventricular cavities while the ventricular muscles are relaxing. Even the muscles surrounding the intestine are expanded when digesting food is forced through the intestinal lumen space by other intestinal muscles in a process called peristalsis. Sphincter muscles close off the end or beginning of a tube and are expanded by material that is pushed through them by other muscles.

Table 5-1   The Principle of Opposition in Skeletal Muscle

MUSCLE	ACTION PRODUCED	OPPOSER	ACTION PRODUCED
Abductor	Move limbs away from axis of body	Adductor	Bring limbs in line with axis of body
Levator	Raise part	Depressor	Lower part
Rotator	Turn part (e.g., head)	Rotator	Turn part in opposite direction
Flexor	Bend joints	Extensor	Straighten joints

AGING MUSCLES
Aging Visceral and Cardiac Muscle

Not as much is known about the aging of smooth (visceral) muscle as about skeletal muscle. Certainly there is a decrease in peristalsis and in the strength of the tissues surrounding the stomach and intestines as a person ages. This contributes to constipation and diverticulosis among some elderly people. The loss of peristalsis may be partially due to failure of nervous stimulation of these muscles as well. This will be discussed in greater detail in Chapter 7. Loss of elasticity and strength in the muscular lining of the arteries and arterioles contributes to one kind of high blood pressure (systolic hypertension) seen in the elderly. This deterioration seems more related to extrinsic factors such as solidification of elastic or collagenous fibers or deposition of calcium and fatty materials in and around the arterial muscles than it does to the muscle itself. This too will be discussed in Chapter 9.

Cardiac muscle deteriorates with age. This seems to result primarily from a decrease in oxygenated blood flow through the coronary arteries to the heart muscles. Coronary artery heart disease is very common in the Western hemisphere. It progresses with age, but is not a result of age alone. Heart damage from coronary artery disease, therefore, is not a result of the age of the heart but of another age-related disease. Heart muscle cells do seem to lose some of their ability to use oxygen as they age but this seems minor when compared to the damage resulting from blockage of the coronary vessels. Problems associated with heart failure will be discussed in Chapter 9.

Aging Skeletal Muscle

Aging skeletal muscle usually causes few problems if the person remains normally active and healthy. There is a progressive loss of muscle mass manifested by a decrease in the number and diameter of muscle fibers. This is particularly evident in the clefts that appear between the long bones of the hands (Figure 5-5) and in the flabby appearance of the lower leg muscles. The loss of mass varies with the muscle and with how much it is used. Most muscle loss in the elderly appears to be the result of disuse. Atrophic aging muscle fibers are gradually replaced with fat and collagen. The fibers that are retained accumulate lipofuscin granules. The impact of accumulation of this pigment on muscle is unknown but it is known to also accumulate in aging cells of other tissues (see Chapter 2).

There appears to be an age-associated decline in the enzymes involved with release of energy for muscle action, especially ATPase. This does not appear to affect the overall muscle performance of most people significantly, and may be another manifesta-

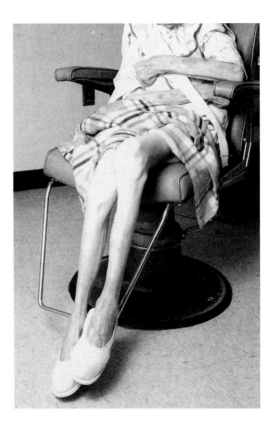

Figure 5-5   Muscle loss causes the joints to appear larger and the areas between the joints to appear thinner.

tion of disuse atrophy. A decrease in the number of capillaries in a given area of muscle has also been observed. Some neuromuscular disorders, along with the aging process and disuse atrophy, result in a loss of muscle tone. These factors slightly reduce the speed (about 13% by age 75) and strength of skeletal muscle contraction.

Muscle Weakness

Elderly people frequently complain of muscle weakness. This, however, is usually not the result of muscles aging. Weakness is most often caused by disuse. It may also be caused by problems with blood constituents, potassium or calcium, or by hormonal imbalance, especially thyroxin. Sometimes it results from interference with the innervation of muscles caused by the impairment of nerves in the motor (anterior) root of the spinal cord or by disturbances deep in the central nervous system. Gradual development of atherosclerosis, chronic pulmonary disease, or declining cardiac output, in addition to psychologic factors such as depression or boredom, can produce complaints of weakness. Atherosclerosis often results in claudication, intense muscle pain and weakness during or after exercise. This condition is caused by ischemia resulting from inadequate circulation. The sudden onset of weakness can also be caused by anemia resulting from undiagnosed internal bleeding or myocardial infarction, which can be painless.

Cramping

Decreased circulation and decreased movement combine to increase the frequency and severity of muscle cramps in elderly people. This may be complicated by deficiency in calcium, sodium, or stored energy. The presence of toxins such as tetanus can also cause cramping. Cramps in the elderly occur most frequently at night, often following attempted movement of a limb that has been still and in which circulation has been impeded because of compression or bending of blood vessels during sleep. These cramps may be frequent and painful but can often be stopped by passive stretching. A hot bath before bedtime will dilate blood vessels and may be helpful in preventing cramps. Various drugs are available that either reduce the tendency of muscles to contract or decrease the likelihood of unwanted nervous stimulation of muscles.

Problems with Nervous Stimulation

Many severe muscle problems result from improper nervous stimulation, either of entire muscles or of individual groups of fibers. Diseases involving the failure of nervous stimulation of muscles include poliomyelitis, multiple sclerosis, amyotrophic lateral sclerosis, cerebrovascular accident (CVA or stroke), and trauma to the brain or spinal cord. Polio has been almost eradicated in the United States by immunization and when it occurs it most often affects young people. However, we are beginning to see a post-polio phenomenon that occurs as people age. All of these are diseases or problems of the nervous system and they will be discussed in Chapters 13 (nervous system) and 14 (sense organs). Diabetes causes muscle failure due to damage to the nervous system both directly and as a result of related circulatory problems. These effects are discussed in both Chapter 10 (cardiovascular system) and with the nervous system. Other problems such as the tremors and eventual paralysis associated with Parkinson's disease result from brain-originated, excessive, uncoordinated, nervous stimulation of muscles. These too will be covered with the nervous system. Table 5-2 lists examples of these types of failures. See the appropriate chapter for more information.

Table 5-2   Failures of Muscle

APPARENT MUSCLE CONDITION	PRIMARY CAUSE OR AREA DAMAGED
Poliomyelitis	Motor nerves of CNS
Multiple sclerosis	Myelinated nerves of CNS
Cerebrovascular accident (CVA; stroke)	Brain damage (hemorrhage or embolism)
Amyotrophic lateral sclerosis	Motor nerves of brain and spinal cord
Parkinson's disease	Brain, basal ganglion
Diabetic amyotrophy	Peripheral diabetic nerve damage
Atrophy of disuse	Failure of muscle use for any reason
Senile myosclerosis	Muscle disuse atrophy with collagen replacement and contraction
Muscular dystrophy	Primary muscle degeneration
Myasthenia gravis	Loss of acetylcholine receptors in neuromuscular function
Polymyositis	Inflammatory muscle disease
Polymyalgia rheumatica	Muscle and artery connective tissue
Calcific tendonitis	Tendon inflammation
Muscle tumors	Abnormal muscle growth

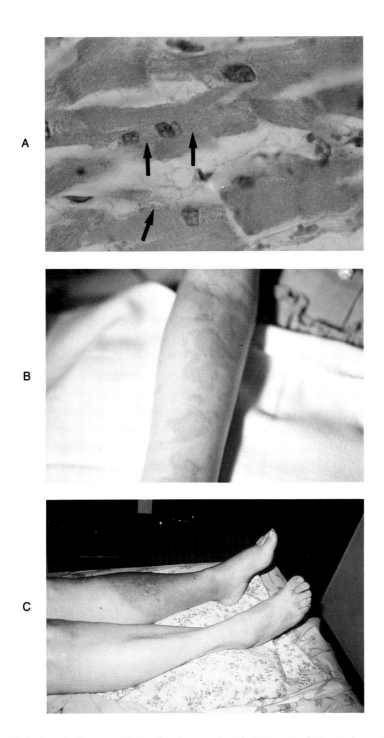

Plate 1   **A,** Tissue with lipofuscin granules. **B,** Urticaria. **C,** Stasis dermatitis.

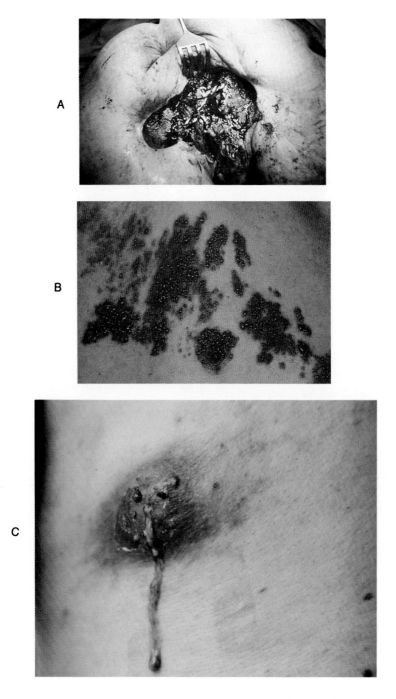

Plate 2 **A,** Decubitus ulcer. **B,** Herpes zoster. **C,** Infected sebaceous cyst.

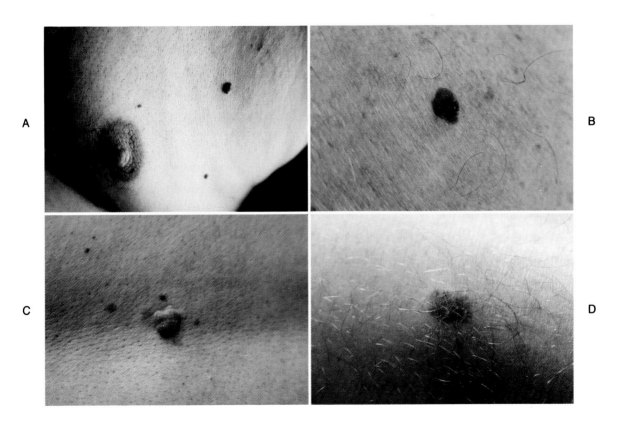

Plate 3 **A**, Cherry angioma. **B**, Seborrheic keratosis. **C**, Fibroepithelial polyp. **D**, Junctional nevus.

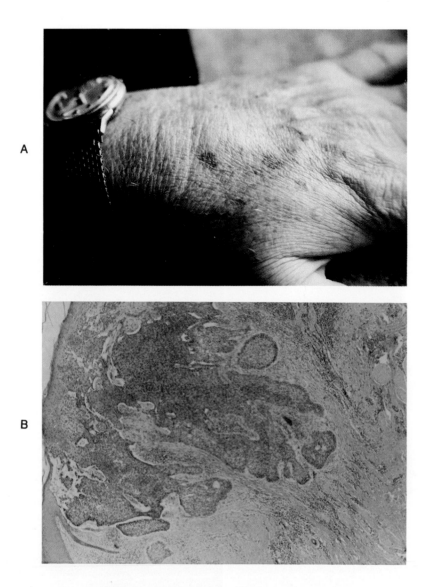

A

B

Plate 4 **A,** Solar lentigo. **B,** Basal cell carcinoma.

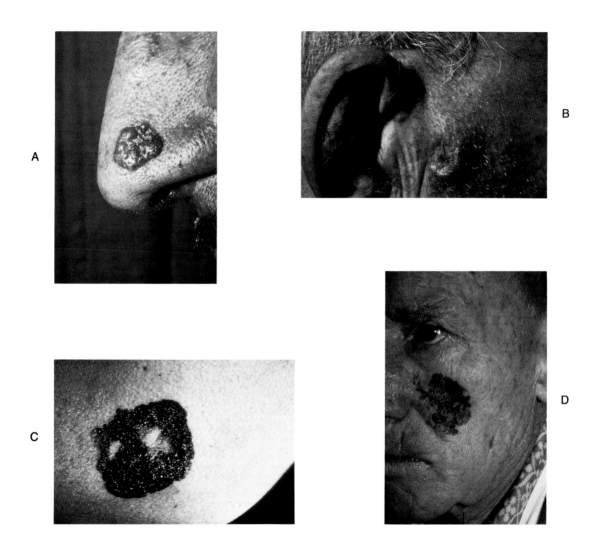

Plate 5 **A,** Basal cell carcinoma. **B,** Squamous cell carcinoma. **C,** Superficial spreading melanoma. **D,** Lentigo maligna.

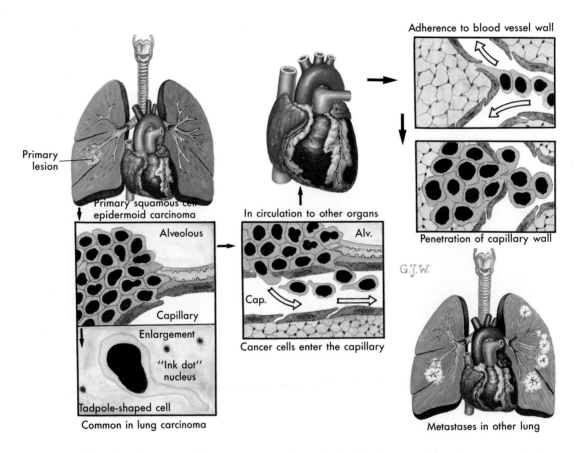

Primary lesion

Primary squamous cell epidermoid carcinoma

Alveolous

Capillary

Enlargement

"Ink dot" nucleus

Tadpole-shaped cell

Common in lung carcinoma

In circulation to other organs

Alv.

Cap.

Cancer cells enter the capillary

Adherence to blood vessel wall

Penetration of capillary wall

G.J.W.

Metastases in other lung

Plate 6   Metastasis of lung carcinoma through the bloodstream. Note that metastasis is commonly through the lymphatic channels and lymph nodes are usually checked for evidence that metastasis has occurred.

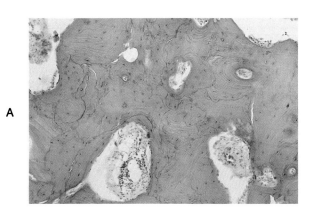

A

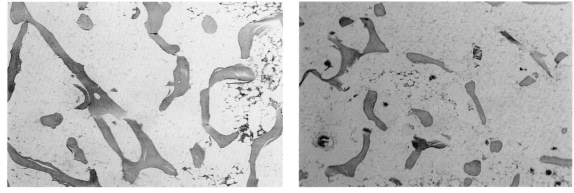

B

C

Plate 7 Photomicrographs of **A,** Paget's disease; **B,** osteoporotic bone; and **C,** normal bone.

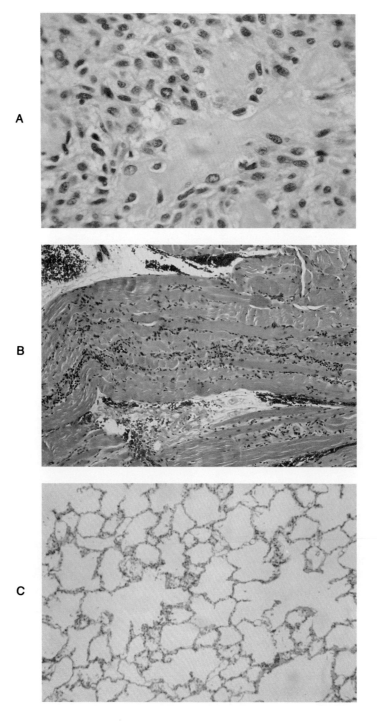

Plate 8 Photomicrographs of **A**, Osteogenic sarcoma; **B**, polymyositis; and **C**, normal lung tissue.

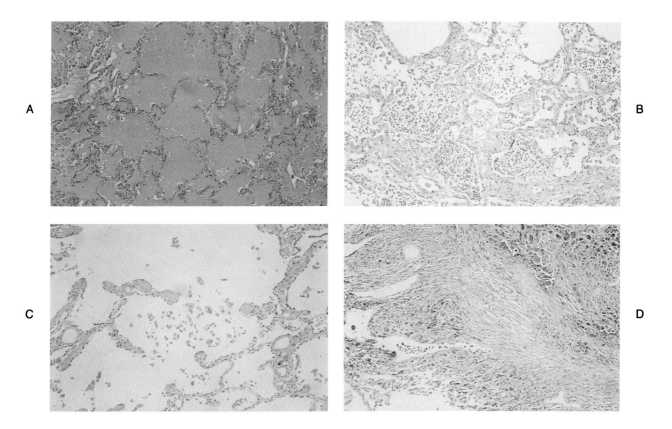

Plate 9   Photomicrographs of **A,** Edema; **B,** pneumonia; **C,** emphysema; and **D,** pneumoconiosis.

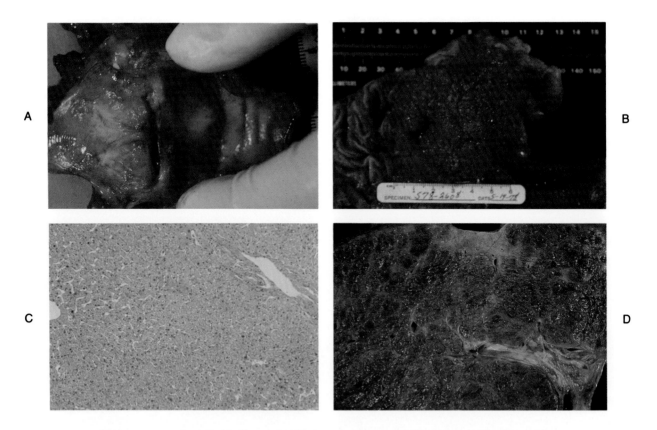

Plate 10 **A,** laryngeal carcinoma. **B,** Villous adenocarcinoma. **C,** Photomicrograph of normal liver. Cirrhotic liver, gross photo.

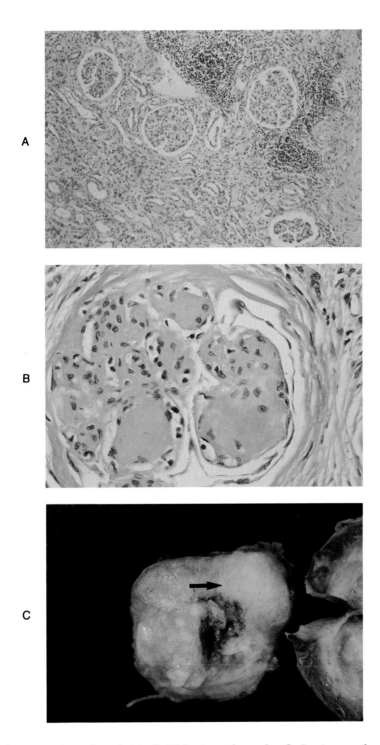

Plate 11 **A,** Pyelonephritis. **B,** Diabetic nephropathy. **C,** Carcinoma of prostate.

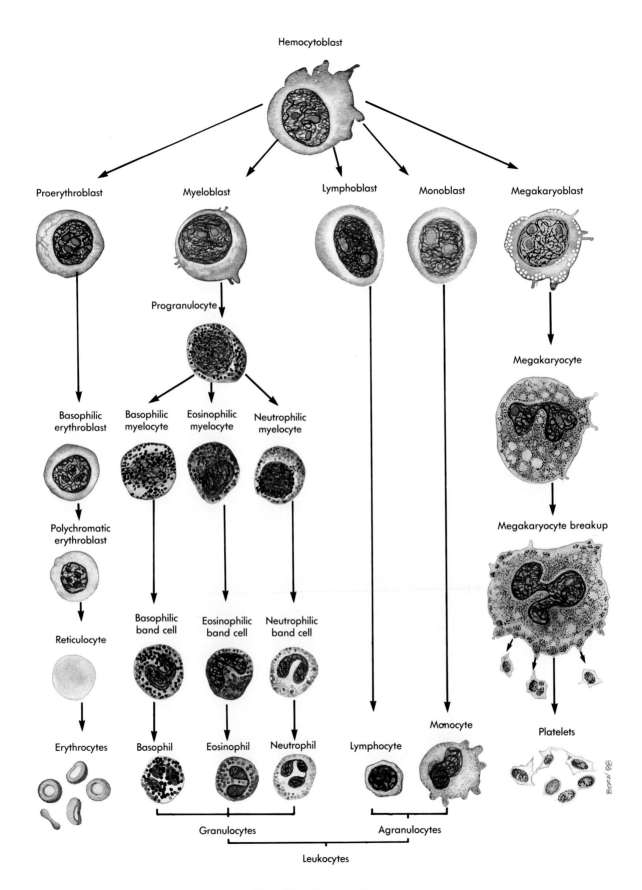

Plate 12   Hematopoiesis.

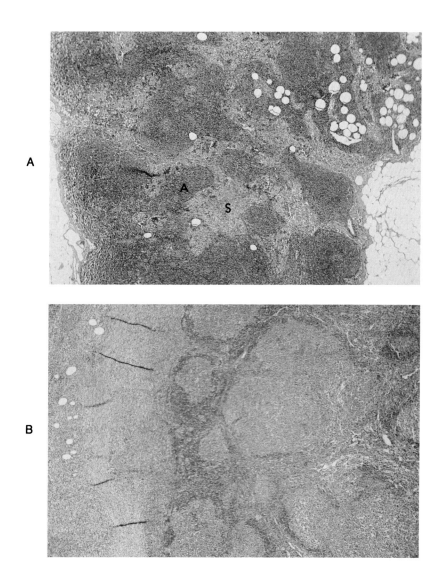

Plate 13   **A,** Normal lymph nodes showing sinusoids *(S)* and aggregates of lymphoid tissue *(A).* **B,** Lymph node with nodular lymphoma. The sinusoids have been obliterated by the malignant proliferation.

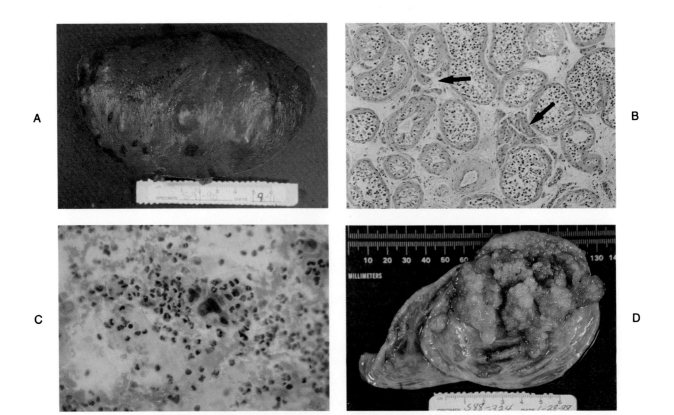

Plate 14 **A,** Spleen infiltrated with malignant lymphocytes as found in some lymphomas and leukemias. **B,** Section of testis showing clusters of interstitial cells of Leydig between the seminiferous tubules. **C,** Positive Pap smear. The presence of cervical cancer is indicated by the shed large orange-staining malignant cells. **D,** Bisected ovarian cyst showing a papillary carcinoma covering the inner lining of the cyst.

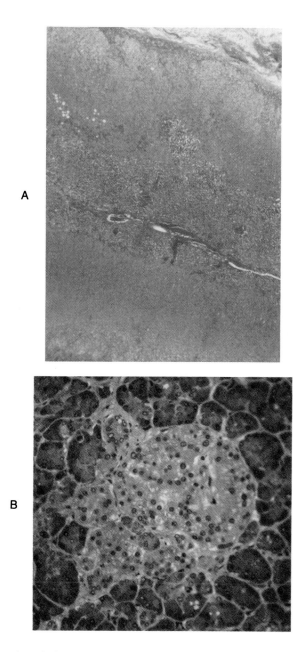

Plate 15 **A,** Adrenal gland. Adrenal cortex *(C)* secretes steroid hormones, such as corti-sol, and the medulla secretes epinephrine and related compounds. **B,** Pancreas. Endocrine Islet of Langerhans resides in a sea of pancreatic acini that secrete digestive enzymes.

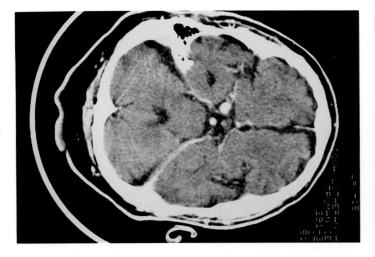

A

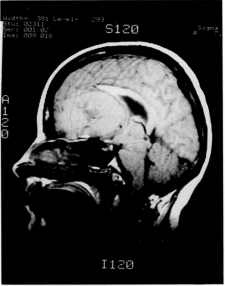

B

Fovea centralis   Macula lutea

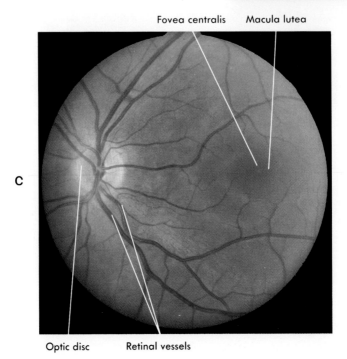

C

Optic disc   Retinal vessels

Plate 16   **A**, Magnetic resonance imaging (MRI) scan showing berry aneurysm (arrow). **B**, MRI scan showing brain tumor (astrocytoma) in the frontal lobe (arrows). **C**, Retina as seen through a fundoscope.

Posture Changes and Exercise

The muscular and skeletal systems work together to produce the posture changes seen in many elderly people. As discussed in Chapter 4, the stance becomes less upright, with the head and neck thrust forward. In addition, the thoracic curvature becomes more pronounced and the lumbar curvature less so. The shoulders may be hunched or rolled forward, and the angle at which the leg articulates with the pelvis changes, leading to a broader based gait with the legs spread farther apart and bent slightly at the knees. Most of this is caused by bone and joint changes. However, in the very old, there is a generalized flexion of the large joints because the flexors are larger and more powerful than the extensors. With age, collagen contraction in the muscles causes the larger muscles to predominate. In addition, weakness and atrophy of rarely used muscles contribute to poor posture.

Regular exercise will not regenerate destroyed muscle fibers. However, a sensible muscle-building program can enlarge and strengthen the remaining fibers, and decrease muscle wasting due to disuse atrophy. In the aged, intact muscle cells can regain strength and tone in a period of 6 to 8 weeks with proper exercise. Most mobile elderly people could benefit from a modest and regular aerobic exercise program involving all the muscle groups, such as aikido. Inaction and lack of normal stress are clearly the greatest enemies of aging muscle. Elderly people decrease their physical activity for many reasons but the net result is always detrimental to the muscular system, and frequently to the rest of the body.

Drug Interactions

It is important to be aware of disease and drug interaction when discussing the problems of any organ system. For example, heart disease is extremely prevalent in the aging population. Often the treatment of heart disease and related disorders involves the use of diuretics to decrease body fluid. Many diuretics, in addition to promoting excretion of water and salt, also promote the excretion of potassium. This may cause muscle weakness that, combined with age-related or other causes of weakness, can have a severe effect on activity and patient attitude.

Conditions and Diseases Associated with Aging Muscle

Posture Changes. Flexors tend to dominate over extensors in most very old people, producing a posture in which the head is bent forward, with the eyes upraised, and the knees, arms, and wrists are slightly bent. As discussed earlier, these changes are caused partly by skeletal changes, especially in the intervertebral disks of the spinal column and the actively used joints of the limbs.

Atrophy of Disuse. This can be observed in young people who have had a leg in a cast. The amount of muscle mass declines rapidly when use of the muscle has been prevented (Figure 5-5). This process is evident also in elderly people who have been confined to bed for even short periods. Some of the weakness associated with bouts of influenza or influenza-like illnesses and other disorders results from atrophy caused by bed rest. Rest in bed for even a few days is detrimental for many elderly persons. Muscular atrophy continues unabated unless some physical therapy is initiated. How severely the immobilization affects an individual will depend on how much muscle mass exists at the onset of the process, how long the immobilization lasts, and what preventive measures are taken.

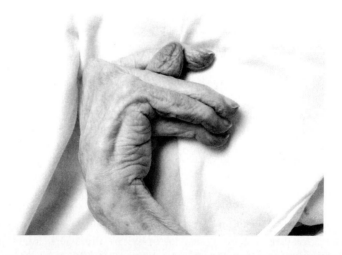

Figure 5-6   Contraction of connective tissue in atrophied muscle may fix the muscle with the joints in a flexed state. In time the angle of flexion becomes more acute and the joints become immovable.

Senile Myosclerosis. This condition occurs in elderly patients who have been bedridden for a very long time. There is considerable muscular atrophy; the limbs become flexed, with the stronger, flexing muscles prevailing over the weaker extensors. Ultimately, muscular atrophy and deposition of collagen fibers result in rigidity (sclerosis) and the limbs can be straightened only with great pain and possible injury. This condition is sometimes called "contractures" before the limbs have become permanently fixed in position (Figure 5-6).

Muscular Dystrophy. This disease is often inherited and usually appears early in life. However, one of its forms may appear as late as the fortieth year. This disease results in the degeneration of individual muscle fibers and their replacement by fat and connective tissue. While this is going on, adjacent muscle fibers may temporarily enlarge. Weakness and muscular atrophy involves the skeletal muscles of specific body areas. Some forms of the disease are selective toward very specific muscle groups such as the ocular or the facial muscles. Ultimately, all of the affected muscles will atrophy but the patient may live a normal life span. The majority of these patients, will eventually require long-term nursing care.

Myasthenia Gravis. This disease results from the loss of functioning acetylcholine receptor sites at the neuromuscular junctions. Although the motor nerve is releasing enough acetylcholine to stimulate a normal muscle, the acetylcholine is not bound to the receptor sites in large enough quantities to start the series of reactions that leads to muscle contraction. The problem appears to be not with the nerve but with the receptive ability of the muscle. The disease is rarely found in people younger than 30 years old and may develop, especially in men, at much later ages. It produces attacks of fatigue of the high use skeletal muscles, especially those of the eyes, face, tongue, and extremities. Severity of the attacks increases as the disease progresses, but the episodes may be separated by periods of spontaneous remission. Death may occur as soon as 6 years after the first symptoms as a result of involvement of vital muscles such as those

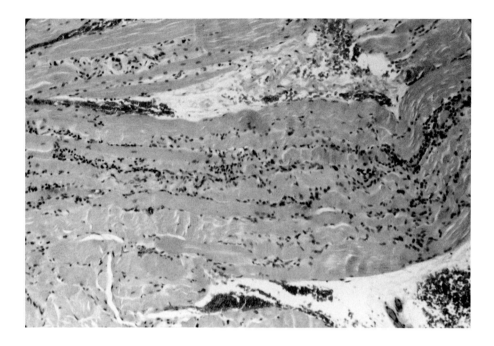

Figure 5-7   Eosinophils congregate around the muscle fibers inflamed with polymyositis. (See also Color Plate 8, *B.*)

used for breathing. Mortality occurs in 14% of men and in 11% of women and is highest in older patients. The generalized form of the disease is more common among women, but a form limited primarily to the muscles of the eyes is more common in men.

Some cases of myasthenia gravis occur in association with thymic problems. In such cases thymectomy, removal of the thymus, may produce improvement. About 80% of people with myasthenia gravis have thymic hyperplasia; about 15% of patients have thymic tumors. About one-fourth of these are invasive but the remainder are benign. Metastasis is rare. Recent studies suggest that the immune system, which involves the thymus, produces antibodies that are responsible for the damage to the acetylcholine receptor sites, thus making this an autoimmune disease.

Polymyositis. This is the most common primary muscle disease of adults. In one-third of the patients the disease occurs alone. In about 40% of them polymyositis is accompanied by skin changes, producing a combination known as dermatomyositis. About 35% of the cases occur in association with lupus erythematosus, polyarteritis, scleroderma, or rheumatoid arthritis. Of patients whose disease begins after 40, 30% have or will have a malignancy.

The disease usually begins with weakness in the legs. Eventually the upper arm muscles become weak in most patients. Muscles of the neck and distal muscles may become involved, but muscles of the face and eyes are usually unaffected. Figure 5-7 shows the accumulation of eosinophils associated with the inflammation of polymyositis. About half of the patients experience pain or tenderness in the affected muscles; about 40% accumulate a pink rash on the face or upper trunk with edema. Erythema is increased redness resulting from expanding capillaries. The dermatitis may involve the hands, feet, and nail beds. Dermatitis precedes the weakness in about half of the patients. The disease may be progressive or broken by spontaneous remissions. In advanced stages, the disease may produce a posture similar to senile myosclerosis. Treatment usually involves corticosteroids or immunodepressant drugs.

Polymyalgia Rheumatica. This and a related condition, giant cell arteritis, occur almost exclusively in the elderly, usually those aged 65 or 70. Giant cell arteritis is an inflammation of the arteries, especially the temporal arteries, causing pain and spontaneous blindness. It is often found in association with polymyalgia rheumatica. Because of this association it is difficult to determine which condition is producing the symptoms observed. Polymyalgia rheumatica is basically a connective tissue disease centered in muscle. The patients usually complain of severe aching and stiffness of the neck, upper arms, shoulders, pelvic region, and thighs. The pain increases with movement and may be accompanied by fever and weight loss. A genetic factor is suspected; it may be that this disease results from an autoimmune problem.

Calcific Tendonitis. This is a condition affecting tendons, not muscles. Pain caused by this inflammation may be interpreted as muscle pain however, and the disease is best covered in this section. Calcium is deposited in the tendons that attach muscles to bones, producing rigidity of the tissue. The calcium deposits may become irritating. Especially when bursa are involved, movement causes pain. This condition is a common complaint of elderly people.

Muscle Tumors. These are generally uncommon and when they occur, they are usually benign. A dangerous but fortunately very rare malignancy of striated muscle is rhabdomyosarcoma.

CASE HISTORIES

❖ Senile Myosclerosis

M.J. is an 86-year-old widow who entered a nursing home after having a massive cerebrovascular accident (CVA) at the age of 69. Her stroke left her unable to move or speak. In the nursing home she was turned on a regular schedule to prevent development of pressure ulcers and was conducted through passive motion exercises every day. Feeding was accomplished through a feeding tube since she was unable to coordinate swallowing.

At about age 78, after 9 years in the nursing home, she began developing flexion deformities (contractures) of the elbows, knees, and back; eventually these became permanent. She is now in a fetal position and her limbs cannot be extended. Any attempt to do so causes obvious signs of pain and distress.

Comment. Senile myosclerosis results from muscle disuse leading to atrophy and replacement of the atrophied muscle fibers with collagenous material, scar tissue. As with all scar tissue, the collagenous fibers contract as they get older. This results in flexion of the joints because the larger flexor muscles are usually replaced with more scar tissue and therefore dominate the smaller extensors. The fact that she remained without contractures most of the nine years post-

CVA can be attributed to the passive range-of-motion exercises performed by the staff of the nursing home. The total disability occurring after a massive stroke requires continuous therapy to prevent the disuse atrophy that occurred with M.J. Even this nonmuscular movement stimulates and preserves muscles and slows the fixation of scar tissue. Often this therapy is not available for economic or other reasons.

❖ Muscular Dystrophy

T.J. first noticed weakness when he was 19 years old. He went to his physician when difficulty in walking became more pronounced. In the last few months he had experienced several instances of stumbling. These falls caused little harm but much embarrassment. T.J.'s initial physical examination showed him to be healthy-appearing except for decreased strength in his arms and legs. This decrease was most pronounced in the large muscles of the upper arms and upper legs, the proximal muscles of his limbs. He had difficulty keeping his arms over his head and he had to use his arms to go from a sitting to a standing position. He walked with a waddling gait. His DTRs (deep tendon reflexes elicited by tapping a tendon with a small mallet) were intact and symmetrical, indicating normal function of the peripheral nerves. All sensory function was normal. The Babinski reflexes, movement of the big toe when the sole of the foot is irritated, were down. This indicated normal control of motor function.

Blood chemistry examination revealed elevation of a muscle enzyme, CK (creatine kinase), to 350 IU/L (normal is 245 or less) and normal levels of other cellular enzymes, aldolase, serum glutamic-oxaloacetic transaminase (SGOT), and lactic dehydrogenase (LDH). A normal WBC and erythrocyte sedimentation rate (ESR) indicated no inflammation or long standing bacterial infection. Examination of a biopsy of a deltoid (shoulder) muscle revealed a slight increase in fibrous tissue without muscle necrosis or inflammation. A diagnosis of muscular dystrophy, limb girdle type was made. There was no history of muscle disease in the family.

During the next 10 years, despite physical therapy, exercises, and splints, his muscle weakness worsened, resulting in confinement to bed. The deterioration involved the muscles of his limbs and trunk, but did not involve his face. At the end of the 10-year period, his face was therefore of near normal contour but his arms and legs appeared to be composed of only skin and bone.

Comment. It should be understood that muscle is a very active organ and as such contains large amounts of enzymes such as CK and SGOT used in the production and storage of energy. Muscle also contains large amounts of aldolase. When muscle fibers are damaged, they release these substances into the bloodstream. The early muscle degeneration in T.J.'s disease led to a mild elevation of CK. The degeneration was not advanced enough to produce elevation in the other enzymes tested.

Although T.J. had lost too much muscle mass to remain functional from a physical standpoint, his mental function was unimpaired. Given the usual development processes that occur in an individual between 19 and 29 years of age, it is not hard to imagine the psychologic impact of this disease on him. Until a way to arrest the physical progression of this disease is found, any treatment program must have a significant psychologic component. (This case is included here because muscular dystrophy patients are often a significant component of a nursing home population; it also provides contrast between a severe primary muscle disease and the muscular atrophy associated with senescence.)

❖ Myasthenia Gravis

G.W. is a 38-year-old housewife who experienced passing episodes of double vision. Her examination revealed nonconjugate gaze (one eye would be focused on an object while the other was pointed toward something off to the side). This lack of normal coordination suggested either neurologic or muscular problems. When her eyes were focused on an object they remained steady with no rapid back and forth movements, therefore showing no signs of nystagmus. Nystagmus is often a sign of neurologic problems. Within 10 days her symptoms spontaneously abated.

Three months later G.W.'s symptoms recurred with the addition of ptosis, or drooping, of the left eyelid. She returned to her physician. Again, her neurologic examination was within normal limits. This time her problem persisted, leading to a more complete evaluation.

The physical examination was normal with the exception of nonconjugate gaze and weakness of her left upper eyelid. As a part of the examination, she was given 6 mg IV edrophonium (an antiacetylcholine esterase agent). This resulted in dramatic improvement in both of her eye problems. Her improvement indicated that her muscles and nerves were normal but that a problem existed with the reception by the muscles of the neurotransmitter acetylcholine. Laboratory examination revealed normal serum enzymes, including CK. However, circulating antibodies to acetylcholine receptors were found and a chest x-ray demonstrated an upper anterior mass believed to be a thymic enlargement. A diagnosis of myasthenia gravis was made.

She was treated with pyridostigmine, another acetylcholine esterase inhibitor, and prednisone, an immunosuppressant intended to reduce the amount of acetylcholine receptor antibody present. This produced only temporary improvement. As her disease progressed, she developed weakness in her upper arms and upper legs, making it difficult for her to rise from a chair or climb stairs. Her eye problems also worsened, leading to bilateral ptosis and continuous double vision. After 9 months of medication, her thymus was removed and found to be hyperplastic with numerous germinal centers. The thymectomy did not improve her clinical course; she developed severe generalized muscular weakness and required placement in a nursing home. In time she became bedridden and began having difficulty with swallowing and speech, necessitating the placement of a feeding tube.

Comments. One of the diagnostic tests for this disease involves injection of an anticholinesterase drug. Since this drug blocks the breakdown of acetylcholine, it allows the acetylcholine to build up and stimulate the muscle even when there are fewer receptor sites available. Her improvement with this drug was conclusive evidence that she had myasthenia gravis. The tests for muscle enzymes such as CK were run to determine whether or not there was rapid muscle breakdown as occurs with some other diseases involving eye muscle symptoms. The levels of these enzymes were normal because in cases of myasthenia gravis the only muscle damage is atrophy.

Myasthenia gravis is an autoimmune disorder of unknown cause that results in reduction of acetylcholine receptor sites by antiacetylcholine antibodies, leading to muscle weakness. This disease is variable and most patients do not become as debilitated as G.W. did. Many respond well to chemical therapy without progression of symptoms. Those who fail drug treatment may benefit from removal of their enlarged thymus gland. In 80% of the cases the gland is tumorous or overactive.

❖ Polymyositis

J.S., a 52-year-old housewife, went to her local physician because of severe weakness and muscle pain of a few months duration. The illness began with mild muscular aching and fatigue, progressively intensifying until she could no longer work. The pain could be elicited by muscle palpation (touching or pushing on it) and was made worse with movement. She was taking no medication. Besides her symptoms she had been healthy and remembered no recent viral infections.

She was hospitalized for a detailed evaluation. Chemical examinations of blood revealed a tenfold elevation in her CK; an SGOT elevation of 5 times normal, and a threefold increase in aldolase, all suggesting a rapid breakdown of muscle tissue. Her CK-MB band (an electrophoretically separated heart fraction of CK) was normal, as was her electrocardiogram (ECG). Her WBC was 18,000 (5,000 to 13,000 is normal) with 36% eosinophils (2 to 4% is normal), and her ESR was 40. The high eosinophil count and erythrocyte sedimentation rate are indications of inflammation. A fluorescent antinuclear antibody (FANA) test for autoimmune diseases such as systemic lupus erythematosus was normal. A workup for malignancy was negative. A biopsy of her deltoid muscle revealed fiber necrosis and a heavy infiltration of eosinophils. A diagnosis of polymyositis was made.

She was placed on high-dose corticosteroids and rapidly improved. The corticosteroids were slowly decreased and eventually stopped. Follow up in 3 years revealed no evidence of malignancy and no further muscular problems. She was carrying a normal work load.

Comments. The high muscle enzyme level in this case indicated fairly rapid muscle deterioration. The muscle damage was caused by inflammation as indicated by the high WBC, disproportionately high eosinophils, and high erythrocyte sedimentation rate. The excellent response to corticosteroids may be the result of both the antiinflammatory and the immunosupressant properties of these drugs, because polymyositis is believed to be an autoimmune disease. About 17% of polymyositis is associated with malignancy; the workup suggesting that malignancy was not involved here was very important.

BIBLIOGRAPHY

Cheshire CM, Cumming WFK: *The musculoskeletal system—skeletal muscle.* In Brocklehurst JC, ed: *Textbook of geriatric medicine,* ed 3, New York, 1985, Churchill Livingstone.

Denham M: *Diseases of muscle and connective tissue.* In Pathy MJS, ed: *Principles and practice of geriatric medicine,* New York, 1985, John Wiley & Sons.

Goldman R: *Decline in organ function with aging.* In Rossman I, ed: *Clinical geriatrics,* ed 2, Philadelphia, 1979, JB Lippincott.

Gorb D: *Common disorders of muscles in the aged.* In Reichel W, ed: *Clinical aspects of aging,* Baltimore, 1983, Williams & Wilkins.

Lifschitz ML, Harman GE: *Musculoskeletal problems in the elderly.* In Schrier, ed: *Clinical internal medicine in the aged,* Philadelphia, 1982, WB Saunders.

Subbarao K: *Radiological aspects of aging.* In Rossman I, ed: *Clinical geriatrics,* ed 2, Philadelphia, 1979, JB Lippincott.

6 *The Respiratory System*

RELEVANT ANATOMY AND PHYSIOLOGY
Respiration

The respiratory system absorbs oxygen from the air and excretes carbon dioxide produced by the body's metabolism. In general usage, the term *respiration* refers to breathing. Respiration is divided into two phases. The first is pulmonary ventilation, or moving air in and out of the lungs. The second phase involves exchange of oxygen and carbon dioxide between the air in the lungs and the blood in the pulmonary capillaries.

Internal respiration refers to the complex chemical process by which sugar is oxidized. (The diagram in Figure 6-1 shows how both external and internal respiration work.) During this process the inhaled oxygen is combined with the carbon atoms of sugars, yielding carbon dioxide, some water, and a great deal of energy. This energy is essential for the thousands of tasks each cell in the body must perform to live. Consequently, when the respiratory system and associated mechanisms fail to deliver adequate oxygen to the tissues for biochemical respiration, the body is stressed. When it fails completely for more than a few minutes, the cells with the highest energy requirements (such as brain cells), and the organs and tissues they make up, begin to die.

Many factors may prevent the body from receiving enough oxygen. These include failure of the respiratory system to deliver enough oxygen to the blood; failure of the blood to absorb enough oxygen; failure of the heart to pump the oxygenated blood; or failure of the blood vessel network to deliver the oxygenated blood to the tissues. In many elderly people, a combination of these factors results in decreased energy and overall decreased viability.

Structure and Function of the Air Passageways

The human respiratory system is composed primarily of the lungs and their conducting system, the collection of tubes and passages leading into the lungs (Figure 6-2). The conducting system is typically divided into the *upper* and *lower airways*. The upper airway is the combination of the nose, pharynx, larynx, and epiglottis. Air normally enters through the nose where it is filtered, warmed, and moistened by the membranes around the coiled nasal conchae. It passes through the pharynx and through the opening called the glottis into the larynx, or voice box. The glottis is protected by the epiglottis, a flap of tissue that normally prevents solids or liquids from entering the larynx when eating. Each day, approximately 10,000 liters of air pass over the vocal cords of the larynx and into the trachea.

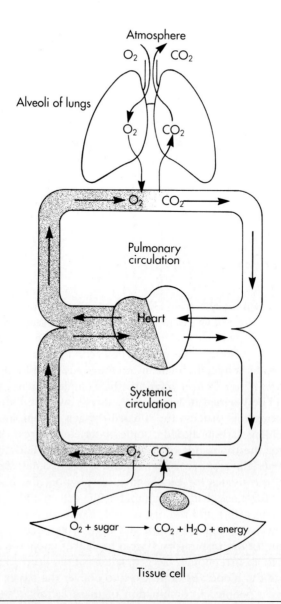

Figure 6-1   External and internal respiration.

The lower airway is composed of the trachea, right and left primary bronchi, secondary (lobar) and tertiary (segmental) bronchi, bronchioles, and terminal bronchioles. This system is also referred to as the *tracheobronchial tree.* The trachea is a large tube supported by cartilaginous C-shaped rings that prevent it from collapsing when the neck is bent or moderately constricted. The trachea divides into a right bronchus and a slightly smaller left bronchus, each of which enters a lung. These bronchi subdivide into progressively smaller bronchi, still supported by cartilaginous rings. These divide to

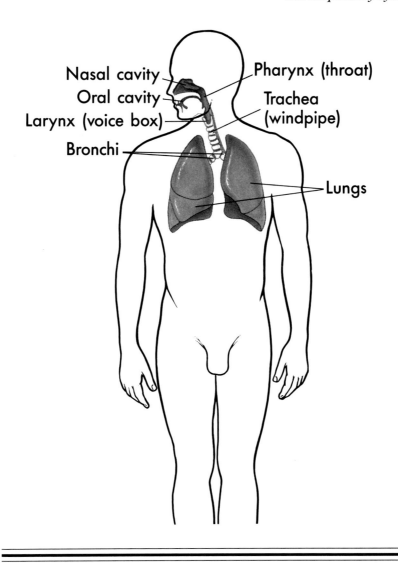

Nasal cavity
Oral cavity
Larynx (voice box)
Bronchi

Pharynx (throat)
Trachea (windpipe)
Lungs

Figure 6-2   The air passages and lungs make up the respiratory system. Some gas exchange occurs across the skin as well.

form bronchioles, which in turn also branch, forming terminal bronchioles. The trachea, bronchioles. The trachea, bronchi, and bronchioles are lined with mucus-secreting and ciliated epithelial cells. The mucus keeps the membranes moist, further moistens the inspired air, and traps bacteria and other particulate matter that has escaped the moist surfaces of the nose and throat. The action of the cilia continually moves the debris-laden mucus toward the trachea and glottis. When it reaches the pharynx, having passed through the glottis, it is usually swallowed. This action of the cilia, along with an occasional cough, and the air currents created by deep breathing, usually keep the lungs free of foreign matter.

The Respiratory Bronchioles and Alveoli

The air passes from the last (terminal) bronchioles into the respiratory bronchioles (Figure 6-3). These structures have irregular, vascular walls containing capillaries which perform the first significant exchange of gas. The air passes from each respiratory bronchiole to several alveoli. Although the alveoli are called air sacs, they are actually complex structures with many folds and projections that provide a large surface area. Most gas exchange occurs here between the air in the spaces inside the alveoli and the blood in the extensive network of capillaries in the alveolar walls. This system of air tubes and alveoli, along with the network of pulmonary blood vessels and the associated connective tissue, makes up the lungs.

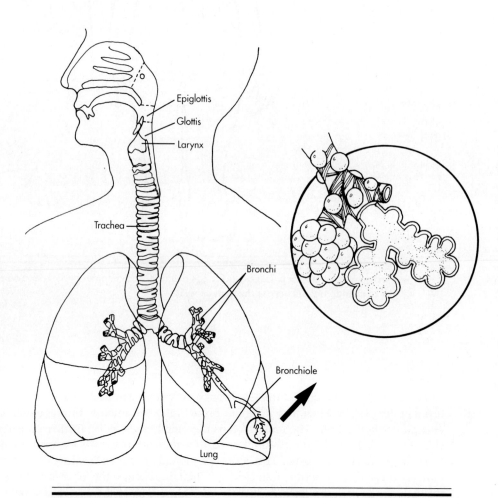

Figure 6-3   Under magnification it becomes obvious that the alveoli are not simple, smooth-walled sacs, but are complex structures set in close association with other alveoli.

The Tissue Barrier to Oxygen and Carbon Dioxide Passage

Closer examination of the alveolar wall at a capillary contact shows that an oxygen molecule in an alveolus must pass many barriers to become attached to a hemoglobin molecule in a red blood cell, or *erythrocyte*. These barriers are illustrated by Figure 6-4. The oxygen must first dissolve in the fluid, mostly water, that covers the alveolar epithelium. This fluid is maintained in a thin film by the surface tension action of phospholipids called surfactant, a natural detergent that prevents the molecular forces of capillary action from "gluing" the walls of the alveolus together.

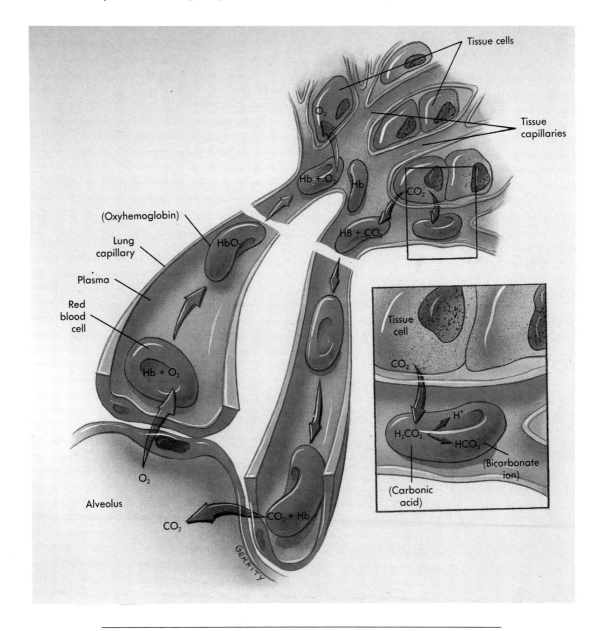

Figure 6-4   The alveolar-capillary interface. Note the number of membranes that an oxygen or carbon dioxide molecule must cross to pass between an erythrocyte and the air in an alveolus.

After the oxygen has dissolved in the water-surfactant layer it must diffuse through a thin, flattened alveolar epithelial cell, the alveolar basement membrane; a thin layer of interstitial fluid; the capillary basement membrane, a thin, flattened capillary endothelial cell; a layer of blood plasma; and finally the membrane of an erythrocyte. Once the oxygen passes these barriers it can dissolve in the cytoplasm of the erythrocyte and attach to a hemoglobin molecule. It is important to recognize that while oxygen is crossing the alveolar membranes in one direction, carbon dioxide is passing in the opposite direction. Both of these substances move by diffusion from areas of high concentration to areas where the concentration is lower. This process does not require cellular energy. When some factor such as abnormal thickness of the alveolar/capillary tissue barrier increases the resistance to gas diffusion, the effect may be both oxygen deficiency and carbon dioxide retention.

Hemoglobin and the Transport of Oxygen and Carbon Dioxide

Hemoglobin consists of four iron-containing molecules, called heme groups, attached to a single globin molecule, a protein. A single oxygen molecule (O_2) can be attached to each of the heme groups. Bear in mind that hemoglobin is three-dimensional although it is illustrated in two dimensions in Figure 6-5.

When hemoglobin acquires its first oxygen molecule, that negatively charged oxygen is attracted by the combined force of all the positively charged iron atoms. This oxygen is therefore bound tightly and is not released when it encounters oxygen-deficient tissues in normal life. When a second oxygen is attached, both oxygens will be

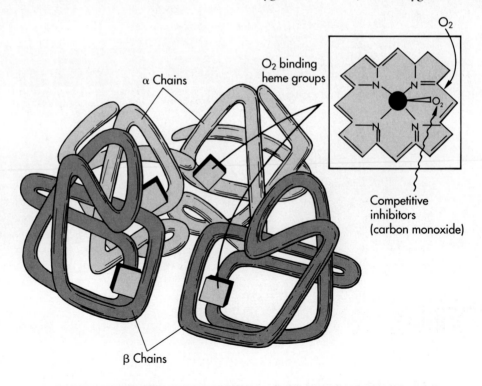

Figure 6-5   Note that oxygen molecules are attracted by all the iron (heme groups) in the hemoglobin molecule but are repelled by each other.

bound a little less tightly, but even these are not normally given up when the hemoglobin passes by an unoxygenated tissue. In the same way the attachment of the third and fourth oxygens progressively cause all the oxygens to be held even less tightly. The attachment of a fourth oxygen results in all the oxygens being held very loosely; any one of them can be easily lost—even to slightly oxygen-depleted tissues. Some oxygen is carried in the cytoplasm of the erythrocytes and some is moved in solution in the water of the plasma, but both of these transport mechanisms are minor in comparison with the amounts carried through attachment to hemoglobin in the erythrocytes.

The hemoglobin molecule's affinity for oxygen changes according to various environmental factors. When the blood is more acidic, oxygen is held less tightly. In the lungs, the high oxygen concentration of the alveoli and low concentration in the blood results in diffusion from the lungs into the blood while the opposite is happening with carbon dioxide. Since the pulmonary blood is less acidic than it is elsewhere in the body, oxygen binds to hemoglobin easily in the lungs. The reverse becomes true when this oxygenated blood reaches the other tissues of the body. There the environment is more acidic because of carbon dioxide being released by the tissues. The oxygen becomes less firmly bound to the hemoglobin and is released to the tissues where the concentration of oxygen is lower.

While oxygen is moving from the blood to the tissues, carbon dioxide is moving in the opposite direction. The cells are manufacturing carbon dioxide, so the concentration of carbon dioxide in the tissues is higher than in the blood. When it enters the blood, carbon dioxide is carried primarily within the erythrocytes and, to a lesser degree, in solution in the plasma. In the erythrocytes it is rapidly converted to carbonic acid in the presence of the enzyme carbonic anhydrase. Some of the carbonic acid may return to the plasma as a carbonate and be transported as bicarbonate. As much as 30% of the carbon dioxide transported can be carried in association with proteins such as hemoglobin.

A lack of available hemoglobin can interfere with delivery of oxygen to the tissues. The body may fail to produce enough hemoglobin because of iron deficiency or other reasons or the hemoglobin may be made unavailable as a result of tight, difficult to reverse combination with outside agents such as carbon monoxide.

Pulmonary Ventilation

Pulmonary ventilation is the process of breathing. *Inspiration* is accomplished by a bellows action involving lowering the diaphragm and raising the front of the rib cage. The diaphragm is a flat, muscular sheet forming a dome as it passes between the liver and the cavities containing the heart and lungs. (See Figure 6-2.) When the diaphragm is lowered it increases the space, thereby reducing the pressure in the chest, or pleural cavity. Air rushes into the lungs, allowing them to expand and fill the relative vacuum. When the front of the rib cage is raised, the angles of the ribs are changed, creating additional space for the lungs with the same effect. Exhaling, of course, is the reverse of this process and is more passive. An individual breath, a single cycle of inhaling and exhaling, is called a *respiration*. A resting person normally has 12 to 18 respirations per minute.

The usual stimulus for taking a breath comes from sensors located in the respiratory center in the medulla and lower pons of the brain. Nerves arising in the respiratory center extend to the muscles concerned with inspiration and expiration. The respiratory impulses are normally rhythmic, with each inspiration and expiration lasting about 2 seconds. Respirations are stimulated primarily by the increased acidity that results when carbon dioxide from the tissue's metabolism accumulates. The carbon dioxide reacts with water to form carbonic acid, which disassociates as follows.

$$CO_2 + H_2O = H_2CO_3 = H^+ + HCO_3^-$$

Carbon dioxide + Water = Carbonic acid = Hydrogen ions + Bicarbonate ions

The hydrogen ions excite the respiratory center, creating the desire to take a breath. This stimulation is much greater from carbon dioxide than from a simple increase in blood acidity. This is because carbon dioxide easily passes through the blood-brain tissue barrier and forms carbonic acid on the inside. However, diseases that result in acidosis, an excessively acidic blood, do cause the patient to hyperventilate, or breathe excessively. A very high level of carbon dioxide (15% or more) may cause suppression of breathing and coma.

Other chemical receptors important in breathing are located in the carotid arteries in the neck and the aorta. These respond both to decreased oxygen and increased carbon dioxide levels. Decreased oxygen concentrations are usually of lesser importance than carbon dioxide levels as a stimulus for breathing. It should be noted that with patients who have chronically high carbon dioxide levels, the acidic stimulus for breathing ceases to function and breathing occurs only in response to hypoxia, or lack of oxygen. Administering high oxygen levels to such a patient can cause death by removing the hypoxic drive to breathe. Pressure receptors present in the carotid arteries increase the rate and force of the heartbeat when the blood pressure suddenly drops.

Lung Capacity

Lung capacity is measured with a spirometer. A normal resting person breathing into a spirometer will exchange about 500 ml of air with each breath. This is known as the *tidal volume.* If a resting person takes a normal 12 breaths per minute, each minute he will exchange about 6000 ml of air. We are capable of inhaling about 3000 ml of air beyond this (called our *inspiratory reserve*) and we are capable of exhaling about 1100 ml beyond our resting expiration (our *expiratory reserve*). The sum of these volumes, about 4600 ml, is called our *vital capacity.* This still does not add up to the total lung capacity. When we have exhaled all the air we can force out of our lungs, about 1200 ml remain. This is the *residual volume.* This air does not remain entirely stagnant. Especially when we breathe deeply, residual air mixes with the newly inspired air. A portion of it will be exhaled with the next expiration. Lung volumes are shown in Figure 6-6.

Lung volumes vary considerably with the age, health, and size of the person being evaluated. It is normal for a person's vital capacity to decrease with age, with a concurrent increase in his or her residual volume. This may result primarily from a loss of the lungs' elastic recoil, or compliance. Studies have not shown change in total lung volume with age.

THE AGING RESPIRATORY SYSTEM
General Comments

The respiratory system is subject to a variety of insults throughout life that take a varying toll. It is difficult to separate damage caused by differing degrees of exposure to cigarette smoke and other damaging substances from changes that would be considered the result of the aging process of a person who has lived a "normal" life. It is interesting to note that nonsmokers often show accumulations of carbon in their lungs that are similar to those seen in smokers. Even primitive people in their "natural" environment are

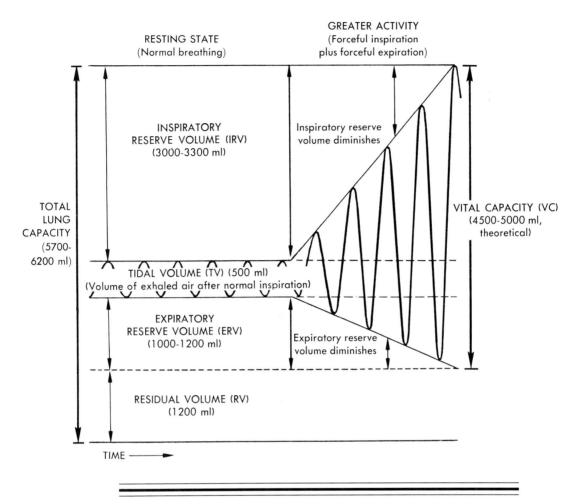

Figure 6-6   Lung volumes.

exposed to large amounts of carbon from wood fires. "Normal" aging changes include changes within the breathing apparatus and tissues of the lungs that are not due to disease or factors arising outside the body.

Threats to the Larynx

The changes that occur in the larynx with age are noticeable primarily for their effect on the quality of voice. An aged voice is typically weak and somewhat intermittent and rasping. The vocal cords may be damaged by misuse or, in the elderly who are isolated or are in a situation where they rarely speak, lack of use. It has been pointed out that the nodules and polyps that destroy or alter voice result from tissue reactions to chronic stress or vocal overuse. These forms of stress include such things as screaming, excessively harsh talking, and faulty singing technique.

Effects on Breathing

Several other changes that may impede respiratory function occur with advancing age. Changes in the vertebral column may cause kyphosis, increased thoracic curve;

scoliosis, a lateral curvature; or the combined condition kyphoscoliosis. In cases of vertebral osteoporosis, compression fractures can decrease the height of the thorax (see Figure 4-6). This decrease is accented by the intervertebral disk compression that occurs with age. Osteoporosis may affect the ribs and the vertebrae; costal cartilages may become calcified and inflexible. As a result the chest cavity will be distorted and relatively rigid, usually in a partially expanded mode. Such conditions allow little opportunity for deep breathing. In support of this, measurements have shown that the anterior-posterior diameter of the chest increases with age, although others report that this is inconsistent in women. Other "external" changes that can affect respiratory performance include shallowness of breathing resulting from obesity, chronic poor posture, or weakness of the breathing muscles as a result of inactivity often associated with advanced age and disease. The speed of muscle contraction (the period between stimulus and relaxation) increases with age and results in more conscious effort being applied to breathing.

Changes within the lung affect breathing as well. All elderly people will have lost some alveolar efficiency because the average size of the alveoli and respiratory bronchioles increase with age. Because the surfaces of the alveoli are gas exchange sites, efficiency is lost. As the alveolus expands, its surface area decreases in relation to its volume. This decreases the functioning absorptive area of the lungs and increases the residual volume.

Scar tissue accumulation in the lungs resulting from past infections and collagen cross-linking decrease the elasticity of individual alveoli and eventually, the entire lung. This loss of elastic recoil causes the lungs to expand less readily to fill the chest cavity when the ribs are raised and the diaphragm is lowered. When the diaphragm is raised and the rib cage is lowered in exhaling, the lungs empty less readily, trapping air that otherwise would have been exhaled. Loss of elastic recoil is a major contributor to the decreased vital capacity seen with advancing age. Bronchitis of various degrees occurs in most elderly patients and causes obstruction of the airways, resulting in further difficulty in exhaling. This too will decrease vital capacity, with an equivalent increase in residual volume.

In summary, decreased capacity for breathing in the healthy aged is attributed to such factors as increased alveolar size; decreased elastic recoil of the lungs; increased stiffness in the framework around the lungs; increased resistance during exhalation; and decreased muscle strength. All of these factors make the treatment of pulmonary diseases even more difficult, with greater morbidity and mortality in the elderly than in younger populations.

Decreased Resistance to Infection

The lungs, skin, and digestive system are all exposed to a continuing onslaught of infectious agents. Skin has a mechanical barrier that, along with its relative coolness and dryness, gives it protection. Protection is afforded the digestive system by its continual flushing action and the exposure of infectious organisms entering with food to tremendous variations in pH. In contrast, the warm, moist surfaces of the lungs provide an ideal environment for bacteria and fungi, especially in the deepest, most poorly ventilated alveoli. Table 6-1 lists the effects of aging on the lung's defenses.

Dyspnea, Hemoptysis, and Chest Pain; Symptoms

Among physical complaints present in elderly people, dyspnea, the inability to breathe enough air, is among the most frequent. A little more than half of such people

Table 6-1   Effects of Aging on Lung Resistance to Infection

DEFENSE	CAUSE OF IMPAIRMENT
Cough reflex	Loss of elastic recoil
	Chest stiffness
	Decreased muscle strength
	Neuromuscular problems
Cilia in bronchi	Chronic bronchial irritation leading to paralysis or loss of ciliated epithelial cells
Immunoglobulin produced on mucosal surfaces	Decline in amount
Alveolar macrophages	Alteration (see Chapter 9)

Adapted from King and Schwartz, 1983

have identifiable cardiopulmonary problems. Many of the remaining complaints can be attributed to the early stages of chronic pulmonary disease, lack of muscular conditioning, or simply an above-average number of insults accumulated in an aged lung. When hemoptysis, the coughing of blood, occurs, one can suspect severe bronchitis, bronchiogenic carcinoma, tuberculosis, or pulmonary embolism. Chest pain is also a common complaint. Although patients often interpret this as heart pain, it may be caused by many other factors. These include muscle pain from severe coughing; pulmonary embolism; costochondritis, the inflammation of a rib or its cartilages; a rib fracture; upper gastric irritation; and pneumonia with pleurisy, inflammation of the membranes surrounding the lung.

Diseases and Conditions Associated with Age

Pulmonary Edema. Pulmonary edema refers to the accumulation of fluids in the tissues and alveoli of the lungs. Specifically, this accumulation starts in the interstitial spaces between the capillaries and the alveolar epithelium and spills into the alveolar lumen. This spillage prevents gas exchange in some alveoli and makes other less efficient.

Pulmonary edema may be caused by allergy, infections such as those producing pneumonia, or cardiopulmonary problems such as congestive heart failure. In the latter instance, a damaged left ventricle of the heart fails to pump blood away from the lungs as rapidly as the right ventricle delivers blood to the lungs. This results in increased blood pressure in the pulmonary veins and capillaries of the lungs. The capillaries begin to leak plasma filtrate into the interstitial spaces, widening the alveolar walls. Eventually the fluid leaks into the alveolar lumen, where it leads to complete alveolar failure. As this fluid finds its way into the bronchi, it triggers the coughing reflex and in severe cases, may be expelled through the mouth and nose as a frothy pink liquid. Figure 6-7 compares a normal alveolar membrane *(A)* with a badly inflamed one *(B)* showing considerable edema.

Interstitial Pulmonary Fibrosis. This condition usually results from prolonged inflammation of the alveolar walls. The onset of interstitial lung disease occurs most frequently between ages 45 and 60 and is often idiopathic, that is, of unknown cause. When the alveolar wall is inflamed, interstitial edema develops in conjunction with an inflammatory cell infiltrate. This may heal by fibrosis and thickening of the alve-

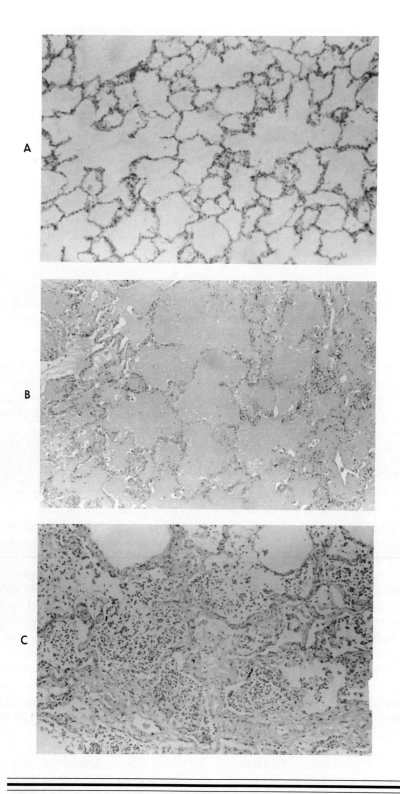

Figure 6-7   For legend see opposite page.

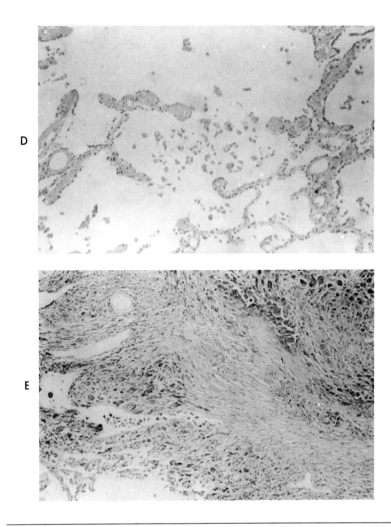

Figure 6-7 A normal lung **(A)** and lung with pulmonary edema **(B).** Note the fluid within the alveolar walls and in the alveoli with edema. **(C)** Pneumonia. **(D)** Emphysema. **(E)** Pneumoconiosis. (See also Color Plates 8, *C* and 9, *A* to *D*.)

ola and walls. Eventually cystic fibrotic lesions develop that are easily seen on x-ray. The connective tissue deposits create an even more severe barrier to gas exchange. The treatment usually involves corticosteroids in an attempt to reduce the inflammation. The overall survival rate for idiopathic pulmonary fibrosis is only 4 or 5 years. About 10% of these patients die from lung cancer.

Bronchial Asthma. Bronchial asthma may occur at any age. When it begins during the early years, it usually disappears by age 40. Like other lung problems, it is usually more dangerous when it occurs in the elderly. These patients may have difficulty both inhaling and exhaling and will exhibit prominent wheezing, especially when exhaling.

This occurs because the bronchioles involuntarily constrict, making it difficult for the air to pass. The severe effects of a single attack can usually be alleviated with bronchodilators.

Pneumonia. Anatomically, pneumonia is diagnosed when the alveoli of a part of a lung have become filled with inflammatory material. In the early stages of the disease the alveoli become congested, usually with proliferating bacteria, plasma exudation, and white blood cells. Later, small capillary hemorrhages occur and the alveoli fill with erythrocytes and plasma proteins, especially fibrin (Figure 6-7, *C*). As the disease progresses, pressure, bacterial enzymes, and enzymes from dying white blood cells may destroy some of the alveolar walls and create abscesses, areas of necrotic tissue infiltrated with bacteria and inflammatory cells. Rarely, an abscess may penetrate the outer wall of the lung and allow air to escape into the pleural cavity which surrounds the lung, causing the lung to collapse. This accumulation of air in the pleural cavity is called *pneumothorax*. The condition of having a collapsed lung is known as *atelectasis*. Atelectasis is more commonly caused by blockage of the bronchioles by mucus. This action results in the complete obstruction of an airway and the collapse of the alveoli as the gas within them is destroyed. The pneumonia begins to resolve when the irritating agent is destroyed or neutralized. The debris in the alveoli are then cleared by the action of macrophages, coughing, and absorption of fluids. The undamaged alveoli will eventually return to their job of gas exchange. Pneumonia can produce permanent damage by causing destruction of some alveolar walls and their replacement by nonfunctional scar tissue. For this reason, people who have had pneumonia several times may have reduced respiratory capacity.

Pneumonia may occur in geriatric patients as a result of normally contagious respiratory infection, or as a result of infective agents that would not create a problem for a healthier, younger person. Pneumonia and influenza are collectively the fourth leading cause of death for people over 65. Pneumonia is often a complication of other serious illnesses and may end a life seriously weakened by other terminal disease. It is the primary killer of patients who are bedridden for prolonged periods. Pneumonia has been termed the "old man's friend," because it has ended the pain and struggle of many hopeless cases.

Chronic Bronchitis, Chronic Obstructive Pulmonary Disease, and Emphysema. These diseases have been traditionally described as separate conditions. New information suggests that they are often phases of the same disease entity. When chronic bronchitis has developed to the point that there is continual obstruction of the bronchioles and smaller bronchi by mucus, the condition is properly called chronic obstructive pulmonary disease, or COPD. When the damage to the alveoli is so severe that the alveolar walls are destroyed either as a result of continuation of this condition or as a result of other causes, the condition is known as emphysema (Figure 6-7, *D*).

Chronic bronchitis develops when the bronchi have suffered an extended period of irritation. The term describes a clinical entity in which an individual has had a productive (sputum-producing) cough for more than 3 months yearly for 2 successive years, unexplained by another disease. In the development of COPD the bronchiole epithelium usually has hypertrophied (thickened). The thickened bronchiole walls and increased mucus block expired air and cause dilation and destruction of the alveolar walls, leading to emphysema.

Emphysema is often associated with exposure to bronchial irritants such as smoke. The passively destroyed alveolar walls are replaced by nonvascular scar tissue that does not aid in gas exchange. This results in larger alveoli with less vascular gas exchange area. Compare the normal lung tissue (Figure 6-7, *A*) with tissue removed from a patient with emphysema (Figure 6-7, *D*).

Patients with severe emphysema are typically anxious due to the unrelenting dyspnea. The work of breathing requires the conscious attention of the patient; the energy required for this extra work also increases. These patients are typically thin because the breathlessness interferes with both the ability to eat comfortably and the ability to shop for and prepare meals. The patient will generally have a barrel-shaped chest with an increased anterior-posterior diameter. The posture typically assumed by someone with emphysema is one that supports maximum ventilatory effort—with the clavicles and shoulders elevated to increase the vertical space in the thoracic cavity. Because of the loss of much alveolar tissue, the lungs on x-ray are more radiolucent (permeable to x-ray) than normal. The diaphragm is flat and the lungs are maintained in a state of partial inflation even after exhaling. Spirometry often shows an increase in total lung capacity. Emphysema can be a crippling and fatal disease, especially if it is associated with chronic bronchitis. Hypoxia, or deficiency of oxygen reaching the tissues, or inadequate oxygenation of the tissues, and cor pulmonale, a condition involving hypertrophy and damage to the right ventricle and sometimes distention of the pulmonary arteries, are frequent partners of advanced emphysema.

Typical pulmonary senescence with enlarged alveolar spaces and postural changes can have the appearance of emphysema. In one study, one half of the individuals over 70 years of age showed mild centrilobar emphysema on autopsy. This is sometimes called senile or benign emphysema and does not cause significant disability.

Patients with COPD usually breathe at a slow rate with an increased tidal volume. They may have developed a tolerance to high carbon dioxide levels with blood that is poorly oxygenated, and the chest may be continually expanded. The forcing of air past inflamed, partially plugged bronchi may produce a wheezing sound. COPD usually involves frequent episodes of bronchitis with bouts of pneumonia; the resulting damage leads to further compromise of pulmonary function.

Several complications (other than the development of specific diseases) arise from COPD. These include cor pulmonale and hypoxia. Cor pulmonale results from increased resistance to blood flow through the lungs caused by scarring of the alveolar tissues or by direct damage to the pulmonary arterioles. Persistent high blood pressure in the right ventricle and the pulmonary arteries (pulmonary hypertension) may eventually contribute to tricuspid valve failure, cardiac arrhythmia (alteration of heart rhythm), and distention and failure of the right ventricle. Periods of cessation of breathing during sleep increase with COPD. This condition, known as sleep apnea, also produces cardiac arrhythmias and may result in sudden death. With COPD there is also a higher incidence of pulmonary embolism (the obstruction of a pulmonary artery or arteriole) usually with a blood clot arising elsewhere in the body.

Pneumoconiosis. Continual irritation of the alveolar walls by foreign matter may result in pneumoconiosis, reactive fibrous thickening and inflammation (Figure 6-7, *E*). This makes gas exchange much more difficult because of the increased thickness of the respiratory membrane. This disease often has an obstructive component, with chronic bronchitis and emphysematous changes. Excessive exposure to mineral dust, especially silica, asbestos, or coal dust, is a common cause of pneumoconiosis. Black lung disease (coal dust pneumoconiosis) has received publicity lately because of its high incidence among coal miners. Pneumoconiosis patients commonly develop dyspnea and become cyanotic, or blue, as a result of low blood oxygen saturation.

Pulmonary Aspiration. Elderly people can develop several conditions that contribute to accidental inhalation of foreign material. Altered consciousness levels and reduced gag and cough reflexes are natural conditions contributing to pulmonary aspiration. Anesthesia, alcoholism, use of endotracheal tubes, and resuscitation procedures may also result in foreign, especially gastric material, finding its way into the lungs. There is a tendency to aspirate into the right lung. This occurs because of the more vertical position of the right mainstem bronchus, resulting in a greater continuity with the trachea.

Aspiration of highly acidic gastric juice is a moderately common and dangerous problem. It occurs when vomitus is aspirated from a nearly empty stomach and may be associated with poor coordination of swallowing and breathing. It produces immediate respiratory distress that worsens over time as the acid damages portions of the lung. The alveolar walls are partially dissolved, resulting in necrosis of the wall, focal hemorrhage, and an outpouring of plasma-like fluid (pulmonary edema). In time the alveoli may become filled with inflammatory cells, necrotic debris, and fluid. This material may form a nutrient medium for a superimposed bacterial infection occurring a day or so later. Aspiration of gastric contents such as partially digested food leads to a more chronic pneumonia that is usually less dramatic and is often associated with bacterial infection. Either form of aspiration, if massive enough, can lead to death.

A different form of aspiration occurs in some elderly and debilitated patients. It results from a combination of poor dentition and unchewed food fragments, recumbent or semirecumbent body position, decreased gag reflex, and an overall weakened condition. Under these circumstances, a large food fragment can lodge in the larynx and result in sudden death from choking.

Tuberculosis. This disease is increasing in incidence in the elderly population. Researchers report that 60% of the cases now occur in people 45 years old or older. It is caused by a bacterium *(Mycobacterium tuberculosis)* and is usually acquired from inhaling the organism or by reactivation of a prior contained infection, with age-associated decline of the immune system (Figure 6-8). In nursing homes or residential institutions, the combination of close quarters, an aged population, and the large number of aerosoled organisms shed by infected people can result in major epidemics.

The disease often occurs in elderly adults who carry latent infection from exposure in their childhood. It erupts when the immune system begins to decline or in the presence of immunodepressant drugs, renal dialysis, alcoholism, or diabetes. Interestingly, up to 20% of patients with active tuberculosis have negative tuberculin tests, demonstrating a decreased immunologic reaction to the organism. Many infected elderly patients fail to show the normal symptoms of high fever, night sweats, and hemoptysis. Because a common observable symptom is loss of appetite, most will have experienced some recent weight loss.

After entering the tissues, tuberculin bacteria are actively consumed by macrophages. However, instead of dying, the bacteria happily reproduce in these white blood cells and may be considered symbiotic. After several weeks have passed, the body mounts an intense immune response against the organisms. Macrophages proliferate and soft tubercles, aggregations of macrophages, develop at the points of infection and in the lymph nodes draining the area. Later, the immune process enables the macrophages to

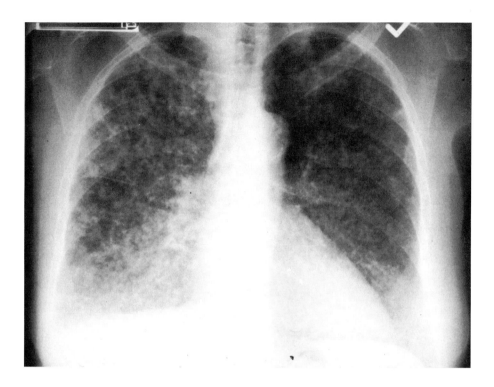

Figure 6-8   This x-ray of the lungs shows the calcified, bacteria-containing tubercles of tuberculosis.

inhibit reproduction of the bacteria, and fibrous tissue walls off the infection. The tissue in the center of this tubercle dies and may become calcified but the bacteria may remain alive. If the bacteria break out of the tubercle and enter the blood stream, the many resulting secondary infections (miliary TB) will produce a severe response by the now hypersensitive body. Virtually any organ can become infected. In addition, the immune system's overactive response to the organisms can cause extensive damage to the surrounding tissues. The disease can be controlled by antibiotics today, but infected people must be educated about the importance of correct disposal of sputum and avoiding contamination of others by coughing. The discovery of active tuberculosis in the close quarters of a nursing home poses obvious problems.

Bronchogenic Carcinoma. This common lung cancer is most often diagnosed in late middle age or later. The disease begins earlier but early tumors are often slow growing and may take up to 20 years to become symptomatic. Bronchogenic carcinoma accounts for 90% of lung cancers. The rate of lung cancer increased 50% between 1960 and 1967, at which time it accounted for 3% of all deaths and 19% of all cancers. It is primarily caused by cigarette smoking and is uncommon among nonsmokers, except those who work with certain carcinogens such as asbestos or some petroleum products.

The four forms of bronchogenic carcinoma are *squamous cell* (Figure 6-9, *A* and 6-10), *adenocarcinoma, large cell carcinoma* (an undifferentiated form of the first two), and *oat cell carcinoma* (Figure 6-9, *B*).

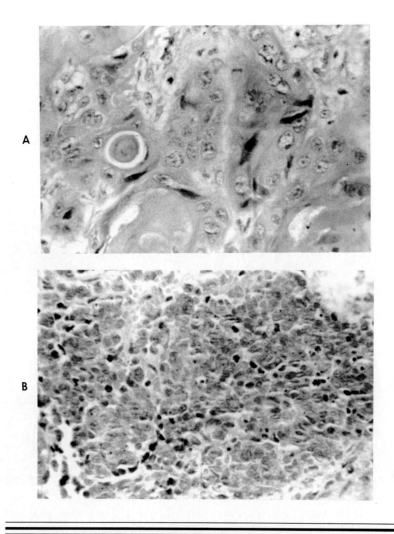

Figure 6-9   Squamous cell (**A**) and oat cell (**B**) carcinomas.

One of the dangers of lung cancer is that its symptoms usually do not appear until extensive metastasis or invasion has occurred. A productive cough with hemoptysis (blood in the sputum) and irritating pain may be present if the lesion develops in the epithelium of the bronchial lumen. This is often masked by a cigarette cough or by chronic bronchitis. If the lesion develops away from the bronchial lumen, a large growth may be produced before symptoms appear. These tumors are often discovered with routine chest x-rays. Many lung cancers, especially the aggressive oat cell form, are found because of symptoms produced by a metastatic lesion, often in the brain, bone, or liver, or because of invasion of the mediastinum, the collection of tissues and organs (including the heart) between the lungs. Sometimes the lesion invades the outer surface of the lung, the pleura, allowing fluid, tumor, and debris to accumulate in the pleural cavity. The prognosis for bronchogenic carcinoma is poor. The average patient survives only 6 to 8 months after the disease is diagnosed. See Color Plate 6 for an illustration of vascular metastasis.

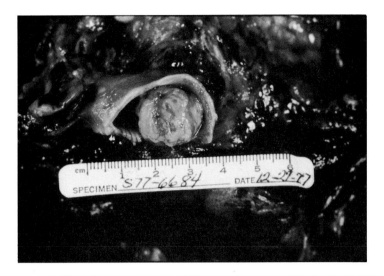

Figure 6-10   Squamous cell carcinoma of the lung projecting from a bronchus.

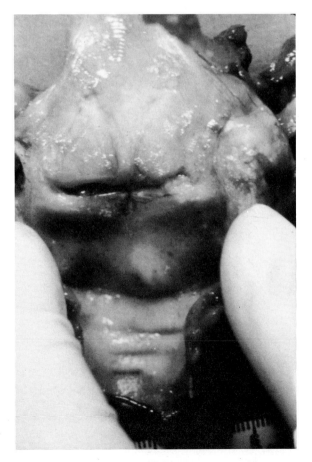

Figure 6-11   Carcinoma of the larynx. In the early stages this cancer may be difficult to distinguish from benign leukoplakia.

Carcinoma of the Larynx. Laryngeal cancer comprises 2% to 5% of all malignancies diagnosed annually. The incidence is higher in men, in smokers and drinkers, and among people in the fifth, sixth, and seventh decades of life. The tumor involves the voice box and often interferes with the function of the vocal cords, resulting in voice changes such as hoarseness (Figure 6-11). The prognosis for this carcinoma is good when the disease is diagnosed early or when the vocal cord can be stripped of the lesion. Hyperkeratosis of the vocal cords may develop, producing a white, lichen-like growth called leukoplakia. This is difficult to distinguish from carcinoma. Benign growths on the larynx are more common than malignancies.

Pulmonary Embolism. The lodging of a blood clot (embolism) in an artery of the lungs is a common cause of death or illness in inactive or postoperative patients. Prolonged inactivity of the lower extremities, due to bed rest or prolonged sitting, can result in the formation of blood clots in the deep veins of the legs. These clots may break loose (becoming emboli), pass through the right side of the heart, and lodge in the lungs. If a large enough embolism occurs, the entire blood flow through the lungs can be blocked, resulting in sudden death. Smaller emboli may go undetected or cause chest pain, shortness of breath, and necrosis of tissue.

CASE HISTORIES

❖ COPD

The first symptom noticed by C.O. was shortness of breath while climbing stairs. This was 9 years ago when C.O., a salesman, was 43 years old. He had been smoking 2 packs of cigarettes a day since the age of 14 and during the past 9 years had experienced extreme intolerance to exercise, a chronic cough most bothersome in the morning, and production of a thick clear to white sputum. He had been advised to quit smoking but all of his attempts to quit failed.

When fully evaluated by his physician at the age of 52, C.O.'s symptoms had progressed to the point of chronic health problems even when resting. Recently he had required yearly hospitalization for pneumonia or bronchitis. His chest had a barrel-like appearance, the diaphragm was depressed, and on percussion (thumping on the chest with the fingers) it produced excessive resonance. With each breath rhonchi (the rattling of thick mucus in the bronchi) could be heard.

On laboratory evaluation, C.O. was noted to have decreased oxygen levels in his blood with increased levels of carbon dioxide and was suffering from acidosis (acidic blood). This, combined with his history and physical findings, resulted in a diagnosis of chronic obstructive pulmonary disease (COPD).

C.O. was again advised to quit smoking and was treated with bronchiodilators, expectorants, and antibiotics as needed. Despite this advice he continued to smoke. Over the years he became more disabled, eventually requiring continuous oxygen administration. For these reasons he became homebound. The respiratory distress worsened until even nasal oxygen was not sufficient. He developed pulmonary hypertension and pulmonary edema. At that time he was offered a tracheostomy with permanent ventilatory assistance but declined. He died a few months later of pneumonia at the age of 61, after a period of right-sided heart failure.

Comments. Eighty percent of the COPD cases are associated with smoking and most have an emphysematous component combined with chronic bronchitis. The bronchitis is responsible for the cough, productive sputum, and airway obstruction, resulting in wheezing and difficulty exhaling. It is often complicated by episodes of acute bronchitis and pneumonia, resulting in acute respiratory distress. The emphysema causes chest expansion with poor respiratory excursions, pulmonary distension and increased unusable breathing space in the lung. With C.O. the increased resistance to blood flow through the lungs resulted from loss of alveoli and their associated capillaries. This created back pressure in the pulmonary arteries (pulmonary hypertension) and contributed to his pulmonary edema and right-sided heart failure.

❖ Pneumoconiosis

J.S. was a 63-year-old retired coal miner who had developed severe respiratory distress. His breathing problems began at the age of 55 years, after 37 years of coal mining in West Virginia. At the onset of symptoms he noticed difficulty in catching his breath while working. This symptom rapidly progressed over a few years to shortness of breath while sitting. A presumptive diagnosis of black lung disease was made and he was released from work. Since that time his health has deteriorated; initially he needed continuous supplemental oxygen to be comfortable but then progressed to needing a respirator and a permanent tracheostomy. While on his respirator he has experienced multiple episodes of bronchopneumonia treated by hospitalization, pulmonary toilet (a combination of chest pounding, deep breathing exercises, and sometimes suction), and antibiotics. Recently he had again developed a fever, become less responsive, and began producing more tracheal secretions tinged with blood. This study begins with his hospitalization for the treatment of pneumonitis.

On physical examination he was noted to be an elderly, emaciated, male with a tracheostomy and on mechanical ventilatory assistance. The vital signs were: temperature 39.6° C, blood pressure 108/64, pulse 118, and controlled respirations of 16/min. His lips and nail beds were cyanotic (blue) and his face was flushed. His chest was barrel shaped with poor respiratory excursions with dullness to percussion in both lung bases. There were decreased breath sounds and rales (a rasping sound indicating alveolar collapse) in the same area. His blood gases revealed profound hypoxia and marked acidosis (blood pH was 7.02; normal is 7.42). There were 26,400/mm^3 white cells (4500 to 11,000/mm^3 is normal) with some immature granulocytes. X-ray confirmed the diagnosis of lower lobe pneumonia and blood and sputum cultures contained *Streptococcus pneumonia*. A CT scan of the brain showed diffuse cortical atrophy.

J.S. was treated with tracheal suctioning, IV penicillin G and replacement fluids but died on his second hospital day. The autopsy revealed bilateral lower lobe pneumonia in the stage of red hepatization superimposed on severe emphysema. The lungs contained large amounts of anthrocotic (coal) pigment and silica.

Comments. Pulmonary emphysema secondary to particulate matter is usually progressive, resulting in death. Often the final event is a superimposed viral, fungal, or bacterial infection as in this case. The physical manifestations of emphysema result from the destructive processes that convert the lungs into large, air-filled cavities. These empty sacs lack the surface area to be effective mechanisms of gas exchange and result in overinflated, noncompliant lungs that cannot move much air. When the process becomes severe enough to require long-term oxygen therapy, it is not unusual for the patient to experience short episodes of hypoxia resulting from inattention to detail or from infection. These episodes are sufficient to cause brain damage and a chronic encephalopathy.

❖ Tuberculosis

P.N. is a 58-year-old epileptic single woman who was brought in for treatment with hemoptysis (blood in the sputum) of recent onset. Her past medical history includes severe epilepsy and she has been in a county care facility all her life because of the seizures and retardation. At the age of 18 she became skin test positive for tuberculosis and has had yearly chest x-rays since then that have been negative. During the last year she has lost 28 pounds, some while dieting, but has continued to lose even after discontinuing her diet. She recently began spitting up blood and has noticed easy fatigability.

Physical examination showed P.N. to be a chronically ill-appearing pale woman who appeared to be of normal intellect. She was mildly anemic with a hemoglobin of 11 mg/dl (normal 12 to 16). Her chest x-ray showed numerous calcified nodules, presumably granulomas, with a superimposed right lower lobe infiltrate suggestive of pneumonia.

Examination of her sputum revealed many white blood cells with normal oral flora grown on cultures. She was treated empirically with various antibiotics over the next 2 months without clearing of her infiltrate. To rule out a tumor, a right thoracotomy was performed with a biopsy of the right lower lobe. Multiple granulomata were found in which tubercular organisms could be seen. Cultures from the lung tissue grew *Mycobacterium tuberculosis* sensitive to all antibiotics generally used for TB. Sputum stains also revealed multiple organisms. She was treated with INH and rifampin and recovered.

Comments. This is an example of *reactivation* tuberculosis, which is a more common clinical problem than the primary infection. Often the first exposure leads only to skin test positivity and perhaps a pulmonary x-ray abnormality. The body limits and contains the organism and no acute disease is seen. In time, and especially if there is a lowering of the patient's resistance, the still-living organism again becomes a problem. During the entire time the patient was treated with ineffective antibiotics, she was spreading infectious organisms. Although we have no knowledge of an epidemic, others could have been infected. Her lifetime in a care facility reflects the historic lack of understanding of epilepsy by the general public and some medical personnel.

❖ Squamous Cell Carcinoma of the Lung

H.R. is a 62-year-old male stockbroker who came to his physician with intense right-sided headaches of a few weeks duration. They began 3 weeks earlier when he fainted, fell, and hit his head. The resulting laceration required sutures but neurological evaluation including a CT scan of the head was negative at that time. Since that episode the frequency and severity of his headaches have been increasing.

H.R.'s medical history was significant for atherosclerotic heart disease (ASHD) requiring coronary bypass surgery 6 years ago. In addition, he was a heavy smoker (75 pack years) until his open heart procedure. For the last 20 years he has had mild COPD.

His physical examination found him to be an obese white man with slow mentation. His skin was flushed, especially over the face, and sweating. The pulse was 88 (normal) and regular and the blood pressure was 170/80 (mildly elevated). The remainder of the examination was negative, including the neurologic portion. The admission diagnosis was "suspect pheochromocytoma" (epinephrine-producing tumor of the adrenal medulla that produces episodes of hypertension, sweating, and headache).

Laboratory tests were inconclusive. The serum epinephrine was normal but the norepinephrine was elevated at 1340 pg/ml (normal up to 750). The tests for excreted adrenaline and its metabolites were normal.

Radiologic evaluation showed a normal chest x-ray. An abdominal CT scan showed a 4.5 cm solid right adrenal mass suggestive of an adrenaline-producing tumor. CT of the head revealed a space-occupying lesion of the cerebellum with obstructive hydrocephalus. A CT scan of the chest showed a 1 cm endobronchial lesion involving the medial and anterior aspect of the bronchus intermedius on the right. Bronchoscopy with biopsy revealed a poorly differentiated squamous cell carcinoma. A diagnosis was made of bronchogenic squamous cell carcinoma metastatic to the adrenal gland and brain.

During the medical examination, H.R. became progressively more confused, agitated and began hallucinating. Palliation was attempted with 3000 rad of radiation therapy over 12 to 15 weeks to the head and chest, but H.R.'s downhill course continued to death.

Comments. Squamous cell carcinoma of the lung is most often associated with inhalation of cigarette smoke. It is usually found centrally in the lung near the bronchi, which react to noxious stimuli by undergoing squamous metaplasia. Many of these tumors may metastasize (spread through the body) prior to diagnosis, making the therapy palliative (designed to provide symptomatic relief). In this case the combination of hypertension, headache and flushing originally led the physicians into a mistaken diagnosis of pheochromocytoma. It is not uncommon to have a primary lung cancer present with a smaller lesion than the metastases.

BIBLIOGRAPHY

Campbell EJ, Lefrak SS: How aging affects the structure and function of the respiratory system, *Geriatrics* 33:68-74, 1978.

Cox DJ, Yeonor RA: Adenocarcinoma of the lung: recent results of the Veterans Administration Lung Group, *Am Rev Respir Dis* 120:1025-1029, 1979.

Freeman E: *The respiratory system.* In Brocklehurst JC, ed: *Textbook of geriatric medicine and gerontology,* ed 3, New York, 1985, Churchill Livingstone.

Harris R: Evaluation of chest pain in the old age patient, *Internal Medicine* 10:65-74, 1980.

Hugh-Jones P, and Whimster W: The etiology and management of disabling emphysema, *Rev Respir Dis* 177:343-378, 1978.

Iseman MD: Tuberculosis in the elderly: treating the "white plague," *Geriatrics* 35:90-107, 1980.

King TE, Schwartz MJ: *Pulmonary function and disease in the elderly.* In Schrier, ed: *Clinical internal medicine in the aged,* Philadelphia, 1983, WB Saunders.

Knudson RJ, and others: Effect of aging alone on the mechanical properties of the normal adult human lung, *J Appl Physiol* 43:1054-1062, 1977.

Sparrow D, Weiss ST: *Pulmonary system.* In *Geriatric medicine,* ed 2, Boston, 1988, Little, Brown.

Stark FE, Dent RG: *Chest diseases.* In Exton-Smith AN, Weksler ME, eds: *Practical geriatric medicine,* New York, 1985, Churchill Livingstone.

Zavala DC: The threat of aspiration pneumonia in the aged, *Geriatrics* 32:46-51, 1978.

7 *The Digestive System*

RELEVANT ANATOMY AND PHYSIOLOGY
Composition and Function

The digestive system is broadly viewed as the system of organs that takes food into the body physically, and chemically reduces it to absorbable form, absorbs it, and transports the remains out of the body. The process begins in the mouth where food is chewed, or *masticated*, mixed and lubricated with saliva, and positioned by the tongue for swallowing. The esophagus conducts the food to the stomach, where digestion of most of the food begins. Food passes from the stomach to a part of the small intestine, the duodenum, where it is mixed with secretions from the pancreas and where most digestion takes place. The remainder of the small intestine is concerned mostly with absorption of nutrients. The large intestine, or colon, absorbs water and compacts waste material for excretion through the muscular rectum and anus. Consult Chapter 15 for a more complete discussion of nutrition.

Digestive glands are located in both the stomach and duodenum. The pancreas, the body's largest digestive gland, empties its digestive enzymes into the duodenum. The liver produces bile, a waste product that also helps to emulsify fat. Bile also is released into the duodenum. The digestive system, therefore, is composed of the tubular system that conducts food through the body and absorbs the nutrients, and the glands that provide the enzymes that digest the food. Most of the digestive system (the lining of the esophagus, stomach, intestines, ducts and secretory tissues of the pancreas and liver) arises from endoderm. The vascular and muscular supporting tissue is mesodermal in origin. The only ectoderm component is the enamel of teeth.

The Teeth

Mastication is accomplished by the teeth. There are four types of teeth; each type is specialized for a different job. Cutting and biting are accomplished by the eight *incisors,* two upper and two lower on each side. Puncturing is the job of the four *canines,* one in each quarter of the mouth. The eight *premolars* have both a cutting and grinding function in the adult mouth and the 12 *molars* are exclusively for grinding. The last molar to come in, the third molar in each line of teeth, is sometimes called the wisdom tooth. Because of the small size of the human jaw, this tooth often has difficulty erupting and may become a source of problems. With the possible exception of the third molars, the loss of a significant percentage of our 32 teeth must result in mastication difficulties.

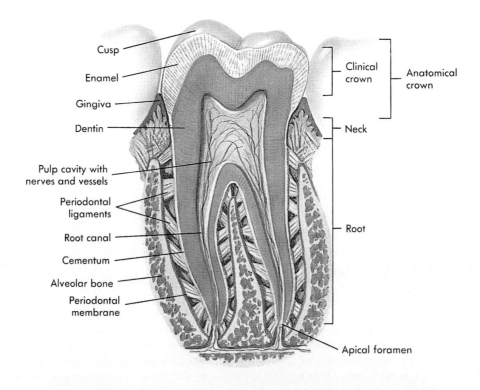

Figure 7-1 Diagram of a tooth. The gingivitis of periodontal disease will eventually cause regression of the gingiva and exposure of the root.

If enough teeth of any one class are lost, they must be replaced with prosthetic teeth (dentures), or a change in diet will probably occur.

Teeth are subject to decay as a result of acids produced by bacterial action. This produces *caries* (dental decay) that penetrates the outer enamel and expands in the softer underlying dentin (Figure 7-1). When decay approaches the pulp cavity, the nerve residing there will register pain. If the infection enters the pulp cavity, it may follow the root canal to the bone of the jaw. From there the infection may spread resulting in an abscess, and may damage the bone, causing the loss of other teeth. An equally important cause of tooth loss today is *periodontal disease.* This involves the establishment of bacterial colonies on the tooth at the gum, or gingiva. These bacteria protect themselves from being washed away by surrounding themselves with a calcareous deposit called *plaque.* The combination of bacterial action and irritation caused by the plaque cause the gingiva to become inflamed. This inflammation, regardless of the cause, is called *gingivitis.* The gingiva eventually responds to this by receding, exposing progressively more of the tooth until, aided by the infection, the tooth falls out.

The Salivary Glands

Only a small amount of digestion takes place in the mouth as a result of salivary enzymes. Protein-digesting and starch-digesting enzymes are present in the watery secretions of the parotid and submandibular salivary glands. These enzymes kill bacteria, digest carbohydrates, and wash and clean the mouth and teeth. The thicker secretions of the sublingual and numerous buccal glands keep the mouth moist, lubricate the food, and glue the particles together to form a bolus for swallowing.

Saliva, or at least a film of water, is necessary for most taste receptors to function and a dry mouth can result in loss of much taste reception. Saliva is normally secreted in response to brain perception that the mouth is dry or that food is about to be offered or at least is desired. Between 500 ml and 1500 ml of saliva is produced each day. Decline in salivary secretion with age is a common problem.

The Tongue, Swallowing Process, and the Esophagus

The tongue positions food under the teeth for mastication, forms the bolus of food, and initiates the swallowing process. Covering the tongue are hundreds of little bumps or papillae, many containing taste buds. Taste buds are capable of being stimulated by sugars (producing a sweet sensation), salts (more or less salty), acids (sour), and alkaline materials (bitter). What we actually taste is a combination of these four basic sensations enhanced by the smell of food, registered by the olfactory nerves of the nose.

Swallowing is a coordinated action of the tongue, the soft palate and the uvula that projects from the roof of the mouth, the muscles of the pharynx, and the peristaltic movements of the esophagus. Normally, humans do not pay attention to the swallowing process because its activities appear to be almost automatic. In fact, the process requires the coordination of complex neuromuscular factors. *Deglutition,* or swallowing, is much more complicated than it appears. It can be divided into two phases: (1) a voluntary phase where the tongue moves upward and backward forcing a bolus of food into the pharynx, and (2) an involuntary phase, which involves transfer and transport. Transfer involves involuntary changes in the pharynx and upper esophagus, starting the involuntary swallowing process. Transport involves the involuntary movement of the bolus through the middle and lower esophagus.

Isolated esophageal contractions occur in some individuals, particularly after middle age. These are nonperistaltic contractions. They do not assist in transport of food and are often painful. The esophagus as a whole conducts food from the pharynx to the stomach. At the lowest segment of the esophagus is a segment called the esophageal or cardiac sphincter. While no specific esophageal sphincter muscle is known to exist in humans, a narrowing is present here, in part due to pressure from the diaphragm. This constriction has a higher resting pressure than either the resting esophagus or the stomach. This high pressure area prevents reflux of the acidic gastric contents into the esophagus. Under normal circumstances, the lower esophageal sphincter relaxes during peristalsis, allowing the bolus of food to pass into the stomach. The movement of food through the digestive system can be followed in Figure 7-2. If mechanical obstruction of the esophagus or paralysis of the swallowing muscles occurs, this process may have to be bypassed. Either a *nasogastric* (NG) *tube* is placed through the nose into the stomach or a surgical opening, or *feeding gastrostomy*, is created directly into the stomach.

The Stomach

The stomach is composed of four parts: (1) a cardiac region where the esophagus enters, (2) an upper leftward projecting pouch called the fundus, (3) the large central body, and (4) a lower pylorus. The digestive tract (including the esophagus, stomach,

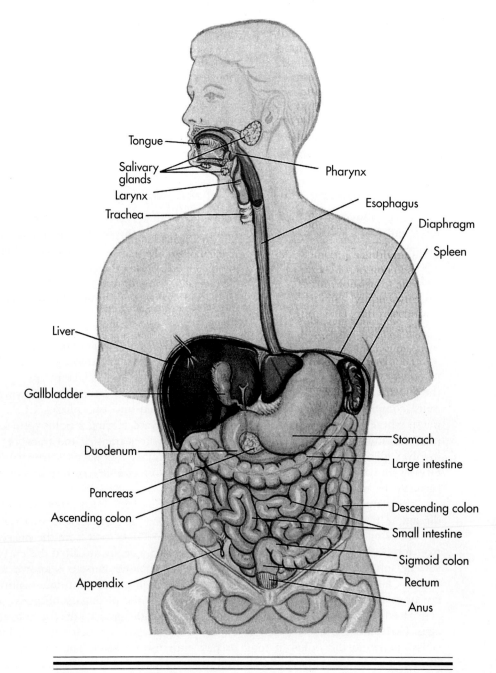

Figure 7-2 The digestive system.

small intestine, and colon) is composed of the same basic tissue layers (Figure 7-3 and 7-4). The inner layer is called the mucosa and is covered with an epithelium rich in mucus-secreting cells. In the stomach this layer contains the *gastric glands*. Beneath this is a thin layer of muscle, the muscularis mucosa, separating the mucosa from the submucosa, a loose collection of fibrous tissue, vessels, and nerves. The stomach is surrounded

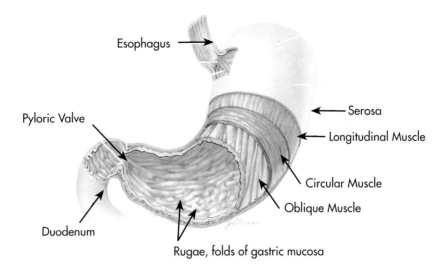

Esophagus

Pyloric Valve

Serosa

Longitudinal Muscle

Circular Muscle

Oblique Muscle

Duodenum

Rugae, folds of gastric mucosa

Figure 7-3 Cross-section of stomach wall. The gastric glands are located in pits in the gastric epithelium. Most of the tissues of the digestive system are arranged in the same way as those of the stomach.

with oblique, circular, and longitudinal muscle. The outside is covered with a loose connective tissue layer, the serosa. The mucosa of the stomach is thrown into large folds called *rugae* that increase the surface area and allow the stomach to stretch.

When it enters the stomach, food causes release of the hormone gastrin from the stomach wall into the bloodstream. (General information on hormones is found in Chapter 12). Both this hormone and stimulation by the parasympathetic nervous system cause the gastric glands to release two digestive compounds, pepsinogen and hydrochloric acid. Pepsinogen is produced by the "chief cells" of the glands and hydrochloric acid is produced by the parietal cells. Pepsinogen is activated by the hydrochloric acid and becomes the powerful proteolytic enzyme pepsin. Together these compounds begin the process of splitting the proteins in food into their component amino acids. Because gastric glands are linked to the central nervous system, stress can cause them to receive excessive parasympathetic stimulation. This can result in excessive gastric secretions and peptic ulcers (discussed later in this chapter).

An essential function of the parietal cells of the stomach is the production of *intrinsic factor*. Intrinsic factor binds with vitamin B_{12} and enables it to be absorbed in the small intestine. A person unable to produce intrinsic factor will be unable to absorb vitamin B_{12} and oral B_{12} will be of little value. Vitamin B_{12} is necessary for DNA synthesis; its absence results in pernicious anemia, (inability to produce red blood cells). Failure to secrete intrinsic factor is associated with achlorhydria, or failure to secrete hydrochloric acid, and the loss of parietal cells. This condition exists in 20% of men and 10% of women over age 60.

In the stomach food is reduced to a partially digested, acidic slurry called *chyme*. Chyme is allowed to pass through the pyloric sphincter into the duodenum, the uppermost portion of the small intestine.

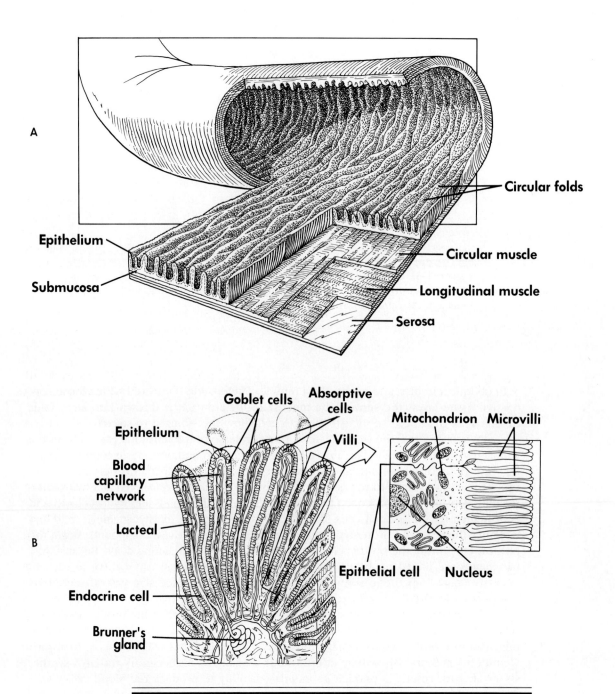

Figure 7-4   The duodemum (**A**) with a villus enlarged (**B**). Note the Brunner's glands located below the muscularis mucosa in **B**.

The Duodenum

The duodenum and the rest of the small intestine differ from the stomach in several respects. Although the arrangement of the tissue layers is the same, the duodenum contains ruga-like folds that form rings called plicae circularis. These plicae help to ensure that the intestinal contents always move toward the colon. Covering the plicae circularis and the spaces between them are thousands of finger-like projections called villi (Figure 7-4). These greatly increase the absorptive area of the intestine.

The portion of the duodenum between the pyloric valve and the ampula of Vater (the entrance of the combined pancreatic and bile ducts), is subject to digestion from the extremely acidic chyme from the stomach. Located in the mucosa and submucosa in this area of the duodenum is a concentration of mucus-secreting glands that produce a large volume of mucus in response to parasympathetic nervous stimulation and contact with chyme. This mucus normally protects the vulnerable part of the duodenum. When it fails to do so, duodenal peptic ulcers result. Approximately 50% of peptic ulcers occur at this site.

Even though the duodenum is only about 10 inches long, it is apparent that the vast majority of digestion takes place there, not in the stomach. In addition to all of the enzymes secreted into the duodenum by the pancreas, and bile from the liver, the crypts of Lieberkuhn (thousands of small intestinal glands) secrete into the duodenum large volumes of fluids that dissolve digested material. All of this is reabsorbed by the intestinal epithelium. This fluid contains small amounts of intestinal amylase. The mucosal cells contain additional enzymes that complete digestion, especially of sugars, as they are absorbed.

The mucosa of the duodenum reacts to the acidic chyme by releasing the hormones *secretin* and *cholecystokinin* into the bloodstream. These hormones act on the pancreas and gall bladder, respectively, to cause the release of their secretions.

Operation of Pancreatic Digestion and the Gall Bladder

The pancreas produces amylase, splitting carbohydrates into their component sugars; lipase, breaking down fats; trypsinogen, the active form trypsin attacking proteins; and bicarbonate to neutalize the gastric acid. The portions of the pancreas responsible for digestive enzyme production are glands called pancreatic acini associated with a network of pancreatic ducts. Trypsinogen is secreted along with a peptide called trypsin inhibitor. This prevents trypsinogen from being activated while still in the pancreas. Because trypsin is the activator for the other pancreatic enzymes, trypsin inhibitor effectively inhibits all of the pancreatic proteolytic enzymes. Sometimes inflammation or duct obstruction promotes release of pancreatic enzymes in excess of inhibitor. The resulting acute pancreatitis is caused by activated enzymes destroying some or all of the pancreas. When severe the condition is often lethal. Under normal conditions when the pancreatic enzymes reach the duodenum, trypsin is rapidly activated by the enzyme enterokinase, which is produced by the crypts of Lieberkuhn.

The hormone cholecystokinin strongly stimulates the gall bladder to contract, emptying bile into the duodenum. The liver produces up to 700 ml of bile per day and while some of it passes directly into the duodenum, much is stored in the gall bladder and secreted in response to a meal. Bile salts do not participate in digestion; they emulsify and increase the solubility of fat globules so they can be digested by lipases and absorbed by the intestinal epithelium.

The Jejunum and Ileum

The digesting chyme moves on through approximately 8 feet of jejunum and 12 feet of ileum, the remainder of the small intestine. Throughout this approximately 20 feet of small intestine, the crypts continue to secrete digestive fluids and absorb the digested nutrients. Here, as in the duodenum, the villi increase the absorptive area of the intestine. At the end of the ileum the remaining undigested material passes through a sphincter muscle, the iliocecal valve, into the cecum, a pouch that forms the beginning of the large intestine.

The Large Intestine

Many herbivorous animals have a much larger cecum than humans do. It is called the pyloric cecum. Humans have the remnants of that large cecum in the form of an appendix. This thin-walled vestigial organ may accumulate bacteria and undigested debris. It may then become infected and inflamed, resulting in appendicitis. So far as is known today, the human appendix has no function.

The large intestine (or colon) has the primary function of absorbing water and compacting the undigested material into feces for convenient excretion. In addition, bacterial action, particularly on the cellulose of plant cell walls, releases vitamins and sugars that are absorbed with the water. Vitamin K, necessary for the clotting of blood, is produced by certain colon bacteria. The crypts of Lieberkuhn are still present in the large intestine, but here they secrete mucus instead of watery material. This lubricates the feces, provides a protective covering on the intestinal epithelium, and glues the fecal particles together. The large intestine lacks villi but the composition of its wall is otherwise similar to that of the small intestine.

When the slurry of undigested material leaves the cecum, it passes into the ascending colon, then to the transverse colon, and into the descending colon. By the time it has reached the distal end of the descending colon most of the water in this slurry has been absorbed and it has become nearly solid. Gradually the sigmoid colon fills and the feces or stool passes into the muscular rectum. Distension of the rectum causes the urge to defecate. Defecation in a continent person includes increase in intraabdominal pressure, tensing of the pelvic floor, rectum contraction, and relaxation of the anal sphincters. Defecation can be the simple emptying of the rectum or it may stimulate mass propulsion and empty the distal half of the colon.

Nondigestive Functions of the Liver and Pancreas

Two organs mentioned earlier have functions beyond digestion. These are the liver and the pancreas. Many of the nondigestive functions of these organs are discussed elsewhere but for sake of convenience their functional anatomy will be covered here.

The liver is the largest single organ in the body. Bile secretion is relatively minor when compared with the liver's other functions. A small portion of the liver is illustrated in Figure 7-5, and a photomicrograph is shown in Color Plate 10, *C*. Bile is actually a waste product, containing the breakdown products of hemoglobin and other ingested and metabolic wastes. All material absorbed into the body from the intestines passes through the liver by way of the portal veins. An exception is fat that may bypass the liver through the lymphatic system. The functions of the liver are summarized in Table 7-1.

The pancreas is both an exocrine, or duct-bearing, and endocrine, or ductless, gland. The exocrine portion is composed of glandular cells arranged around ductules, the acini. This part of the pancreas is concerned with manufacture of digestive enzymes.

Table 7-1 Functions of the Liver

FUNCTION	SUBSTANCE AFFECTED	RESULT OF ACTION OR DYSFUNCTION
Removal	Bacteria and other foreign matter	
	Bilirubin	Failure to function results in accumulation of bilirubin in blood; jaundice
Metabolism	Carbohydrates	Converts sugars to glycogen (the form in which carbohydrates can be stored in the body); releases them as glucose if blood sugar levels fall
	Fats	Converts proteins and carbohydrates to fat; fat to carbohydrates
	Proteins	Synthesizes plasma proteins; interconverts many amino acids for necessary balance.
Formation	Blood proteins	Using Vitamin K, forms fibrinogen, prothrombin, and clotting factors, as well as albumin, some globulins, other proteins
Formation/excretion	Cholesterol	Maintains low level of cholesterol by excreting excess; this is probably an inherited trait. Also forms cholesterol.
Conversion	Ammonia to urea	Failure to function results in accumulation of ammonia in blood; hepatic coma can result
Storage	Vitamins, iron	Stores vitamins A, D, B_{12}, as well as iron (as ferritin). Iron can be released to replenish levels in the extracellular fluid

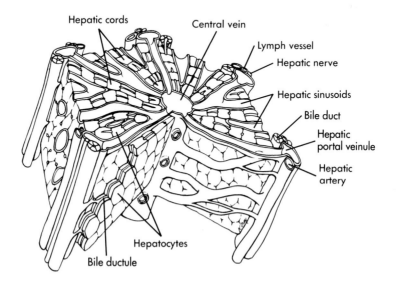

Figure 7-5 Liver lobule. Bacteria are removed from the blood in the hepatic sinusoids. Waste products are excreted from the lobules into the bile ductules.

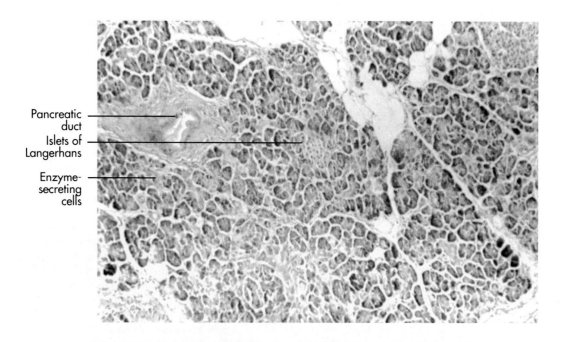

Pancreatic duct

Islets of Langerhans

Enzyme-secreting cells

Figure 7-6 Note the digestive-enzyme–secreting cells and the large pancreatic ductules. The other masses of cells are the islets of Langerhans and pancreatic duct.

The small ductules connect to the pancreatic duct, which often combines with the bile duct and empties into the duodenum through the ampulla of Vater.

The endocrine portion of the pancreas is composed of the islets of Langerhans (Figure 7-6). It forms two major hormones, insulin and glucagon. Insulin increases the permeability of cell membranes to glucose, enabling the sugar to pass from the blood into the cells to be metabolized. Glucagon stimulates the release of glucose from the liver when the blood sugar is excessively low. When the endocrine portion of the pancreas fails, the patient has diabetes mellitus, which will be covered in more detail in Chapter 12.)

THE AGING DIGESTIVE SYSTEM
Changes in the Teeth and Mouth

Until the twentieth century it was common for most people to lose most of their teeth by age 40 as a result of periodontal disease or infection. Today modern dentistry can preserve enough teeth to maintain normal eating habits through most of the lifetime of many people, despite diets high in sugar and carbohydrates. Still, the loss of teeth resulting from these causes is common. The effects of these losses are cumulative. The teeth of elderly people are often darker or more yellow than in the young. The chewing surfaces may have worn down so that they do not occlude well, although the teeth may

Table 7-2   Some Causes of Xerostomia

1. Mouth breathing causes drying and sometimes cracking and bleeding of the oral surface. Xerostomia is more likely among emphysema patients or those with allergies or infections that cause increased nasal resistance to air flow.
2. Decreased salivary secretion. As much as 25% of the secretory cells of the salivary glands are replaced by fibrous or fatty material.
3. Drug side effects. Antihypertensive, antihistamine, and antidepressive drugs are commonly prescribed for the elderly. When prescribed for patients with already reduced salivary secretions or nose-clogging allergies, they may produce xerostomia.

appear longer because of gum tissue withdrawal at their base. The expression "long in the tooth" truly characterizes some elderly people. Loss or removal of teeth aggravates the process of bone loss in the jaw as a result of lost beneficial chewing stress. Even without tooth removal, this bone loss contributes to loosening of otherwise healthy teeth.

Malfunction of the temporomandibular joints is a common cause of eating difficulties experienced by the elderly. Such malfunctions are thought to result from years of stress and trauma (for example, malocclusion or the loss of certain teeth), rather than from the aging process. Dentures provide a means of maintaining a normal diet in spite of extensive tooth loss. However, many elderly people prefer to leave their dentures out, possibly because of mouth irritation or poor fit. The net effect of tooth loss and associated jaw and denture problems is an altered diet and probably reduced pleasure from eating. This may result in eating less, eating less frequently, eating less appropriate food, and ultimately, inadequate nutrition. For this reason helping an elderly person to obtain proper tooth care and proper denture maintenance and adjustment can be one of the most useful things someone working with the elderly can do.

A condition causing concern and discomfort in many elderly is xerostomia, or dry mouth. This may be caused by any of the situations listed in Table 7-2 or may be the result of their combined effects:

Xerostomia may also be a side effect of radiation to the head region, or of some chemotherapies used to treat malignancies. Sjögren's syndrome (discussed later in this chapter) produces severe xerostomia. Sometimes people complain of xerostomia even though their salivary secretions are normal. This xerostomia of unknown cause is most common in postmenopausal women. Suggestions for managing xerostomia include ensuring adequate fluid intake and, sometimes, frequent rinsing of the mouth with water. It is also possible to stimulate salivary secretion by chewing gum or sucking on tart, hard candy.

The Aging Esophagus

Failure of the esophagus to transport food adequately occurs more often in the elderly than in the young. The swallowing process may be interrupted by inadequate coordination of the cricoid cartilage and associated pharyngeal muscles. There also appears to be an age-related decrease in the strength of peristaltic contractions and an increased incidence of nonperistaltic contractions, often with inadequate relaxation of the lower esophageal sphincter. Changes in esophageal motility may be responsible for older people having difficulty emptying the esophagus while lying on their backs. This may contribute to both poor nutrition and aspiration of esophageal contents into the lungs. Any problems in esophageal mobility, tone, or strength of contraction, may result in esophagitis, diffuse esophageal spasm, or gastric esophageal reflux. Both cardiac and esophageal disease are common in the elderly. Therefore it is important to remember that distinguishing between lower esophageal pain and pain from the heart may be difficult.

Changes in the Aging Stomach and Small Intestine

The stomach undergoes substantial change with age. By age 64 the gastric mucosa has usually undergone some degree of atrophy. It is thinner and has fewer active cells. The most observable result of this is decreased hydrochloric acid production. In cases of severe atrophic gastritis, hydrochloric acid may be eliminated entirely (achlorhydria), along with other gastric enzymes. The production of intrinsic factor also decreases or fails entirely, with gastric atrophy resulting in failure of vitamin B_{12} absorption, producing pernicious anemia.

Some atrophy is seen in the small and large intestine but often less than in the stomach. In the large intestine there is atrophy of the mucosa and circular and longitudinal muscles, with increased deposition of connective tissue and inflammation of the lamina propria. There is an increase in the prevalence of diverticula, herniations of the mucosal layer of the intestine through the intestinal wall. Even though the elderly commonly complain about constipation, frequency of defecation is not significantly different after age 60 than before. This ranges from as many as 3 defecations per day to as few as 3 per week. People sometimes confuse the production of hard feces with constipation. It has been shown that the volume of rectal distention necessary to cause discomfort is larger in elderly people.

Although absorption is the most critical function of the gastrointestinal system, food must be digested before it can be absorbed. In aging persons, the rate and degree of digestion, the integrity of the absorbing surface, the adequacy of the transporting mechanisms, and gastrointestinal motility all decline, influencing the absorption of nutrients.

The Aging Liver

Liver weight decreases with age; the number of hepatic cells (hepatocytes) in the lobules also declines. There is an increase in fibrous tissue and the size of the hepatocytes (liver cells). The size of the nuclei of the hepatocytes increases with age. Protein synthesis by the hepatocytes declines, accompanied by a decrease in enzymatic activity. These changes are usually quite minor and can be observed only with careful statistical studies. Overall, the liver has enough reserve capacity to outlast the rest of the body, if it has not been damaged by alcoholism or disease.

Conditions and Diseases of the Aging Digestive System

Sjögren's Syndrome. Most elderly patients with this relatively rare condition have had symptoms since middle age. Patients generally have at least two of the following ailments: dry eyes with inflammation of the conjunctiva, called keratoconjunctivitis; xerostomia; or atrophy of the dorsal surface of the tongue, so that it presents a cobblestone appearance. In some patients there is total replacement of the salivary glands with inflammatory tissue. The gingiva may have receded from the incisors. Caries are abundant and in unusual places. The only treatment for this possibly inherited disease is designed to alleviate the symptoms. Sjögren's syndrome is believed to be an *autoimmune disease;* some suspect that it may be more common in its milder forms than is generally believed.

Reflux Esophagitis. Reflux of gastric contents into the esophagus is a common problem at any age. It produces heartburn or occasional regurgitation. The epithelium becomes inflamed and may ulcerate if the problem persists. With long-standing inflammation, the epithelium eventually responds by producing microscopic columnar cells that are similar to the lining of the stomach or intestine. This condition is called Barrett esophagus. Esophageal reflux may result from gastritis, inadequate function of the lower esophageal constriction, esophageal motility disorders, or hiatus hernia. Mild cases are treated with antacids and elevation of the head of the bed at night. More severe cases may require surgery either to correct the muscular tone of the cardiac sphincter or to remove a stricture created by the chronic inflammation.

Esophageal Carcinoma. Men are twice as likely to develop esophageal cancer as women. Its incidence increases steadily with age. The cancer is usually a squamous cell carcinoma, and is one of the most common causes of esophageal obstruction. About 10% of the cases are adenocarcinoma, which is especially common in patients with Barrett esophagus. Esophageal cancers tend to be locally invasive, often extending into the pericardium, aorta, mediastinum, or respiratory tree. They are sometimes discovered after sudden aspiration of gastric contents through a fistula into the trachea. More frequently they are found because of difficulty in swallowing caused by the neoplasm forming an obstruction that prevents food from passing. The prognosis for these cancers is poor, with a 5-year survival rate of less than 6%.

Hiatus Hernia. Herniation of a part of the cardiac and sometimes fundal portions of the stomach through the diaphragm at the hiatal aperture occurs in 69% of adults over 70 years old, but in only 9% of adults under 40. While most hiatus hernias are asymptomatic, those that cause symptoms usually do so by causing esophageal reflux, creating heartburn (esophagitis) and sometimes refluxing gastric juices into the mouth. Esophageal strictures may be present as well, producing difficulty in swallowing.

Atrophic Gastritis. This is a progressive, chronic condition of the stomach involving atrophy of the glandular epithelium, loss or flattening of the rugae, and degeneration of the gastric glands. The latter results in loss of hydrochloric acid and pepsin, with an obvious effect on digestion. Intrinsic factor cannot be formed to absorb vitamin B_{12}, sometimes resulting in pernicious anemia. Iron deficiency anemia also may develop as a result of poor absorption or iron loss through gastric or intestinal bleeding. Patients with atrophic gastritis are typically elderly. The atropic stomach is easily irritated, and reflux of gastric and sometimes intestinal contents is common. Very elderly patients may have such a problem that vomiting occurs at the slightest provocation, causing loss of appetite among others sharing the same dining room. There is good support for atrophic gastritis being an autoimmune disease.

Adenocarcinoma of the Stomach. While stomach cancer may occur at any age, it is particularly a disease of the elderly. It has shown an absolute decline in recent

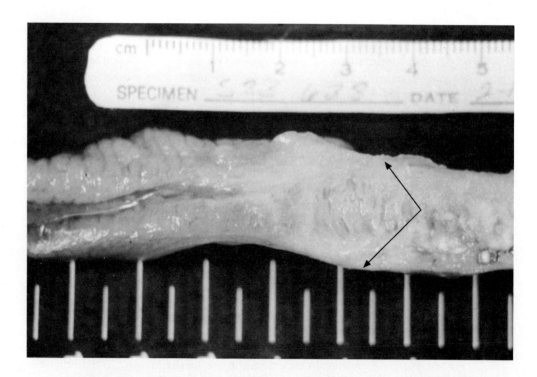

Figure 7-7 Adenocarcinoma of stomach. Stomach cancers may hemorrhage and give patients the impression they have ulcers. Arrows demonstrate a diffuse thickening of the stomach wall by carcinoma. Compare with the normal stomach on far left.

years for unknown reasons. There is a strong familial tendency toward the disease; a lifetime of eating smoked foods has been shown to contribute. People with pernicious anemia, atrophic gastritis, and polyps are at increased risk for this disease. Because the symptoms of adenocarcinoma of the stomach are usually minor or nonexistent, metastatic lesions are often found before the primary gastric lesion. Metastasis usually involves the liver and abdominal lymph nodes. These cancers frequently invade adjacent organs. Overall, the 5-year survival rate is only about 9%. However, patients with well-differentiated polyploid, in situ lesions have a 5-year survival rate of better than 95%. A gastric adenocarcinoma is shown in Figure 7-7.

Peptic Ulcer Disease. Despite the fact that hydrochloric acid secretion is usually lower in elderly patients, peptic ulceration may occur in as many as 9% of those who are chronically ill. In one study, nearly half the patients who had peptic ulcer were over 60 years old. These ulcers occur in the stomach and duodenum and occur equally among elderly men and women. Activated pepsinogen must be present to produce a peptic ulcer. This requires both hydrochloric acid and pepsinogen, so the disease is absent in patients with totally atrophic stomachs. About 85% of patients with peptic ulcer disease secrete excessive amounts of pepsin and hydrochloric acid. The symptoms are often atypical in the elderly, who may have become more accustomed to gastric irritation. Gastric or intestinal bleeding is often the first indication of the disease. Because of this, surgical intervention is more frequently necessary than with younger patients. Non-peptic gastric ulcers can occur in patients with achlorhydria; these have a higher incidence of malignancy.

Appendicitis. The incidence of acute appendicitis is somewhat lower in the elderly than in younger people. A cause of concern is that the symptoms may be masked by other problems. Pain may not be noticed or if described as in the stomach, may be assumed to be gastritis often seen in the elderly. Perhaps for this reason, the diagnosis of appendicitis may be delayed and this delay may contribute to the increased likelihood of perforation (rupture). A recent study of people over 60 years old showed that 37% to 71% of acute appendicitis patients had perforated appendixes by the time of surgery. Mortality from acute appendicitis is reported as 4% for people 60 to 69 and 20% for people 70 years old or older.

Diverticular Disease. Diverticula become increasingly common in the colon with age; they are present in 5% of the population older than 50 years. Diverticula are herniations of the mucosa and submucosa through the muscularis of the large intestine. They appear as flask-shaped sacs, usually full of debris and usually in the sigmoid colon (Figure 7-8). Inflammation (diverticulitis) occurs when feces become impacted in the diverticula. Ultimately the diverticula and adjacent walls of the colon undergo fibrotic thickening. This may produce a stenosis (sten-O-sis), a narrowing, and there may be intestinal bleeding. The patient suffers abdominal pain and changes in bowel habits (with either constipation or diarrhea). Approximately 22% of those with symptoms have blood in the stools. Most patients with diverticulitis can be managed without surgery.

Vascular Disease of the Small Intestine and Colon. The development of atherosclerosis with age sometimes results in occlusion of the mesenteric arteries. This results in obstruction at the point of the atheromas, or downstream as a result of emboli arising as thrombi in the atheromas or elsewhere. The reduced blood flow can cause ischemic damage in the intestine. Acute ischemia of the small intestine typically produces abdominal pain, distention, and fever. It may also produce diarrhea, usually with blood and leucocytosis. When the condition is chronic, the patient typically avoids eating because the pain is worse after meals. Ischemic colitis is more likely to produce abdominal cramping and rectal bleeding, but otherwise the symptoms are little different from those of intestinal ischemia.

Fecal Impaction. The formation of a large, firm, immovable mass of stool in the rectum is a serious problem in bedridden patients in nursing homes and occasionally with more mobile patients. The condition most often results from slowing or inhibition of the rectal muscles. For this reason, impaction occurs in narcotic addicts, patients taking large doses of prescribed narcotics or tranquilizers, and debilitated patients. Patients with a history of constipation are at increased risk. As the mass enlarges and the rectum expands, the mass becomes too large and solidified to be forced through the anus. Sometimes more fluid feces may flow around the impaction and present as diarrhea or fecal incontinence. While urinary incontinence can generally be concealed from the public, this is not true with fecal incontinence because of the odor produced. Therefore this is one of the most devastating problems an active person can have.

The physician or nurse can often break up low-lying impactions by using a finger. A combination of enemas, oils, and mechanical breaking up of the impaction may be required for impactions farther up in the sigmoid colon. Only rarely is surgery required.

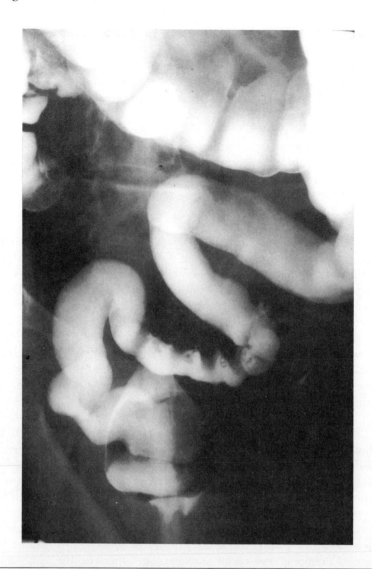

Figure 7-8   Barium x-ray of diverticula in colon. Diverticula occur when the colon herniates because of high internal pressure.

Adenocarcinoma of the Colon. This disease, most common between the ages 50 and 70, is the second leading cause of death from cancer in men. The symptoms usually include abdominal cramps, a change in bowel habits, and blood in the feces. Sometimes there is an excess production of mucus. Many authors believe that the malignancy usually begins in a benign polyp, such as a villous adenoma. Many of these do show malignant changes (Figure 7-9).

Colon adenomas are now removed as early as possible because they are difficult to differentiate from small cancers. Both can ulcerate and bleed. When a growth removed by colonscopy is found to be cancer with involvement of the intestinal wall, it becomes necessary to remove the invaded portion of the colon along with a safety margin of normal tissue. If the removed growth is found to be a benign polyp, no further surgery is needed. Prognosis for cancer of the colon are given in Table 7-3.

Table 7-3   Prognoses for Colon Cancer (%)

STAGE	EXTENT	5-YEAR SURVIVAL RATE
1	No spread outside the rectum, no lymphatic metastasis;	85-100
2	Invasion of extrarectal tissue, no lymphatic metastasis;	30-70
3	Lymphatic metastasis	45

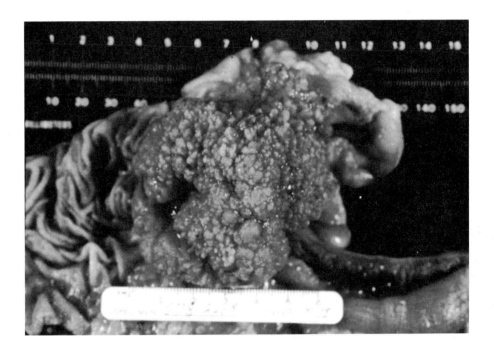

Figure 7-9   Villous adenoma from the colon, gross view.

If it becomes necessary to remove a large portion of the colon or rectum, making it impossible for fecal matter to continue to pass through the anus, a colostomy will be performed. This procedure creates a temporary or permanent opening between the colon and the skin of the abdomen. Depending on the amount of colon that remains functional, the feces flowing from the opening may be wet or relatively dry. A device is now available that greatly reduces the problems associated with this, making it possible for many colostomy patients to continue a normal social life (Figure 7-10).

Cirrhosis of the Liver. *Cirrhosis* properly describes a condition involving liver cell necrosis during which the destroyed tissue is replaced with nonfunctional connec-

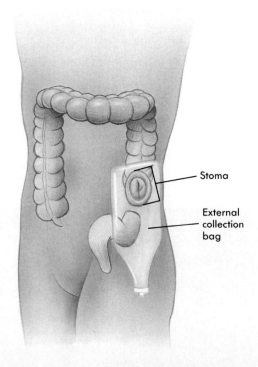

Stoma

External
collection
bag

Figure 7-10   When a section of colon is removed, a colostomy is sometimes necessary. This requires that the colon contents exit from the body through the abdominal wall. With the help of the appropriate apparatus colostomy patients can live a normal life.

tive tissue. As the disease progresses, the symptoms of decreased liver function develop. Patients with the disease show jaundice, enlargement of the liver and spleen, and increased abdominal girth secondary to fluid accumulation, or ascites. As the condition becomes more severe, the patient develops ammonia poisoning and passes into hepatic coma. Many develop portal hypertension leading to distension of blood vessels in the stomach, ulceration, and life-threatening bleeding into the stomach. Even though the prognosis for alcohol cirrhosis is poor, few alcoholics with the disease give up drinking. Therefore 80% to 90% of alcoholics with the disease die within 5 years (Color Plate 10, *D*).

Obstruction of the common bile duct and congenital problems, as well as infection, can produce cirrhosis. The former condition, biliary cirrhosis, can occur in the later years of life as a result of gallstones but is not particularly a disease of old age. Sections of normal and cirrhotic liver are shown in Figure 7-11.

Hepatotoxic Injury. New drugs and other chemical agents are developed every day, so it is not surprising that the number of agents found to be toxic to the liver is increasing. The elderly are particularly vulnerable to drug toxicity because their ability to detoxify and excrete drugs is usually diminished. Therefore, drug dosages that are not toxic to younger people may be toxic to the elderly. A side effect of this is that drugs may have the desired effect in much smaller dosages than usually prescribed.

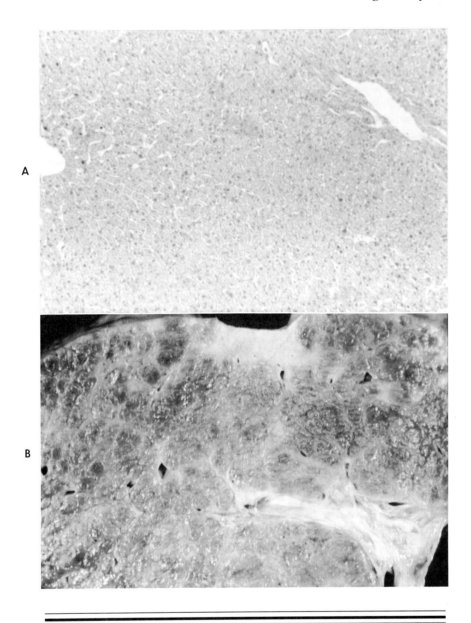

Figure 7-11 Photomicrograph of normal liver (**A**) and a gross photo of a cirrhotic (**B**) liver. Connective tissue and fat have replaced the lobules in the cirrhotic liver.

Gallstones. Gallstones usually pose no problem and are often discovered during routine x-ray. Some gallstones can be dissolved but unfortunately, usually not those that show up on x-ray. Approximately 40% of people with asymptomatic gallstones will develop obstruction and require surgery. Incidence of gallstones is higher in women than in men and is higher in patients with cirrhosis of the liver. Most physicians today avoid surgery with asymptomatic gallstones in elderly people, except with diabetics. In diabetics, serious life-threatening problems may develop. Therefore surgery is often performed.

CASE HISTORIES

❖ Esophageal Carcinoma

L.J. is a 47-year-old accountant who noted the slow onset of pain in his lower abdomen and back. The pain had been present for 2 weeks and seemed to be increasing when he first saw his doctor. His physical examination revealed only a tender mass in the right groin believed to represent a hernia. The physician referred him to a surgeon who was unsure of the nature of the mass but suspected that it could be associated with an infection. A trial period of antibiotic therapy was begun.

L.J. returned for reexamination after 3 weeks of antibiotic therapy, feeling no improvement. After examining him again, the surgeon concluded that the mass probably represented an internal process and was not a simple infection. He conducted a complete radiographic survey of the gastrointestinal system. The barium enema study of the colon was unremarkable. The upper GI study, using oral intake of barium, revealed distortion of a large area of the mid-esophagus. The distortion was believed due to a mass thought to be a tumor. When specifically asked about his ability to swallow, L.J. said that he had noticed slight difficulty since his last visit. It felt like food would get stuck in his throat for a few seconds before continuing down into his stomach.

Prior to this illness, L.J.'s only significant problem was a torn knee ligament that required surgical repair. His personal habits included drinking 8 to 10 cans of beer per week and smoking 1 pack of cigarettes per day. L.J.'s complete physical examination showed him to be a healthy-appearing, middle-aged male in no distress. The only abnormality was the slightly tender swelling in his right groin. A battery of diagnostic tests confirmed the large tumor in the mid esophagus and a biopsy of the area found a moderately differentiated squamous cell carcinoma. A marker for intestinal tumor, carcinoembryonic antigen (CEA), was mildly elevated in C.J.'s blood (CEA was 22; normal would be less than 5).

The groin mass was now believed to be a metastatic esophageal cancer. Both areas were given radiation therapy, and L.J. also received chemotherapy. Both the esophageal tumor and the groin swelling disappeared and the CEA dropped to 4.8. An exploratory surgery was conducted to remove any residual tumor. During the procedure, many obviously malignant lymph nodes were noted in the abdomen and in the tissues near the esophagus. The residual tumor was deemed inoperable, clips were placed about all the identifiably metastatic nodes, the new tumor areas were irradiated, and he was continued on chemotherapy. Despite a good initial response, L.J.'s CEA again became elevated. His tumor continued to spread, and he died of debilitation and pneumonia 2 years after he first sought medical help.

Comment. As in this case, esophageal cancers have often spread throughout the body by the time of diagnosis. This significantly reduces the chance for a cure. Most gastrointestinal malignancies liberate CEA, which is often proportional to the amount of tumor present and can be used to monitor tumor growth. If the CEA had remained normal after treatment, one could have assumed all of the tumor had been eradicated or controlled. Unlike this case, the most common initial symptoms of esophageal carcinoma are difficulty in swallowing and weight loss.

❖ Pernicious Anemia

R.S. is a 73-year-old banker who found that he was having difficulty concentrating. He had been active in the banking business all of his adult life and at the time of his illness was heading a banking division. Over the prior year he had noticed progressive difficulty with simple mathematics and his wife reported that he had been "walking funny." He had recently become forgetful and had been worried about having Alzheimer's disease. He had no fever, weakness, or abnormal sensations. He reported no head trauma or headaches and in general felt healthy.

The physical examination was performed by his family doctor and revealed no medical problems except for subtle neurological defects and a mild anemia. R.S. could not progressively subtract 9 from 100 (a test used to establish reduced *mentation,* or problem-solving ability). He had difficulty sensing vibrations and with proprioception (knowing where his limbs were without looking at them). When the doctor moved R.S.'s toe, he could not tell the doctor whether it was pointing up or down without looking at it. His proprioception failure contributed to his broad-based, waddling gait.

Examination of his blood revealed macrocytic anemia, with unusually large red blood cells. His mean corpuscular volume was 108 (normal is 98) and his hemoglobin was 11 gm/dl (normal 14 to 18). The nuclei of his granulocytic white blood cells were hypersegmented, with some of them having up to seven segments when they would normally have only three or four. In addition, the enzyme lactate dehydrogenase (LDH) was elevated to 240 IU/L (normal to 190). This enzyme is present in red cells; the increase reflected their unusually high rate of destruction. His vitamin B_{12} level was 120 pg/ml (normal 330 to 900). His gastric acid was low and his ability to absorb B_{12} was reduced, reflecting a lack of intrinsic factor. An upper GI endoscopy found no evidence of tumor. A biopsy of stomach wall revealed atrophic gastritis.

A diagnosis of pernicious anemia was made and R.S. was begun on replacement B_{12} injections. During the next 2 months his gait returned almost to normal and his blood count and mentation (mental activity) returned to normal levels.

Comment. Pernicious anemia usually presents with anemia and only the most severe cases show neurologic damage such as degeneration of the spinal cord or dementia. In some elderly patients, the neurologic symptoms predominate over the hematologic and some dementias responding to B_{12} have normal hemoglobin levels. Once recognized, the blood picture can always be successfully treated but often the central nervous system manifestations never return to normal.

❖ Adenocarcinoma of the Stomach

R.N., a 69-year-old postal worker, developed weakness and weight loss that became progressively worse over a 4 month period. He noticed a gradual loss of energy and experienced difficulty getting things done. In a conversation with his family doctor, he reported a decrease in appetite over the same period and his weight had dropped from 180 pounds to 160 pounds. He had not been aware of any pain, change in bowel habits, blood in his stool, or fever.

The physical examination determined that R.N. was a thin, elderly man in no acute distress. A hard, unmovable mass measuring 2 cm by 1.5 cm was felt just above his left collar bone and a second, grapefruit sized mass could be seen just above his umbilicus (navel). The fixed nature and hardness of the upper mass suggested that it was a malignant lymph node. Benign nodes are usually smaller, softer, and slide under the skin. The abdominal mass seemed to be attached to the liver. The chest, heart, and rectal examinations were normal.

Laboratory examination revealed an iron deficiency anemia with a hemoglobin of 9.6 gm/dl (normal 14 to 18). The red cells were pale and small. The chemistry profile suggested liver problems with liver enzymes elevated; alkaline phosphatase 10 times normal, GOT (glutamic oxaloacetic transaminase) 5 times normal, and a mild elevation of bilirubin, a waste product excreted by the liver. There were decreased amounts of substances formed by the liver, including albumen and clotting factors. The CEA tumor marker was elevated to 135 (normal 5.0). X-ray of the stomach and esophagus showed a large mass located predominantly on the anterior wall but filling almost the entire stomach. A liver-spleen scan showed numerous space-occupying lesions in both lobes of the liver. A chest x-ray demonstrated bilateral fluffy, round densities in the lungs that suggested metastatic cancer.

A biopsy of the left supraclavicular lymph node was performed. The results showed moderately differentiated adenocarcinoma with dense fibrosis, probably of gastrointestinal origin. The patient declined chemotherapy and received a feeding jejunostomy before discharge. With the aid of Hospice, he died at his home 5 months after discharge.

Comment. Carcinoma of the gastrointestinal tract usually spreads via the lymphatics to the local lymph nodes and through the veins into the liver, lungs, and bones. Currently gastrointestinal malignancies respond poorly to chemotherapy. Most of the therapy is directed toward comfort and support.

❖ Diverticulitis

W.R. is a 62-year-old man whose illness began suddenly with fever and abdominal pain. He sought medical help after 2 days of increasing pain and illness. He reported to his doctor that 3 weeks ago he had cramping in the lower abdomen before and during bowel movements. Since W.R. has had a long history of "constipation" and the pain was not severe, he was not alarmed. Forty-eight hours ago the pain had become constant and, in addition to cramping, was associated with a dull ache between bowel movements. His physician felt a lower abdominal mass.

W.R. was hospitalized for treatment and further evaluation. On admission, he appeared to be in acute distress. His temperature was 38.1° C, and his pulse was rapid. His abdomen was slightly distended and the bowel sounds were hyperactive and high pitched. There was an ill-defined, tender mass in the lower left part of the abdomen. No other masses or organs could be felt. The distension and bowel sounds were suggestive of a partial bowel obstruction.

Laboratory analysis showed signs of infection with an elevated white blood cell count of 16,400 (normal 4500 to 11,000) with a shift to more immature forms. The urine analysis and chemistry profiles were normal, as was the carcinoembryonic antigen (CEA). The chest x-ray and ECG were within normal limits. A lower gastrointestinal barium enema examination demonstrated a near-total obstruction just above the sigmoid colon with an irregular ragged edge. It gave the impression of a tumor but an inflammatory mass could not be excluded.

W.R. was placed on nasogastric suction to rest his GI tract. He was treated with antibiotics and IV fluids. Over the next 2 days, he passed only minimal gas and no stools but his fever and white blood cell count reverted to normal. The abdominal mass and tenderness remained. On the third hospital day an exploratory laparotomy was performed and revealed a mass in the left descending colon which was removed by segmental bowel resection. The frozen section diagnosis was diverticulitis with near total obstruction. W.R. recovered uneventfully and was discharged on the eighth hospital day.

Comment. Diverticulitis is a more common problem in patients with constipation who "strain at stool." The diverticulum forms when the mucosa of the bowel is forced through the surrounding muscle at weak points, usually associated with penetrating blood vessels. If this outpouching becomes inflamed, often because of fecal obstruction, the infection may spread into the surrounding bowel and mesentery. When this happens, it is often difficult to tell whether one is dealing with cancer or diverticulosis.

❖ Colon Carcinoma

This case concerns a 58-year-old male computer programmer who had gradually developed fatigue and lack of stamina during a 2-year period. At the time of his examination he was overweight and appeared pale. On testing, the hemoglobin was 8.6 gm/dl (normal 14 to 18) and a stool was positive for occult (hidden) blood. He was admitted to the hospital for further (studies).

His physical examination at the hospital revealed no abnormalities but laboratory evaluation showed anemia with a hemoglobin of 7.8 mg/dl and a hematocrit of 25% (normal 42% to 52%). The red cells were small and pale, suggesting iron deficiency. His CEA tumor marker was 1.7 (normal less than 5). The serum iron was low at 25 μ/dl with a high total iron-binding capacity of 430, confirming iron deficiency. A barium enema examination demonstrated a defect in the cecum suggestive of a tumor. Flexible colonoscopy of the cecum found a mass lesion. A biopsy of the mass was performed; the results showed benign villous adenoma with atypia (atypical cells).

The patient was given oral iron and 4 units of packed RBCs and taken to the operating room. A right hemicolectomy (removal of the cecum and ascending colon) was performed. The pathologic analysis found moderately differentiated adenocarcinoma adjacent to areas of benign villous adenoma. The cancer extended through the mucosa into the submucosa but did not extend into the muscularis externa. Four lymph nodes were found but they did not contain metastasis.

After an uneventful recovery, the patient was discharged and followed at 6-month intervals. At 5 years there has been no evidence of recurrence, and he is assumed to be cured.

Comment. Colon cancers are often not diagnosed until they have spread to other organs. Those diagnosed early like this case are either found by routine testing or because of related symptoms. Some colon cancers ulcerate and bleed, causing iron deficiency and anemia. Others are discovered as a result of obstruction. Any adult onset iron deficiency without obvious cause should be evaluated for occult (hidden) bleeding from the GI tract. It may be caused by a curable gastrointestinal tumor.

❖ Fecal Impaction

S.G. is a frail 76-year-old woman whose illness began 1 month ago when she was hospitalized with a fractured right hip. The hip was successfully pinned, but after the surgery S.G. developed a respiratory infection requiring a prolonged recovery period. The patient's past history revealed a long-standing addiction to codeine. Just before her planned discharge, she was noted to have a distended abdomen with a palpable (feelable) colon. A fecal impaction was diagnosed. The impacted stool was mechanically removed and she was given mineral oil and a stool softener and discharged. One week after returning home, she developed severe abdominal pain quickly followed by fever and a marked reduction in awareness.

When admitted to the hospital, she was unconscious with a blood pressure of 58/?, pulse of 134, and temperature of 40.1° C. Her tissue hydration was poor and her mucous membranes were dry. Her abdomen was boardlike without bowel sounds; the merest touch caused her to moan. She was treated for shock with IV fluids, broad-spectrum antibiotics, and a drug to raise her blood pressure until she seemed stable. The admission laboratory work demonstrated an infection with a WBC of 24,000 (normal 4500 to 11,000) with an increase in immature forms. The abdominal x-ray showed a ground-glass appearance with possible free air, an indication of perforation.

In surgery, the abdomen was found to contain a large amount of green, purulent, foul-smelling fluid in which multiple formed stools floated. The colon was distended by impacted feces and the transverse colon had developed a hole through which fecal material escaped. A left transverse colectomy was performed. The peritoneal cavity was emptied and irrigated with antibiotic solutions. In spite of large amounts of fluid, antibiotics, and supportive therapy, she died on the second postoperative day.

Comment. Fecal impaction is usually an innocuous but troublesome condition. It is associated with poor regularity and constipation and can often be found in those patients with decreased sensations. One of the less common causes is narcotic usage. All the opiates decrease peristalsis, cause constipation, and sometimes lead to fecal impaction. In this particular case, the prior hospitalization with its disruption of normal routine led to an increased probability of constipation and impaction. One can also speculate that the narcotic blunted this woman's sensory stimuli, leading to fecal rupture of the colon, peritonitis, septicemia, and death.

❖ Cirrhosis

Z.J. is a 63-year-old bartender who sought medical help because he was concerned about his increasing abdominal girth and lack of appetite. The symptoms had developed so gradually that it was difficult to say when they started. Questioning revealed that Mr. J. was self-employed as a tavern owner for his entire adult life. He drank daily; each day he consumed 10 to 20 boilermakers consisting of a shot of whisky chased by a glass of beer. Despite this consumption, he never appeared inebriated. Over the last years he has eaten less and has developed a persistent diarrhea. Although he has thin arms and legs, his abdomen measures 42 inches.

On admission to the hospital, he appeared to be a chronically ill man, older than his stated age. He had spider angiomas (spider-like abnormally enlarged clusters of blood vessels) on his palms, abdomen, and chest and appeared flushed over the cheeks and palms. His eyes were slightly jaundiced. The chest and heart were normal. His abdomen protruded and a prominent venous pattern was centered on the umbilicus. A percussion wave could be elicited when the abdomen was tapped and shifting dullness was present. These findings indicate ascites, a large amount of fluid in the abdomen. Results of the neurological exam were within normal limits but the spleen was enlarged; there was pitting edema noted in the ankles. *Pitting edema* refers to such excessive fluid in the tissues that small depressions remain when the skin is depressed.

Laboratory examination revealed a macrocytic anemia typical of liver disease. His hemoglobin was 12.2 g (slightly low) and his MCV (mean corpuscular volume) was 108 (normal 82 to 98). The white cell count was normal. His blood chemistry profile showed a pattern suggestive of liver failure: total protein of 4.8 (normal 6 to 8) with 2.1 grams of albumin (normal 3.5 to 5.5), calcium 7.8 mg/dl (normal 8.5 to 10.5), alkaline phosphatase of 176 IU/l (normal to 100), and a total bilirubin of 4.2 mg/dl (normal <1.5). Prothrombin time (a measure of prothrombin and other clotting factors produced by the liver) was 14.2 seconds (normal is 13.0). Abdominal paracentesis (removal of a fluid sample) recovered a faintly turbid greenish fluid with a few nonmalignant cells but an excess of protein. These tests show that the liver is not removing what it should from the blood (bilirubin) or producing as much as it should for the blood (prothrombin and albumin).

During the second day of hospitalization the patient was noted to be agitated, delirious, and feverish and began having hallucinations. These were diagnosed as delirium tremens, typical of alcohol withdrawal, and he was treated with sedation over the next 4 days but still required restraints because of his lack of orientation and combativeness. On the seventh hospital day he underwent a percutaneous liver biopsy that revealed micronodular cirrhosis with a background of a fatty liver. The process was believed to be active and caused by chronic alcohol ingestion.

He was treated with fluid restriction, diuretics, vitamin K, general diet, and IV albumin, all of which led to a 5 kg weight loss. He was instructed to avoid alcohol and join Alcoholics Anonymous and was released from the hospital. He returned to his usual job, continued to drink, and was hospitalized 3 years later in hepatic coma. After a 1-year stay in various institutions, he died.

Comments. This case demonstrates many of the classic findings in chronic alcoholism. These findings include the apparent lack of effect of high alcohol consumption (impression of alcohol tolerance) followed by delirium tremens after abstinence; insidious liver damage which, at the time of diagnosis, reveals too few viable liver cells to cause an elevation in the usual liver enzymes GOT and LDH; poor hepatic function resulting in lack of vitamin K dependent clotting factors with a marked decrease in albumin; and finally, the difficulty in changing a life-long habit.

❖ Hepatitis

P.R., a 68-year-old Cuban physician, requested a physical examination because of increasing fatigability and lack of appetite. The symptoms developed gradually during the last few months. Dr. R. had been a practicing anesthesiologist for the last 30 years. He had habitually sniffed the gases he was working with to make sure they were flowing properly. He knew of no other hepatotoxins or contacts with people with liver disease.

His physical examination was negative with the exception of a palpable, slightly enlarged, tender liver. There was no edema, ascites, icterus (jaundice), or palpable spleen. Laboratory testing showed hepatitis with a serum glutamic oxaloacetic transaminase (SGOT) of 546 IU/I (normal to 36), SGPT of 342 (normal to 40), alkaline phosphatase of 200 (normal to 100), and a bilirubin of 3.1 mg/dl (normal to 1.5). Tests for hepatitis A and B were negative, as was a FANA (fluorescent anti-nuclear antibody, used to test for autoimmune disease). Normal alpha-1 antitrypsin levels were found. A needle biopsy of the liver revealed an irregular, macronodular, active fine septate cirrhosis. The disease was ascribed to chronic exposure to anesthetic agents.

P.R. retired from practice but his liver function test results remained abnormal during the next 6 years. At that time he experienced a rapid increase in abdominal girth and was found to have a palpable liver mass. Biopsy showed a hepatoma and an attempt was made to excise the malignant tumor. Over the next year he was treated with 5-fluorouracil, a chemotherapeutic agent, but showed tumor progression. His liver function continued to deteriorate and he died in hepatic coma.

Comment. This is an unusual case of hepatitis. The common causative agents include the viruses, hepatitis A, B and C. Certain drugs, such as some anesthetic agents, various solvents, alcohol, and occasional medications are hepatotoxins and can cause chronic or acute hepatitis. Autoimmune mechanisms can cause liver damage, and a deficiency of alpha-1-antitrypsin may lead to cirrhosis. When a cause cannot be established one is left with an "idiopathic" (or unknown)" cause. Dr. R.'s hepatitis was either "idiopathic" or anesthetic induced. His terminal complication, hepatoma, is an unusual but not-unheard-of complication of cirrhosis.

BIBLIOGRAPHY

Andrews GR, Haneman B, Arnold BJ: Atrophic gastritis in the aged, *Austral Ann Med* 16:232-240, 1967.

Brocklehurst FC: *The gastrointestinal system—the large bowel.* In Brocklehurst FC, (ed): *Textbook of geriatric medicine and gerontology,* New York, 1985, Churchill Livingstone.

Earnest D: *Other diseases in the colon and rectum.* In Sleisenger MH, Fordtran JS, eds: *Gastrointestinal disease,* vol 3, Philadelphia, 1989, WB Saunders.

Gibbons JC, Levy SM, *Gastrointestinal diseases in the aged.* In Reichel W, ed: *Clinical aspects of aging,* ed 3, Baltimore, 1989, Williams & Wilkins.

Hutter RVP, Sobin LH: A universal staging system for cancer of the colon and rectum, *Arch Pathol Lab Med* vol 110:367-368.

Langman MJS: Epidemiology of cancer of the esophagus and stomach, *J Surg* 58:794, 1978.

Miles EAW: *Aging in the teeth and oral tissues.* In Bourne GH, ed: *Structural aspects of aging,* 1961, Hafner.

Minaker KL, Bonis P, Rowe JW: *Gastrointestinal system.* In *Geriatric Medicine,* ed 2, Boston, 1988, Little, Brown.

Pridie RE: Incidence and coincidence of hiatus hernia, *Gut* 7:188, 1966.

Robbins SL, McAngell: *Basic Pathology.* Philadelphia, 1971, WB Saunders.

Vierling JM, Reichen J: *Physiology and diseases of the digestive system in the aged.* In Schrier, ed: *Clinical internal medicine in the aged,* Philadelphia, 1985, WB Saunders.

Walker DM: *Oral disease.* In Pathy MJS: ed: *Principles and practice of geriatric medicine,* New York, 1985, John Wiley & Sons.

CHAPTER

8 *The Excretory System*

RELEVANT ANATOMY AND PHYSIOLOGY
Functions and Components of the Excretory System

The excretory system eliminates the wastes generated by metabolism while conserving the materials required for life. Both these processes are essential to survival. Metabolic wastes can be toxic when accumulated, although in some cases they may be stored harmlessly. For example, bilirubin from the breakdown of hemoglobin can be stored in the skin, producing a yellowish skin color (jaundice). In most instances, however, metabolic wastes must be removed from the body through the excretory system, consisting of the primary excretory organs (the kidneys and the urinary bladder), and other organs that have excretion as a secondary function.

Organs in other body systems that also have an excretory function include the lungs, which excrete carbon dioxide and, under certain circumstances, other metabolic gases such as ammonia and acetone. The skin excretes metabolic wastes in sweat and in the process of shedding dead cells. The liver excretes many metabolic wastes, including bilirubin and cholesterol, in the bile. (It is when the liver begins to fail that bilirubin is stored rather than excreted.) The secondary excretory organs are discussed more fully elsewhere in this text. This chapter will discuss the kidneys, their collecting system, and the urinary bladder.

Ammonia, a toxic product of the breakdown of amino acids, is converted by the liver to urea, a less toxic compound. The kidneys remove most of the urea from the body, although under extreme conditions the sweat glands may also remove substantial amounts. The kidneys remove other toxic or unneeded compounds from the blood while conserving needed materials. Indeed, more energy may be expended conserving needed elements than excreting unneeded ones. Neither process is completely effective, so small amounts of some needed elements are excreted in the urine. For example, calcium is continually lost through the kidneys and must be replaced by dietary intake. Without sufficient oral intake, the blood calcium level would fall, requiring absorption of calcium from the skeleton to maintain adequate levels in the circulation.

The Filtering of Blood

Approximately 21% of the blood pumped by the heart through the aorta passes into the renal arteries and through the kidneys. This is called the renal fraction of the cardiac output. It may vary greatly with activity, emotional state, or the condition of the

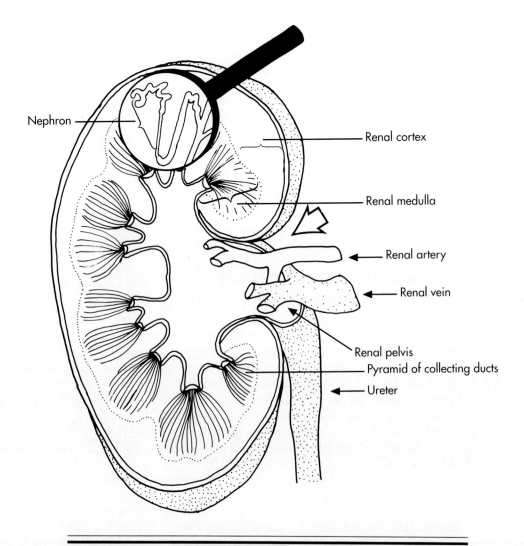

Figure 8-1   Cross-section of kidney. Blood enters the kidney through the renal artery and leaves through the renal vein. In the kidney it is forced through the capillaries of the nephrons located mostly in the renal cortex. Tubules collecting the urine from the nephrons make up most of the renal medulla and conduct the urine to the renal pelvis.

kidneys or renal arteries. Because the blood mixes as it is returned to the heart, ultimately all blood will pass through the kidneys. Figure 8-1 shows a cross-section of a human kidney.

Blood enters the kidney through the renal artery. This artery subdivides many times into sets of smaller arteries that end in over one million afferent (incoming) arterioles in the renal cortex, or outer portion of the kidney. The blood from each afferent arteriole then enters a separate functional unit of the kidney called a nephron, shown in Figure 8-2. Each nephron consists of a glomerulus and its associated Bowman's capsule, the system of tubules leading away from the Bowman's capsule, and the interconnecting

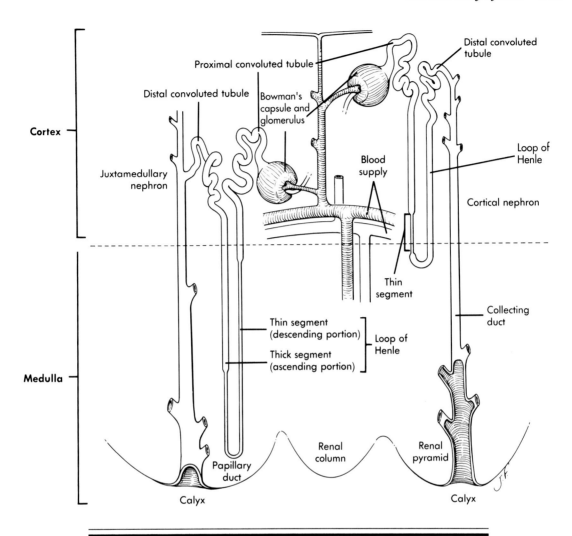

Figure 8-2 Very dilute urine is squeezed out of the blood in the glomerulus of each nephron. About 99% of the water that enters the Bowman's capsule is reabsorbed into the blood by the time the urine reaches the collecting duct.

chain of blood vessels that are in close contact with the tubules. There are approximately 1 million nephrons in each human kidney.

Each of the afferent arterioles deposits its blood in an arteriole ball called a glomerulus surrounded by a capsule of tissue called Bowman's capsule. The glomerulus and part of its associated Bowman's capsule make up the filtering structure of the nephron. The blood pressure at the entrance to the glomerulus is about 70 mm Hg. Pressure is regulated primarily by the size of the lumen of the blood vessel leaving the glomerulus, or the efferent (outgoing) arteriole, in combination with cardiac output and several other factors described later. A small lumen increases back pressure on the blood and increases the amount of plasma filtrate squeezed out of the blood into the nephron. The high pressure combines with the selective porosity of the glomerular walls to filter about 19% of the plasma volume through the glomerular walls into the space of Bowman's capsule. This filtrate is similar to plasma except that it normally lacks most of the plasma proteins and all the cells and platelets of blood. It is usually referred to as glomerular or plasma filtrate. By the time the final urine is formed, about 99% of the water of the plasma filtrate will have been reabsorbed along with many other needed compounds. The urine will also have been augmented with additional materials actively removed from the blood and secreted into the urine by the renal tubules.

Refining Plasma Filtrate into Urine

The Proximal Convoluted Tubule. The glomerular filtrate passes from each Bowman's space into a proximal convoluted tubule. Capillaries carrying the blood from the efferent (outgoing) arteriole form a close association with the proximal convoluted tubule, allowing active and passive reabsorption of many constituents. The blood in these capillaries is concentrated and contains many large proteins and other large molecules that did not pass through the glomerular membrane. The fluid in the proximal convoluted tubule is dilute, containing a proportionately larger volume of water. This results in movement of water by osmosis from the tubule back into the blood. Because the wall of the tubule is more permeable to water than it is to urea, most of the urea remains in the urine. Approximately 60% of the water that was filtered out of the blood in the glomerulus is reabsorbed by the proximal convoluted tubule. In addition, cellular energy is used to actively transport glucose, amino acids, calcium, phosphate, and sodium ions from the glomerular filtrate into the interstitial fluid so that they can be absorbed into the blood.

The Loop of Henle and the Counter Current Mechanism. From the proximal convoluted tubule, the filtrate passes into the loop of Henle. The loop of Henle forms an elongated U that begins in nephrons near the medulla and extends deeply into the medulla of the kidney (Figure 8-2). In outer cortical nephrons, the loop of Henle is short; only the tip penetrates the medulla. The loop of Henle is responsible for increasing the build-up of salt in the medulla of the kidney through which the collecting tubules pass.

The loop of Henle concentrates salt in the interstitial fluid through a multistep process. In the descending portion of the loop, its tubular wall is thin and permits the passive diffusion of water, some urea, and lesser amounts of certain ions. The walls of the ascending portion are very thick and nearly impermeable to water and urea. They prevent these substances from diffusing back into the urine. More importantly, the ascending portion actively transports sodium, chloride, and potassium ions out of the tubule lumen into the interstitial fluid. This leaves the urine with less salt and leaves the interstitial fluid very salty. The direction of flow within the blood vessels associated with the loop of Henle (vasa recta) is opposite to the flow of the urine (Figure 8-2); the blood absorbs and carries with it some of the salts. This, together with the loop of Henle, creates a counter current mechanism that is responsible for the efficient concentrating of salts at the base of the medulla of the kidney.

The Distal Convoluted Tubule. The loop of Henle terminates in the distal convoluted tubule. The blood that has been in association with the proximal convoluted tubule and the loop of Henle flows in close association with the distal convoluted tubule. The distal convoluted tubule, under the influence of the hormone aldosterone secreted from the adrenal gland, actively reabsorbs sodium from the forming urine and with it, considerable water. As the sodium is reabsorbed, large amounts of potassium and hydrogen ions are excreted. This increases the acidity of the urine. The distal tubule strongly influences the pH (acidity or alkalinity) of blood by varying the amount of hydrogen excretion.

The Collecting Tubules and Ducts. From the distal convoluted tubule, the fluid flows into the collecting tubules and on into collecting ducts. These ducts flow through the salty interstitial fluid near the loops of Henle. As they do so they become extremely permeable to water, due to the influence of antidiuretic hormone. The water then rapidly flows by osmosis from the urine into the interstitial fluid and into the blood. Osmosis is the passive movement of water across a membrane from high water concentration (low salt, protein concentration) to low water concentration (high salt, protein concentration). The tubular mechanism is so well-developed in some desert rodents that their urine is concentrated into a paste. Urea is also absorbed from the urine by the collecting ducts. Most of it reenters the urine in the loop of Henle. While in the interstitial fluid the urea further aids water absorption from the collecting ducts by osmosis. As a result, about 39% of the water originally filtered out of the blood by the glomerulus is reabsorbed into the blood by the combination of the loop of Henle, the distal convoluted tubules, and the collecting ducts. Added to the 60% of the glomerular filtrate that was absorbed by the proximal convoluted tubule, this makes up the total of approximately 99% of the filtered water that is returned to the blood.

The Kidneys, Regulatory Hormones, and Homeostasis

The Renin-Angiotensin-Aldosterone Mechanism. The kidneys regulate blood volume and ion balance by using two mechanisms. One of these involves a group of cells called the juxtaglomerular apparatus. It is composed of specialized cells in the walls of the afferent and efferent arterioles, plus another group of cells called the macula densa, located in the walls of the distal convoluted tubule. The distal convoluted tubule is folded back against the glomerulus so that the macula densa and the juxtaglomerular cells form a unit (Figure 8-3). When inadequate filtration occurs, such as would happen if the blood pressure declined, the fluid flowing through the tubules is reduced and most of the salts are absorbed from the tubular fluid. The presence of inadequately salty urine in the distal tubule causes the macula densa to stimulate the juxtaglomerular cells to release the enzyme renin. Renin circulates through the body, causing the conversion of a blood protein, angiotensin I to angiotensin II. Angiotensin II causes constriction of arteries and a consequent increase in blood pressure and glomerular filtration.

Angiotensin II raises blood pressure by other means than arterial constriction. It causes the cortex of the adrenal glands to release aldosterone. This steroid hormone increases the active transport of salt out of the ascending portion of the loop of Henle and the distal convoluted tubule, increasing the concentration of salt in the medulla of the kidney. As the salt moves into the interstitial fluid, water from the distal tubule and collecting duct follows. The salt and water then diffuse into the bloodstream, increasing the blood volume and pressure. As the blood pressure rises, glomerular filtration increases, and renin release ceases. Angiotensin II also stimulates the vasomotor center, which stimulates the sympathetic nervous system. This reaction accelerates the heartbeat, further constricts the blood vessels, and elevates the blood pressure even more.

The Antidiuretic Hormone Mechanism. The second mechanism for regulating blood volume and pressure through the kidneys is through antidiuretic hormone (ADH). ADH, also known as vasopressin, is produced by the hypothalamus of the dien-

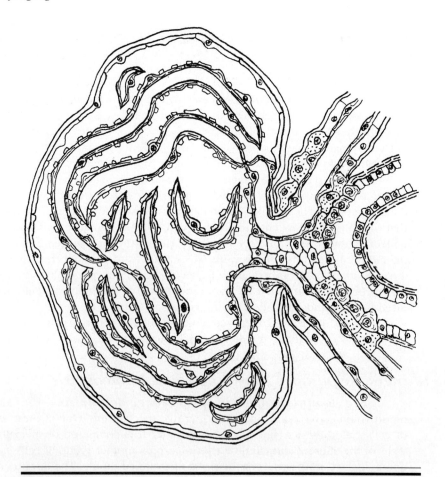

Figure 8-3   A small portion of the distal convoluted tubule contacts the afferent and efferent arterioles near the glomerulus. Here specialized cells that form the macula densa stimulate the juxtaglomerular cells to release renin into the bloodstream. The cells of the macula densa become active when salt concentration in the distal tubule becomes too low. This occurs when flow through the tubules is reduced.

cephalen of the brain and released by the posterior lobe of the pituitary gland in response to increased blood salinity. When water is lost from the blood, or for some other reason blood salinity increases, the posterior pituitary gland releases ADH into the bloodstream. The release of this hormone increases the permeability of the distal convoluted tubules and collecting tubules and ducts to water. This results in much more complete absorption of water from the glomerular filtrate into the blood. This in turn increases the blood volume and pressure, dilutes the salts in the blood, and concentrates the urine. This mechanism prevents the excretion of an abnormally high volume of dilute urine, or diuresis, consequently the name antidiuretic hormone. The disease diabetes insipidus involves inadequate ADH production and tremendous urine volume, along with tremendous thirst to replace the lost blood volume. Note that aldosterone could also be considered an "antidiuretic" hormone but it decreases urine volume by increasing sodium chloride reabsorption from the tubules, a different mechanism.

The ADH and renin-aldosterone mechanisms would not be of much help if they were not supported by an intense thirst mechanism. When blood salinity increases, the thirst center in the hypothalamus responds to the osmotic dehydration of its cells by stimulating the brain to generate a sensation of thirst. It is the resulting intake of water that actually increases the blood volume and dilutes the salt in the blood, because all the renal regulatory mechanisms can do is lower the rate of water loss. It is interesting to note that if the renin-aldosterone salt retention cycle were to increase blood salinity, this then would increase the thirst response, thereby increasing blood volume and decreasing relative salinity. It is easy to see how maladjustment of this mechanism can lead to perpetually higher blood volumes and pressures.

Lowered blood pressure resulting from either hemorrhage or decreased cardiac output also results in a thirst reflex. It does this through the release of renin and the resulting production of angiotensin II. Angiotensin II directly stimulates the thirst center in the hypothalamus. Assuming that the hemorrhage has stopped and that the kidneys are conserving water by decreasing urine output, the water consumed should return the blood volume and pressure to normal. It is important to remember that blood pressure alone is an important regulator of blood volume. As blood pressure increases, glomerular filtration and the tubular flow of urine increases. As fluid moves through the renal tubules more rapidly, efficiency of the urine concentration mechanisms declines. This results in increased rate of fluid loss from the blood and a returning of the blood volume and pressure to normal. Conversely, as blood pressure declines, the amount of water filtered from the blood by the glomeruli declines. It moves through the tubules more slowly so water is reabsorbed more efficiently, and further loss of blood volume is reduced.

Tests of Kidney Function. The kidneys decline in capacity and efficiency with age. Knowledge of the extent of this decline is essential, not just to determine whether the kidney is getting rid of waste, but to determine how well renally excreted drugs are being cleared from the blood. Toxic effects can occur when reduced renal clearance allows normal drug dosages to accumulate to overdose concentrations in the blood. To measure kidney function, two tests may be used, inulin clearance and creatinine clearance. Table 8-1 gives information on common renal function tests.

Inulin is a complex sugar that is filtered out of the blood by the glomerulus but is not reabsorbed by the tubules. A known injected amount in the blood produces a known concentration in the blood. Thereafter, an hourly measure of inulin concentration in the urine gives a precise measure of glomerular filtration rate. A more frequently used but slightly less accurate measure of kidney filtration is creatinine clearance. Creatinine is a natural waste product that is always present in the blood. Using a blood creatinine level and a determination of the amount of creatinine in a timed urine collection, one can estimate the glomerular filtration rate. However, creatinine is both filtered by the glomerulus *and* excreted by the distal tubule, making this test less accurate than inulin clearance.

Some Other Kidney Functions

The functions of the kidneys are more numerous and complex than might be expected, but essentially all of them are involved with the maintenance of homeostasis. Table 8-2 lists the major functions of the kidney.

Table 8-1 Common Renal Function Tests

TEST	KIDNEY DISEASE	AGED KIDNEY	CONFOUNDING FACTOR
BUN (Blood Urea Nitrogen)	Elevated in direct proportion to renal impairment	Slight elevation	Falsely low in protein malnutrition Falsely high in GI bleeding Insensitive for renal disease
Creatinine	Elevated in direct proportion to renal disease	Slight elevation	Proportional to muscle mass. Insensitive for renal disease
Creatinine clearance	Decreased in direct proportion to renal disease	Decreased (see text)	Very sensitive to loss of functioning kidney Measures glomerular flow plus tubular excretion
Inulin clearance	Decreased in direct proportion to renal disease	Decreased (see text)	None Requires injection of inulin.

Table 8-2 Major Functions of the Kidneys

Secretion	of renin (an enzyme) and erythropoietin (involved in the production of red blood cells)
Excretion	of metabolic wastes such as urea, creatinine
Conversion	of 25-hydroxycholecalciferol to active Vitamin D
Maintenance	of proper blood pH (discussed in this chapter) of proper electrolyte balance (discussed in this chapter)
Conservation	of essential elements such as sodium and calcium

Collection, Storage, and Release of Urine

Once the urine has been formed, it collects in the renal pelvis and leaves each kidney through a ureter (Figure 8-4). The ureters carry the urine to the urinary bladder where it is stored until the bladder is full and conditions are appropriate for micturition, or the elimination of urine from the body. When micturition occurs, the bladder contracts and forces the urine out through the urethra. In males, the urethra passes through the prostate gland and becomes a common duct for semen and urine.

The flow of urine into the urinary bladder from the ureters is continuous. The expansion of the bladder to accommodate this input stimulates an increasing number of impulses to be sent to the micturition center in the spinal cord. When the bladder wall is adequately stretched, enough impulses will be received by the micturition center to produce the micturition reflex. This causes simultaneous contraction of the bladder wall and relaxation of the bladder neck and urethral sphincters. Control of micturition is

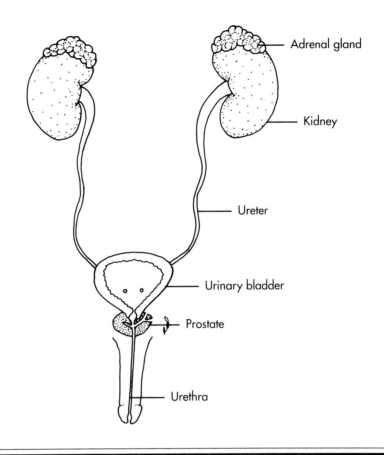

Adrenal gland

Kidney

Ureter

Urinary bladder

Prostate

Urethra

Figure 8-4 Male urinary system. Urine passes from the kidneys through the ureters to the urinary bladder, where it is stored until it is released into the urethra. Note that in men, the urethra passes through the prostate.

learned by young children as a result of impulses being passed up the spinal cord to the region of the hypothalamus and cerebral cortex. People learn to inhibit and facilitate micturition through associations made with repeated stimulation of these higher centers. When these higher centers are damaged, loss of bladder control is a common result. Loss of bladder control is a very common problem that will be discussed in detail later in this chapter.

The urinary bladder is surrounded by a network of smooth muscle fibers. When stimulated by the spinal reflex, these detrusor muscles contract, increasing the pressure on the bladder fluid. A separate set of circular smooth muscle fibers surrounds the bladder neck in men but is poorly developed in women. In men, the primary function of these bladder neck muscles may be to prevent reflux of semen into the bladder during coitus. It is evident that they also provide an additional level of bladder control.

Additional control is provided by a specialized involuntary muscle in the distal portion of the prostatic urethra of males and the middle third of the urethra of females. The voluntary striated sphincter is distal to this and is innervated by the pudendal nerve. This sphincter functions in emergencies but otherwise is probably not involved in bladder control.

THE AGING EXCRETORY SYSTEM
Anatomic Changes

Several anatomic changes develop in the kidney as it ages. Much of the following discussion is drawn from the excellent review of these provided by Bichet and Schrier (1982). The kidney loses about one-fifth of its size, declining from a volume of about 250 ml to 200 ml. There is a decrease in the surface area and number of glomeruli. Only about three quarters of them remain functional. More outer cortical nephrons are lost than inner cortical nephrons. In the medulla there is an increase in fat, interstitial tissue, and sclerosis of the tubules of the pyramids. Amazingly, aging kidneys continue to function normally. They maintain homeostasis of needed elements and compounds, do a fair job of maintaining blood volume and pressure, and still manage to clear the body of most undesirable materials.

Changes in Glomerular Filtration and Vascular Response

The glomerular filtration rate decreases gradually from age 30 until age 45, and more rapidly thereafter probably reflecting decline in number of functional nephrons. Creatinine clearance declines from 140 ml per minute at age 29, to 97 ml per minute at age 80. The determination of inulin or creatinine clearance is an important measure of kidney function and is usually estimated before treatment with nephrotoxic agents. Knowledge of kidney function is useful regarding other drugs normally excreted by the kidney. A toxic drug eliminated by urinary excretion, prescribed in dosages high enough to maintain an adequate level in the blood with normal functioning kidneys, may accumulate to dangerous levels in elderly patients with reduced renal clearance.

The effect of age is evident when one kidney is removed from an elderly patient. The remaining kidney is slow to undergo the normal vasodilation, hypertrophy, and increase in blood flow typical of younger people. The ultimate filtration rate may be only half the already reduced rate for someone this age who had not just had a kidney removed.

Changes in Renal Hormones and the Thirst Response

The distal convoluted tubule and collecting ducts probably lose some of their responsiveness to antidiuretic hormone (ADH), making it more difficult for water to return to the bloodstream. A decrease in sodium transport across the membrane of the loop of Henle has been reported. The latter condition could decrease the hypertonic state of the interstitial fluid in the medulla and thereby decrease the urine concentration mechanism of the kidney, resulting in lower blood volume and pressure. Many elderly people have a decreased thirst response to excess sodium in the blood, or hypernatremia. This, in combination with lower ADH or aldosterone, could result in low blood pressures and even kidney failure. The ability to excrete acidic substances such as ammonium chloride also declines with age.

Several of the hormonal functions of the kidney also have been shown to decline with age. Both plasma renin secreted by the juxtaglomerular cells and aldosterone secretion from the adrenals decrease with age. The increase in renin and aldosterone following sodium restriction has also shown an age-related decline. Hypoaldosteronism and its resultant hyponatremia may be caused by insufficient stimulation of the adrenals by the renin-angiotensin system. Hypoaldosteronism has been related to heart block, ECG irregularities, and other heart problems. Excessive antidiuretic hormone is also seen in the elderly as a cause of hyponatremia. ADH secretion can be stimulated by various conditions such as stroke, or by certain drugs. These drugs also may stimulate the kidneys' responsiveness to this hormone.

Conditions of the urinary system that frequently cause problems in elderly people are discussed in the following section. Problems such as renal and essential hypertension also are discussed in this section although they are vascular in effect, because it seems likely that the kidney or its hormones will be found to be at the root of both.

Diseases and Conditions Associated with Age

Urinary Incontinence. Urinary incontinence is the inability to control the release of urine from the urinary bladder. While this condition is only a symptom of a deeper underlying problem, it is included here because it afflicts a very large number of elderly people and is one of the most common reasons for placing an elderly person in a nursing home. In one study of elderly patients admitted to a long-term care facility, 38% were incontinent. Additionally, 10% approached incontinence with symptoms of frequency, urgency, nocturia (night urination), or had dysuria (inability to urinate without pain). Twenty percent had urinary drainage devices. Of course the rate of incontinence in the elderly population at large is much less than this. However, the manufacture of adult disposable diapers is now a multibillion dollar business. This shows that there is a large number of people with this problem. Although the use of adult diapers carries some risk of enabling a person with incontinence to ignore a serious underlying condition, these products have enabled many elderly people to continue active social lives. The following discussion of incontinence is adapted from Farrar (1985).

Urethral obstruction is most common in men because their urethra passes through the prostate gland. When this gland is enlarged, it encroaches on the lumen of the urethra. Prostate obstruction of the urethra produces symptoms of overflow incontinence, including poor urine stream, continual dribbling, frequency and nocturia, and post-micturition dribble. Retention may also result from urethral stricture or carcinoma of the prostate. A frequent complication of the resulting urine retention is bladder stones. In women, obstruction may occur at the bladder neck or in the distal portion, sometimes as a result of prolapse of the urethra.

Genuine *stress incontinence* results from loss of the bladder neck control mechanism, leaving the urethral sphincters unable to withstand the bladder pressures. Bladder neck weakening may be congenital, or arise from surgery, or other instrumental damage. In women it may arise from childbirth or hormone deficiency, the latter sometimes involving cystocele, or prolapse of the bladder into the vaginal space. Neuromuscular age-related factors can affect either sex.

Unstable incontinence can result from uninhibited detrusor muscle contractions. With this condition, the bladder spasmodically contracts and the sphincters are unable to withstand the pressure. This is the most common cause of incontinence in the elderly and produces both urgency and frequency. It varies greatly in severity. Unstable incontinence may follow prostatectomy when bladder instability persists following relief of prostate obstruction.

Neuropathic bladder refers to bladder dysfunction as a result of neurologic disease. Many forms of age-associated dementia may result in lack of inhibition regarding voiding. Cerebrovascular accidents may produce a lack of conscious bladder control, or they may damage the neural pathways controlling the bladder. Loss of bladder sensation may result from damage to the dorsal roots of the spinal nerves or their ganglia. Diabetes can cause a peripheral neuropathy and result in poor bladder control.

Overflow incontinence results from excessive volumes of urine in the bladder.

Urethral obstruction produced by an enlarged prostate, or stricture resulting from gonorrhea infection may damage the detrusor mechanism so that a persistent daytime dribble is produced. In women bladder obstruction may occur at the bladder neck or in the distal portion, sometimes as a result of prolapse of the urethra. Since the bladder is carried partially full all the time, increasing frequency produced by the limited available bladder reservoir may result in miscalculations of the time necessary to find an appropriate place to urinate.

Renal Calculi. Renal calculi, or kidney stones, develop when salts are concentrated in the urine and precipitate as stones. The stones are very small (smaller than sand grains) when first formed and usually wash out of the urinary system as harmless sediment. However, problems may develop if urine volumes are very low, allowing the stones to grow.

Kidney stones usually become a problem when they become lodged in the ureter while passing from the pelvis of the kidney to the urinary bladder. The ureters narrow as they approach the bladder; the pain produced as the stone irritates the ureter, causing spasmodic peristaltic contractions, is one of the most excruciating pains known. Blockage may occur, expanding the ureter between the stone and the kidney and flooding the kidney with fluid. Blood escaping from the damaged walls of the ureter may appear in the urine.

Most forms of kidney stones are uncommon in the elderly; the peak incidence for calcium stones is between ages 30 and 50. However, staghorn calculi, large branching stones that remain in the renal pelvis, occasionally occur in elderly patients. These are frequently associated with infection. They may form abscesses and sometimes block the ureter. The formation of renal calculi usually can be prevented through maintenance of a normal diet and adequate fluid intake. A normal exercise regimen is also helpful.

Urinary Tract Infections. Bacterial urinary tract infections are common and often asymptomatic. They carry the risk of spreading through the body where, with their toxins, they may produce septicemia and shock. If the infection involves the kidneys, significant renal damage and failure may result. Most such infections are limited to the bladder and lower urinary tract. A recent study found urinary infections in 8% of persons between 25 and 44 years old and in 20% of people from age 65 to 96. It found that more active, noninstitutionalized people generally have a lower infection rate. Urinary tract infections are more frequent with pregnancy, insertion of instruments into the urethra, and advancing age. They are more common in women than men, although infections are increased with urinary obstruction caused by an enlarged prostate, a problem limited to men. Decreased immune response and inadequate blood supply are the two most likely factors contributing to the presence of bacteria in the urine (bacteriuria) of elderly people who do not have a recent history of urinary obstruction or instrumentation (usually catheterization).

Approximately 85% of urinary tract infections are caused by the most common organism found in feces, the bacterial species *Escherichia coli.* After multiple infections have been treated by antibiotics, more resistant organisms such as *Klebsiella, Enterococcus, Pseudomonas,* and *Proteus* are often found to be present. All of these organisms are normal flora of the colon and probably enter the bladder through the urethra. Infections are more common in women, probably because of their shorter urethra. The usual urinary infections are limited to the urinary bladder and urethra and cause fever, lower abdominal pain, painful urination, and the need to urinate more frequently. Because antibiotics tend to concentrate in urine, these infections respond well to antibiotic therapy.

When urinary tract infections are left untreated for extended periods or when especially virulent organisms are involved, the infection may extend to the renal pelvis and kidney. Antibiotics are less concentrated in these tissues, which makes these infections harder to cure. Upper tract infections produce fewer symptoms and are often present for long periods of time, resulting in chronic pyelonephritis.

Urine is a good culture medium for bacteria. For this reason any urinary stasis (abnormal retention of urine) must be eliminated when treating an infection. Paradoxically, indwelling urine drainage systems (catheters) used to relieve prostatic obstruction or in treatment of urinary incontinence, all result in infection within 1 week of placement.

Pyelonephritis. Urinary tract infections may involve the kidney and produce an interstitial inflammation, resulting in destruction of glomeruli and their renal tubules. These interstitial infections and their associated inflammation are known as pyelonephritis. Recently, cases thought to be bacterial pyelonephritis have been shown to result from sensitivity to certain drugs such as sulfonamides, phenytoin, methicillin, and some analgesic drugs. Even ischemia due to renal vascular problems can produce changes that appear as pyelonephritis. The nonbacterial pyelonephritis is now properly termed *nonbacterial interstitial nephritis,* although their histologic and clinical pictures are similar. Drug-induced interstitial nephritis may be associated with allergic symptoms, including skin rash, pruritis, fever, eosinophilia (excessive eosinophils), and azotemia, or urea in the blood. Removal of the drug usually stops progression of the disease.

Two schools of thought exist regarding how pyelonephritis originates. The older view is that bacteria ascend through the urethra, urinary bladder, and ureters and infect the renal pelvis. Infected papillae at the ends of the collecting ducts become inflamed. The bacteria then follow the ducts through the associated nephrons, penetrate the tubules, and enter the interstitial spaces. The more recent view is that infection occurs through septicemia, a body-wide bacterial infection developing from a bladder infection, or by following the lymphatic tracts around the ureter to the kidney. In both instances the infection involves the entire nephron from the glomerulus in the renal cortex to the ends of the collecting ducts in the medulla. The tubules may become distended with acute inflammatory cells and edema develops in the interstitial spaces (Figure 8-5 and Color Plate 11, *A*). Occasionally abscesses develop.

The early form of the disease usually responds to treatment if it is caused by a correctable urinary blockage and appropriate antibiotics are used for the proper period of time. The acute process heals with fibrous tissue; the wedge of scar tissue left in the kidney, where the block of nephrons was destroyed by the infection, is permanent. Unfortunately, chronic pyelonephritis is usually silent and slowly progressive, ultimately resulting in reduced renal reserve capacity. When contributing causes such as urinary tract obstruction can be removed, most patients with acute pyelonephritis recover fully. However, in chronic cases where the infection has developed into chronic pyelonephritis and/or where hypertension, poor urinary concentration, and increased urea in the blood are seen, even removal of urinary obstruction may not help. About 90% of these patients will develop progressive renal failure.

Glomerulonephritis. Acute poststreptococcal glomerulonephritis is often a disease of children and young adults, but a description of this condition provides a good background for understanding glomerular kidney disease. This form of glomerulonephritis normally begins about 10 days after a streptococcal infection, usually a "strep throat." Antistreptococcal antibodies in complex with remains of the streptococcal bac-

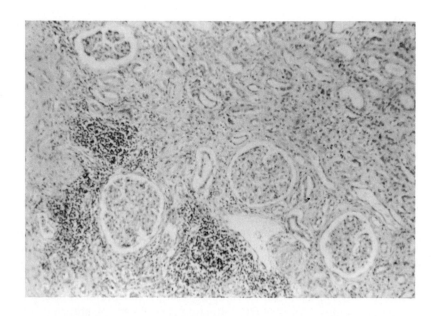

Figure 8-5 The glomerulus here looks relatively normal except for some expansion of Bowman's space. There is considerable interstitial inflammation and some collecting tubules are expanded with fluid. (See Plate 11, *A*, for color).

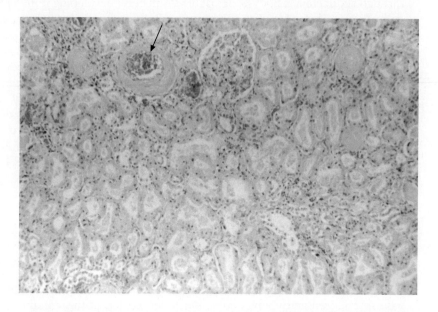

Figure 8-6 A crescent of proliferated epithelial cells and monocytes is shown in the Bowman's space of the affected glomerulus (arrow).

teria deposit on the glomerular membrane. The inflammatory reaction is caused by the effects of the immune system on the antigen-antibody complex. It results in cellular proliferation, edema, and disruption. The disrupted glomeruli leak red blood cells into the urine, producing hematurea. They also leak large proteins, producing proteinurea. The glomeruli and nephrons may then become blocked with swollen tissue and inflammatory cells. If enough glomeruli are shut down, acute renal failure with uremia can develop. About 80% to 90% of children and 40% to 50% of adults recover from this form of glomerulonephritis with little permanent impairment.

Idiopathic Crescentric Glomerulonephritis.

This is the most common cause of nephritis in the elderly. There is no evidence of a streptococcal origin of this disease, but it is also the result of antigen-antibody deposits. In this disease, Bowman's capsules become filled with proliferating epithelial cells and monocytes. These form a crescent around the compressed glomerulus and interfere with the filtering function (Figure 8-6). Diffuse atrophy of the tubules occurs but there is little change in the renal arteries and arterioles. The symptoms of this disease are minimal and similar to those of acute glomerulonephritis. There is often rapid progression of this disease, resulting in death in a few weeks or months.

Membranous Glomerulopathy.

This condition and crescentric glomerulonephritis have been reported as the most commonly diagnosed ailments of elderly patients who have glomerulonephritis. As with the previous form, membranous glomerulopathy is the result of deposition of antigen-antibody complexes with no preceding streptococcal illness. It is found 50% more frequently in men than women. Early in the disease the patient usually develops the nephrotic syndrome: edema with massive proteinurea and microscopically visible hematuria. Gross observation of blood in the urine is uncommon. Unlike many forms of glomerulonephritis, there is no significant glomerular inflammation. There is instead a uniform thickening of the glomerular basement membrane, interfering with the filtering function of the glomerulus (Figure 8-7). This is believed to result from progressive deposition of small soluble immune complexes in the subepithelial spaces of the glomerulus. The antigen is usually unknown but certain drugs, heavy metals, infectious agents, and carcinoma have been implicated. The disease progresses slowly to renal insufficiency at variable rates, producing uremia in 5 to 10 years. Overall adult survival rates are 5-year, 70%; 10-year, 40%; and 15-year, 30%.

Diabetic Nephropathy.

The prevalence of diabetes mellitus increases with age. The effect of the disease on renal function becomes particularly significant in the later years. About 5% of deaths of diabetics are due to diabetic nephropathy. In insulin dependent diabetes, the first indications of nephropathy usually appear 10 to 15 years after the onset of the diabetes.

Diabetes involves decreased insulin production or decreased insulin receptiveness by the body and consequent sugar utilization problems. Widespread basement membrane abnormalities and vascular lesions develop but the reasons for this are unknown. The vascular consequences are seen throughout the body and especially in the kidneys and eyes. These consequences appear to be proportional to the average blood sugar level. Higher blood sugars and poor diabetic control result in accentuation of the vascular problems. Diabetics with renal involvement usually have diffuse glomerulosclerosis.

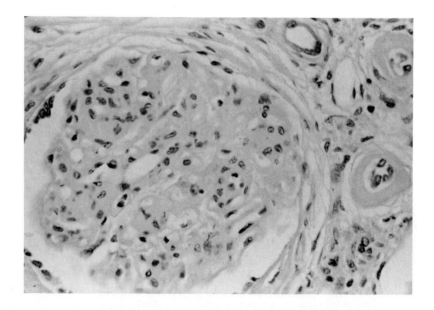

Figure 8-7   In cases of membranous glomerulopathy, the walls of the glomer-ulae are thickened and will not function adequately as filters.

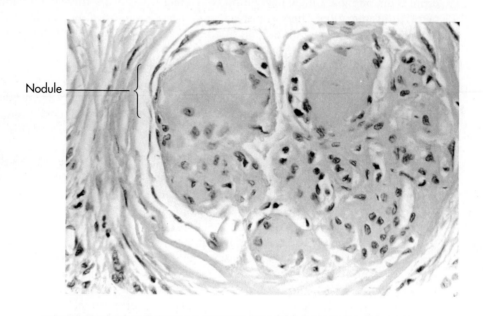

Figure 8-8   Nodules are visible on the glomerular membrane in this example of diabetic nephropathy.

Diabetic glomerulosclerosis may be membranous or may produce a nodular thickening of the glomerulus basement membrane called Kimmelstiel-Wilson nodules (Figure 8-8). Hyalin arteriolosclerosis usually occurs fairly early in the disease and affects both afferent and efferent arterioles. Advanced cases require hemodialysis or kidney transplants. Hemodialysis presents particular problems with diabetics because of clotting and infection along blood access routes, muscle wasting, worsening of vision, fluid overloading, and low sodium complications. Renal transplants have recently shown much more success. Without dialysis or transplant the life expectancy of a patient after the diagnosis of diabetic glomerulosclerosis has been only 6 months to 2 years.

Renal Atherosclerotic Embolism. This renovascular problem occurs with severe abdominal atherosclerosis, primarily in men over 50 years old. With this condition, fragments of atherosclerotic material rich in cholesterol detach from the abdominal aorta or renal arteries and become lodged in the small renal arteries or arterioles, causing renal infarctions. The affected arterioles are usually occluded and heal by fibrous scarring (Figure 8-9). When a large vessel or a critical number of smaller ones have become blocked, symptoms of renal hypertension appear. The resulting decreased tubular flow activates the renin-aldosterone mechanism and thereby increases the blood pressure. Treatment involves anticoagulation therapy and attempting to control cholesterol and hypertension.

Acute Renal Failure. Renal shutdown can result from almost any of the aforementioned internal kidney problems as well as many not discussed in this text. It can result from decreased blood delivery to the kidney (prerenal failure). This is a condition commonly seen in the elderly as a result of decreased blood volume and pressure. De-

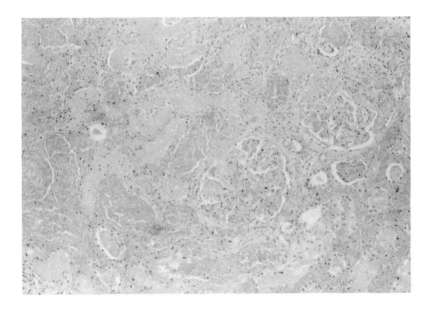

Figure 8-9   The blood supply of this atrophied glomerulus has been interrupted by a renal atherosclerotic embolism.

Table 8-3   Some Factors Involved in Hypertension

Overactivity of the renin-aldosterone mechanism
Arterial constriction resulting in excessive secretion of aldosterone
Increased cardiac output combined with narrowed lumens in arteries and arterioles
Glomerular inflammation preventing adequate blood flow past the juxtaglomerular apparatus

creased thirst reflex, excessive use of diuretics, diarrhea, antibiotic induced renal toxicity, vascular blockage, or decreased cardiac output can produce renal shut-down as well. Post-renal failure normally occurs because of prostate blockage, renal or bladder stones, or carcinoma of the ureters, bladder, or kidney. Elderly patients with poor renal reserves must be treated aggressively, using dialysis before severe uremia develops and the kidneys are irreparably damaged.

Hypertension. Normal blood pressure for a young person is near 120/80 (systolic/diastolic). Definitions of hypertension differ, but a diastolic blood pressure (the minimum pressure attained between heartbeats) of 105 or greater is generally recognized as moderate hypertension. Pressures in the mid to upper 90s are now being treated to prevent more severe hypertension later. Systolic pressure, the maximum pressure achieved immediately after the heartbeat, is usually less important in defining hypertension. The kidneys are implicated in many forms of high blood pressure. In fact, the kidneys, in conjunction with the autonomic nervous system, are the primary regulators of blood pressure. That is why hypertension is discussed in this chapter rather than with the circulatory system. Hypertension may result from several factors, as shown in Table 8-3.

Essential hypertension is hypertension produced by unknown causes. Historically this description has included about 85% of all hypertension cases. Currently more of these are being associated with renal or renovascular abnormalities. Theoretic causes include impaired ability of the nephrons to remove sodium from the blood and the release of excessive renin by the juxtaglomerular cells under conditions when blood flow to the kidney is normal.

Malignant hypertension is life-threatening high blood pressure, often in excess of 200/140. High pressure in the cerebral vessels can cause headache, vomiting, blindness, convulsions, coma, and of course a very high risk of cerebrovascular accident (CVA). The cause of malignant hypertension is unknown, but the following scenario has been suggested.

An extended period of very high blood pressure (diastolic above 140 mm Hg) stresses the renal arterioles, causing a fibrous narrowing of the lumen (renal arteriolosclerosis). This reduces flow through the tubules, stimulates the juxtaglomerular cells, and results in more renin release. This generates more angiotensin II, causing even more arterial constriction and releasing more aldosterone. The high aldosterone causes more sodium reabsorption by the renal tubules and additional water reabsorption from the urine. Reabsorption combines with the thirst reflex and increases blood volume and further increases the pressure. The higher pressure results in even more fibrous response and the cycle repeats.

Renal Cell Carcinoma. This cancer represents 80% to 90% of all kidney malignancies. It most commonly occurs between ages 50 and 70 and is found twice as often in men than in women. The disease is often discovered upon investigation of hematuria, pain, or a large palpable mass. Fever and polycythemia (excessive numbers of blood cells), presumably from excess production of hematopoietin, may also be indications of advanced tumors. Unfortunately, in many cases the primary tumor is silent and is discovered only as a result of symptoms produced by metastases, often in the lungs and bones and less commonly in the regional lymph nodes, liver, and brain.

Renal carcinoma cells are deceptively benign in appearance. The tumor mass, although it is invasive, often gives the false impression of being encapsulated because of its tendency to compress the surrounding tissue. It has been known to enter the renal vein and produce a finger of tumor tissue extending all the way to the heart. The overall 5-year survival rate is about 35%, but the disease is unpredictable. In some cases a metastasis may remain dormant for up to 20 years after a nephrectomy (removal of a kidney) and then suddenly become active. In still others, removal of the kidney containing the primary tumor has resulted in disappearance of other known metastases. Because this tumor exhibits such a variable growth rate it is difficult to say when a patient has been "cured."

Transitional Cell Carcinoma of the Bladder. These cancers are associated with aniline dyes, smoking, and aging. They are indolent neoplasms and are often confined to the urinary bladder. Unfortunately, they tend to be multifocal with multiple areas of tumor development. When removed, the malignancies often grow back or at least redevelop from nearby tissue. Painless hematuria is the usual indicator of a problem leading to discovery of this tumor. When there is no invasion of the bladder wall, the prognosis is good. With invasion, the 5-year survival rate is only 20%.

Nodular Hyperplasia of the Prostate. The prostate normally begins nodular hyperplasia at about age 50. By age 80, about 80% of men have some hyperplasia. The nodularity resulting from glandular hyperplasia occurs primarily in the median lobe and central portions of the lateral lobes, producing pressure on the urethra. This restricts and eventually blocks urine flow, causing difficulty in initiating urine flow and producing post-urination dribble. Despite the abundance of this condition, only about 10% of men require surgery to improve urine flow during their lifetime. Many others will suffer frequency of urination, nocturia, and some difficulty with bladder control. Severely affected individuals may have acute bladder distention and multiple bladder infections resulting from chronic severe obstruction. Back pressure can flood the kidneys, causing hydronephrosis and promoting kidney infection. There is no known association between prostate hyperplasia and prostate cancer.

Carcinoma of the Prostate. Although prostate cancer is the most common cause of cancer in men, most cases are not fatal. It is usually discovered during surgery to relieve the obstruction of nodular hyperplasia, or at autopsy. Some of the cancers, often in younger men, are aggressive and metastasize widely, making prostate cancer the third leading cause of death from cancer in men (after cancer of the lung and colon).

The cancerous growth produces a pattern of densely packed glandular structures penetrating the prostate (Figure 8-10 and Color Plate 11, *C*). Unlike nodular hyperplasia, prostate cancer usually arises on the posterior lobe and immediately beneath the capsule. Metastasis is often to the paravertebral venous plexus and the bones of the axial skeleton. Acid phosphatase levels are often elevated because a normal prostate produces this enzyme. Five-year survival rates range from 70% for non-aggressive prostate cancer (about equal to the survival rate of the normal population of that age) to 33% for aggressive, metastatic disease.

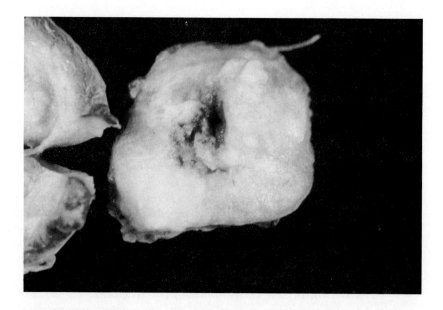

Figure 8-10   The nodular area of the dorsal surface of this prostate gland represents underlying cancer. This can often be felt by the physician during a rectal examination.

CASE HISTORIES

❖ Urosepsis, Urinary Tract Infection

P.C. was a 76-year-old woman who had been in a nursing home for the last 3 years. She needed nursing home care because of a stroke that left her with slurred speech and paralysis of her right side. P.C. was taken by ambulance to the hospital when she could not be awakened one morning. She had a mild fever (temperature 101.6° F) and was hypotensive with no readable diastolic blood pressure. Her blood pressure was recorded as 66/? (Normal 120/80).

Her physical examination revealed the following observations: She was in coma and was responsive only to deep pain. This was indicative of profound brain dysfunction. Her skin was cold and clammy, suggestive of excessive release of adrenaline secondary to shock. Hypersensitive deep tendon reflexes on the right side demonstrated lack of brain input. This was consistent with her right-sided paralysis and positive Babinski's sign (the big toe curls up instead of down when the bottom of the foot is stroked).

The laboratory study's objective was to find the cause of her coma. An x-ray examination revealed no abnormalities in the chest cavity. The ECG showed no evidence of myocardial infarction (death of heart muscle tissue that would have indicated a heart attack that could have damaged the brain). Enzyme studies revealed her CK was three times normal and LDH was six times normal. However, the heart fraction of CK (the CK-MB fraction) was normal. Separation of the LDH fractions revealed a high LDH5, indicating shock. Her urine was cloudy, pink-white, and very concentrated (specific gravity 1.035 vs normal of 1.015 to 1.020). The urine contained red and white blood cells and excess protein. Her complete blood count revealed 18,600 white cells (normal to 11,000) with many "bands" (immature granulocytic white blood cells). This suggested acute inflammation and infection and a presumptive diagnosis of urosepsis was made. *Urosepsis* is a term indicating urinary tract infection that has invaded the bloodstream, causing septicemia. The toxins released in severe septicemia result in shock; the decreased blood flow to the brain often results in confusion, coma, and death. Cultures of bacteria from the blood and urine were obtained to determine resistance to antibiotics. P.C. was started on multiple antibiotics, blood pressure supportive agents, IV fluids, and diuretics to keep her well hydrated with functioning kidneys. In spite of intensive management, her blood pressure and urine output remained low, and she died 36 hours after admission. The blood and urine cultures grew *E. coli,* a common species of intestinal bacteria.

Comment. One of the most common reasons for transfer from a nursing home to a hospital is urosepsis. This entity is one of the major causes of body-wide infection in the elderly. Typically *E. coli* is the infective agent. This bacterial species is found in the large intestine of everyone and is usually not pathogenic until it enters the urinary tract or the tissues of the body. When it enters the bloodstream, endotoxins are released that result in hypotension. The MB band is one of three isoenzymes of CK, a muscle enzyme, and represents the heart fraction. As such it is often elevated in myocardial infarction. LDH has five isoenzymes with a normal pattern of $2 > 1 > 3 > 5 > 4$. Reversal of this pattern is indicative of shock.

❖ Chronic Pyelonephritis

M.P. was an 82-year-old woman who complained to her physician of weakness and sleepiness. The symptoms had gradually increased during the last few months to the point where she was no longer capable of the simplest activity and was sleeping most of the day. She had not noticed any fever or headache that could indicate meningitis, and said she had not suffered any recent trauma or pain. She had experienced no weight loss and had gained 12 pounds despite her poor appetite. To her physician she appeared chronically and acutely ill; she was admitted to the hospital.

Her prior medical history included adult onset diabetes, which was controlled by diet and weight loss. Her only medications were digitalis to control atrial tachycardia, or abnormally rapid heart rate. During the last 15 years her only hospital admission had been for an episode of urinary tract infection 12 years earlier.

Positive findings from her examination included obtundation (dull mentation with increased drowsiness), moderate pitting edema of the ankles, and mild obesity. The laboratory data supported glucose intolerance with an elevated blood sugar of 126 mg/dl and documented chronic renal failure. Blood values for urinary excretory products were high. Blood urea nitrogen (BUN) was 186 mg/dl (normal 22); creatinine was 21.2 mg/dl (normal to 1.4), and uric acid was 15.3 mg/dl (normal to 6.0). She appeared malnourished with a total protein of 4.9 mg/dl and 2.1 g/dl of albumin. Her electrolytes were all altered, with sodium 121 mEq/L, potassium 6.9 mEq/L, chloride 89 mEq/L, and carbon dioxide 15 mEq/L. She was anemic with 8.8 g/dl of hemoglobin (normal 11 to 15) but had a normal (4,600) white blood cell count with a normal differential. Her urine showed a specific gravity of 1.010, pH 7.8, and positive tests for protein, blood, WBCs, ketones, and erythrocyte casts.

An x-ray of her abdomen showed small kidneys with no evidence of kidney stones. A culture of her urine showed a large number of *E. coli:* more than 100,000 colony-forming units per ml. A percutaneous renal biopsy revealed no normal renal tissue. The glomeruli were replaced by fibrous tissue. There were increased numbers of inflammatory cells and the residual tubules were nonfunctional. A diagnosis of chronic pyelonephritis was suggested with end stage kidneys. Within a few days she became comatose. She died after eight days, without regaining consciousness.

Comment. The renal failure of chronic pyelonephritis has an insidious onset due to the slow cumulative effects of the chronic inflammation. Often the insult is bacterial as in this case, but a similar histologic and clinical picture can result from other causes. Prolonged ingestion of antiinflammatory drugs such as phenacitin or aspirin is one such cause. The kidney functions as a homeostatic organ and to remove waste. With failure, waste products normally excreted by the kidney accumulate in the blood. These include urea nitrogen, creatinine, uric acid, and fixed acids such as sulfuric acid. Normal kidneys preserve the body's internal environment by saving sodium and excreting potassium and water. Failure usually leads to water retention with decreased levels of sodium (leading to edema) and increased levels of potassium. Failure to excrete hydrogen ions contributes to blood acidity.

❖ Glomerulonephritis

E.J., a 50-year-old previously healthy man, had a mild viral-like illness with fever, nasal congestion, and cough. Gradually, as the cold abated, he noticed fatigue. Recently he developed gastrointestinal upsets and back pain. The back pain was in the mid-back, bilateral, dull, constant, and not affected by motion or weight bearing. There was no change in bowel habits but he had noticed a less frequent need to urinate. He had gained a small amount of weight despite a decreased appetite.

E.J.'s medical history revealed no significant prior illnesses and no use of medication. His father had died of a heart attack at age 53. His mother is still alive but has adult onset (type II) diabetes. He is an accountant and is not exposed to chemicals or fumes. He does not drink or smoke.

During physical examination he appeared to be his stated age but seemed acutely ill. His face and hands showed edema. His neck veins were distended, suggesting heart failure or increased blood volume. The heart sounds appeared distant as if muffled by a layer of water; a scraping friction rub could be heard after each beat. These findings indicate an inflammatory process in the pericardial sac with fluid accumulation (pericarditis with hydropericardium). The breath sounds included numerous fine crackles (rales), indicating some fluid in the alveolar sacs. Percussion (here, a gentle kidney punch) revealed bilateral flank pain just under the ribs posteriorly (Murphy's sign). He was admitted to a hospital immediately.

Laboratory examination revealed chronic renal failure with many of its complications. These included anemia, uremia with pericarditis, fluid overload, and electrolyte imbalance. The chronic renal failure was seen in the accumulation in the blood of waste products normally excreted by the kidney. These included BUN of 148 mg/dl (normal to 22), creatinine of 12.3 mg/dl (normal to 1.5), and uric acid of 10.6 mg/dl (normal to 8). Urine examination showed the results of glomerular damage with microhemorrhages, increased amounts of protein, and blood. There were casts of the renal tubules composed of protein and blood. The presence of oval fat bodies indicated glomerular damage as the cause. An electrocardiogram (ECG) demonstrated diffuse changes (elevated ST waves) suggestive of pericarditis and tall peaked T waves with prolongation of the P-R interval. These changes are associated with an excessive potassium level.

Emergency hemodialysis was ordered but before it could begin, the patient developed cardiac arrest. Cardiopulmonary resuscitation was not successful. At autopsy the significant findings were acute glomerulonephritis with crescent formations, uremic pericarditis and gastroenteritis, and cardiac tamponade, or acute suppression of the heart by fluid or blood in the pericardium.

Comments. The patient died of heart failure secondary to multiple effects of acute renal failure. The most noticeable change was a large amount of fluid around the heart. It had accumulated as a result of uremia, the toxic effects of waste products that could not be excreted by the diseased kidneys. Although the actual toxic chemicals are not known, they tend to parallel the BUN and creatinine. Excessive potassium is also a waste product and is normally excreted by the kidneys. The high levels of potassium in this case could be expected to produce cardiac arrhythmias. The renal findings of diseased glomeruli with crescents is indicative of rapidly progressive glomerulonephritis. This disorder is believed to be caused by circulating immune complexes in about a third of the cases but its cause is unknown in the other two thirds.

❖ Renovascular Hypertension

L.A. was a 26-year-old mother of two who came to her doctor after 2 weeks of severe headaches. She had been in excellent health and there was no history of medication, migraines, head trauma, or tumors. The family history revealed that her mother had diabetes and was obese. On examination in her doctor's office she was found to have a blood pressure (BP) of 260/178 (normal 120/80). She was sent to the hospital as a semiemergency to attempt to prevent a hypertensive stroke.

On admission to the hospital she was found to be a well-developed, healthy appearing female in no acute distress. Except for her blood pressure, her vital signs (including her pulse) were normal. The eye grounds (the retina) showed slight A-V nicking (due to the tense arteries pressing on the veins) but no evidence of exudates, hemorrhage, or papilledema. Palpation of the flanks did not increase the BP or headache. The kidneys were not tender or palpable. Neurologic examination results were within normal limits (WNL).

Her admission laboratory work, including CBC, urine analysis, chemistry profile, ECG, and chest x-ray was normal. L.A. was started on pain medication and IV antihypertensive drugs. Twenty-four hour urine collections revealed normal amounts of vanillylmandelic acid (VMA) and catecholamines. These agents are breakdown products of adrenaline metabolism and would be expected to be elevated if the high blood pressure were caused by an adrenaline-producing tumor. Urinary aldosterone was 104 µg/24 hours (normal to 33), which indicated that the blood pressure was being elevated through the renin-aldosterone mechanism or that there was an adrenal tumor producing excess aldosterone. There was a slight increase in epinephrine. An abdominal x-ray revealed bilateral renal outlines of normal size.

L.A.'s blood pressure was controlled within 24 hours but remained labile. Peripheral vein renin levels were twice normal, again indicating that her problem was caused by the renin-aldosterone mechanism. Visualization of the renal arteries was undertaken. The vena cava renin level was elevated threefold with a fivefold elevation in the right renal vein and a decreased level in the left renal vein. The aortography revealed a markedly narrowed, smooth-walled, right renal artery with an isolated lesion. These findings are conclusive for renal artery stenosis with hypotension of one kidney that responds by producing large amounts of renin, driving the blood pressure up.

L.A. developed hypotension during the procedure which was quickly followed by oozing from the puncture wounds and spontaneous mucosal bleeding. She was treated over the next 4 weeks for an allergic dye reaction with disseminated intravascular coagulation and acute renal failure. After her kidneys recovered function, the narrowed portion of her right renal artery was removed and the free ends of the artery were connected. Follow-up examination showed that her hypertension had completely disappeared and her renal function was normal. There has been no recurrence for the past 13 years.

Comment. Hypertension caused by renal artery obstruction is rare but very treatable. This cause is most often looked for when the patient is young and the onset of the condition is sudden. It is caused by decreased blood flow to a healthy kidney. The normal response is secretion of renin, which increases blood volume and pressure through the aldosterone mechanism. Unfortunately the stenosed renal artery still does not allow enough increased blood flow to the kidney to stop the secretion of renin, so the condition worsens as malignant hypertension. In this case the artery involved had an idiopathic muscular hypertrophy; most cases are due to atherosclerosis. This case illustrates the importance of the kidneys and the renin-aldosterone mechanism in regulating blood pressure.

❖ Renal Cell Carcinoma

M.B. was a 59-year-old male tire worker who came to his local physician after experiencing 2 weeks of grossly bloody urine. Careful questioning elicited the statement that smokey-appearing urine had been present for about 5 months (a less obvious form of hematurea). Accompanying this symptom was a dull ache in his left flank. He reported no change in his bowel habits, frequency of urination, or pain with urination. He had unintentionally lost 25 pounds during the last 6 months. His medical history revealed multiple episodes of kidney stones, beginning at age 40. M.B. drank an average of two highballs daily and smoked an average of six cigars a day.

When examined, he appeared ruddy (his face and hands were flushed), but in no acute distress. The chest showed dullness to percussion at the left base and a few small rales. The most significant finding was a large palpable mass occupying the upper quadrant of the abdomen, extending across the midline. An enlarged right supraclavicular lymph node was noted. It was fixed to the underlying tissues and firm. These findings suggested a metastatic malignancy.

Radiologic examination showed multiple "snow ball" masses in both lung fields with a pleural effusion on the left. A renal x-ray, intravenous pyelogram (IVP) showed a mass lesion destroying the left kidney. Computed tomography (CT) scans showed a tumor mass obscuring the left kidney, massively extending into the left renal vein and inferior venacava. Laboratory data of note included elevated hemoglobin of 18.8 g/dl with a 59% hematocrit (normal to 50% of blood that is cellular in men), grossly bloody urine, normal BUN and creatinine, and a 320 IU/L LDH (normal lactate dehydrogenase to 200). The pleural fluid was drained and found to contain malignant cells. The supraclavicular lymph node was biopsied and found to contain adenocarcinoma compatible with kidney origin. The patient was deemed inoperable and offered chemotherapy. He refused and died several months later.

Comments. Renal cell carcinoma is a difficult tumor to diagnose during its early stages when it may be curable. As with many of the deep organs, symptoms usually develop late in the course of the illness. Renal cell cancer is unusual because a small primary tumor may give rise to extensive and large metastases. It is also a tumor in which a single metastasis can appear and surgical removal may result in prolonged control of the disease. This tumor demonstrated production of erythropoietin, which is normally made in the kidney and acts to stimulate the bone marrow to produce red blood cells. His polycythemia (high number of blood cells) was most probably caused by tumor production of this hormone. The elevated LDH reflected necrosis of tumor tissue. The normal renal function tests merely reflect the typical reserve potential of his healthy kidney.

❖ Benign Prostate Hyperplasia with Retention

A.N. is a 63-year-old family physician who found himself unable to "pass water" for 24 hours. His present illness actually began 3 years earlier when he noticed changes in his urinary stream. It slowly became difficult to start, was diminished in strength, and dribbled at the end of urination. In time he noticed that he had to urinate more frequently than usual and passed a smaller volume of urine.

When he visited his urologist, he was in acute distress. His lower abdomen was markedly tender and a large suprapubic mass could be palpated. Percussion (tapping on the mass) demonstrated dullness to the level of the umbilicus, indicative of a markedly swollen bladder. With difficulty, a small-caliber straight catheter was passed and about 2500 ml of clear fluid was recovered. Rectal examination revealed no masses but showed a markedly enlarged, soft, prostate.

A.N. was hospitalized for surgery. A transurethral resection of prostate (TURP) produced 76 g of tissue compatible with benign hyperplasia of the prostate. Following the operation he was slightly incontinent for 3 months but then regained full bladder control.

Comment. Prostate problems will affect most men in our society if they live long enough. The gradual enlargement of the gland with age ensures a slow development of problems and these are often ignored until complete obstruction results, as in this case. Because the gland lies at the base of the bladder and surrounds the urethra, its enlargement disturbs the urinary stream. The accumulation of residual urine after "emptying" allows the bladder to develop compensatory hypertrophy so that it can hold progressively larger amounts of urine. Common complications not seen in this case are urinary tract infection and bladder stones.

❖ Prostate Cancer

A.V., a 59-year-old successful small manufacturer, was seen by his local physician for a routine physical examination. The only remarkable finding was a single firm nodule on the left posterior aspect of the prostate. The urologic consultant performed a transrectal needle biopsy that showed a poorly differentiated adenocarcinoma of prostate origin. A.V. was offered a radical prostatectomy but declined further therapy because of his fear of impotence. He visited an immunotherapy institute outside the United States and received various serums. For 2 years he was lost to medical follow-up.

On his next visit, he complained of back pain. The pain was of sudden onset and very severe. Physical examination revealed an overweight male in acute pain. The positive findings were limited to the back and prostate. His second lumbar vertebral body was tender to palpation and the pain in this location limited back movement. A rectal examination showed an enlarged, firm prostate.

The CBC and urinalysis were normal except for microscopic hematuria. A chemistry profile was abnormal for LDH, 240 (normal less than 200); the bone enzyme, alkaline phosphatase was 310 (normal less than 100); and glucose, 125 mg/dl (normal below 110). Acid phosphatase levels are often elevated in prostate cancer but were normal in this case. A spine x-ray showed no pathologic condition. A bone scan revealed multiple areas of alkaline phosphatase-producing osteoblastic activity in the vertebrae and pelvis. A presumptive diagnosis of metastatic prostate cancer was made and the patient was offered radiation therapy. He chose to begin a new total hormonal ablation therapy available in Canada.

Over the next year A.V. continued on antitestosterone therapy plus adrenal gland suppression. He was seen multiple times to monitor testosterone and cortisol levels. His back pain disappeared and the firm, enlarged prostate returned to normal. Follow-up x-rays or bone scans were not obtained. Suddenly his bone pain reappeared and evaluation demonstrated osteolytic and blastic changes in multiple vertebral bodies. The acid phosphatase was still normal but the tumor marker, prostatic specific antigen (PSA), was markedly elevated.

Before a plan of therapy could be agreed upon, A.V. suffered an apparent stroke and died. Autopsy revealed no evidence of residual cancer in the prostate but a poorly differentiated adenocarcinoma compatible with prostate origin was present in bone, brain (no evidence of stroke), and lymph nodes.

Comment. It would be easy in this case to suggest that if conventional medical therapy had been tried, the patient could have been cured. Unfortunately, a small percentage of stage 1A (early) prostate cancers have spread beyond the gland prior to diagnosis and could have shown the same clinical evolution. This case does demonstrate one of the fears of oncologists: The selection of a clone of tumor cells resistant to therapy. It seems that the lack of residual cancer in the prostate demonstrated that the antitestosterone therapy was successful for certain portions of the tumor but not for all of it. Prostate cancer has two serum markers, acid phosphatase and prostate specific antigen. About 90% of these tumors in bone are associated with high acid phosphatase but this case was not. A.V.'s tumor did produce the other marker, PSA.

BIBLIOGRAPHY

Bichet DG, Schrier RW: *Renal function and diseases of the aged.* In Schrier RW, ed: *Clinical internal medicine in the aged,* Philadelphia, 1982, WB Saunders.

Brocklehurst JC: *Textbook of geriatric medicine and gerontology,* New York, 1978, Churchill Livingstone.

Cox JR, Shalaby WA: *Renal disease.* In Pathy MJS, ed: *Principles and practice of geriatric medicine,* New York, 1985, John Wiley & Sons.

Farrar DJ: *The bladder and urethra.* In Pathy MJS, ed: *Principles and practice of geriatric medicine,* New York, 1985, John Wiley & Sons.

Faubert PF, and others: *Medical renal disease in the aged.* In Reichel W, ed: *Clinical aspects of aging,* Baltimore, 1989, Williams & Wilkins.

Feinstein EI: *Renal disease in the elderly.* In Rossman I, ed: *Clinical geriatrics,* Philadelphia, 1986, JB Lippincott.

Jenis EH, Lowenthal DT: *Kidney biopsy interpretation,* Philadelphia, 1977, FA Davis.

Jewett MAS, and others: Urinary disfunction in a geriatric long term care population: prevalence and patterns, *J Am Geriatr Soc* 29:211, 1981.

Moorthy AV, Zimmerman SW: Renal disease in the elderly, clinicopathologic analysis of renal disease in U.S. elderly patients, *Clin Nephrol* 14:223, 1980.

Tanagho EA, McAninch JW: *Smith's general urology,* ed 12, Hertfordshire, 1984, Prentice-Hall.

Weidmann PS, and others: Effect of aging on plasma renin and aldosterone in normal man, *Kidney Int* 8:325, 1975.

Williams GH, Braunwald E: *Hypertensive vascular disease.* In Massry SG, Glassock RJ, ed: *Textbook of nephrology,* Baltimore, 1989, Williams & Wilkins.

9 *The Cardiovascular System*

RELEVANT ANATOMY AND PHYSIOLOGY
General Composition and Function

The circulatory system (Fig 9-1) is made up of cardiovascular and lymphatic components. The cardiovascular component comprises the heart, arteries, veins, and blood. The lymphatic component comprises the lymphatic channels, lymph nodes, and lymph. The components of the circulatory system work together to transport oxygen, hormones, and nutrients to the tissues of the body, and to carry carbon dioxide and other waste materials away. In this text, blood is discussed in chapter 10 along with the lymphatics. This chapter deals only with the heart and blood vessels. Cardiovascular disease remains the leading cause of death in the United States in spite of recent medical advances.

The heart pumps blood to the extremities of the body through arteries. These are muscular, tubular vessels that carry blood under relatively high pressure. Arteries have several structural specializations that enable them to withstand this pressure; these are illustrated in Figure 9-2. The highest pressure, called the systolic pressure, is attained immediately after the ventricles of the heart contract (systole). The largest artery, the aorta, takes the greatest force from the heartbeat. It protects the smaller arteries by dilating and storing up to 50% of the stroke volume (amount of blood pumped out of the ventricles with each heartbeat) of the heart. The elastic, muscular tissues of the aorta then decrease the size of the arterial lumen, forcing the blood through the arterial tree. This maintains continual pressure and movement of blood through diastole, the period between heartbeats and the period of lowest blood pressure.

Arteries branch to form smaller arteries and finally to become small arterioles . These branch into capillaries where gases and nutrients are exchanged with the adjacent tissues. This exchange is possible because capillaries have extremely thin walls, creating the smallest possible barrier to the diffusion of nutrients and wastes. Typically, capillaries are composed of a single stretched layer of endothelium surrounded by a thin outer layer of connective tissue. Because capillary walls are thin and, to a degree, porous, it is normal for them to leak small amounts of plasma filtrate. The transition area from arteriole to capillary is surrounded by a ring of smooth muscle fibers called the precapillary sphincter. The amount of fluid allowed to enter the capillary at a given blood pressure is determined in part by the degree of constriction of the precapillary sphincter and in part by constriction of the smooth muscles in the entire arteriole. Smooth muscle contraction may be regulated either by the autonomic nervous system or by hormones such as adrenaline.

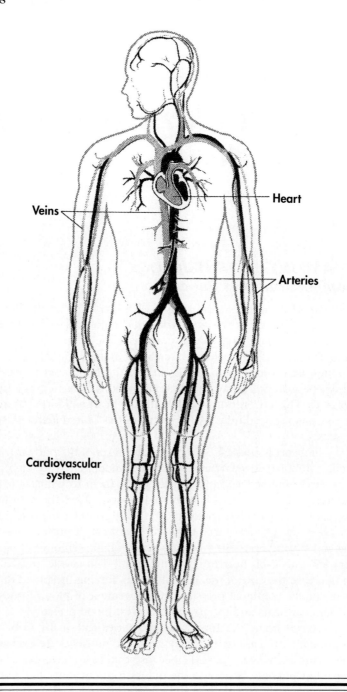

Veins

Heart

Arteries

Cardiovascular
system

Figure 9-1   The cardiovascular system.

The inner cell layer of arteries is called the endothelium. It, along with a layer of elastic and collagenous fibers and a few scattered myointimal cells, make up the tunica intima (See Figure 9-2). Beneath this lies a thick layer of muscle tissue, the tunica media. The outer layer consists of connective tissue and is called the tunica externa. Folded elastic fibers in both the tunica intima and tunica externa allow arteries to be stretched without rupturing. Although arteries usually carry oxygenated blood, passive diffusion of oxygen from this blood is inadequate to maintain all of the thick arterial tissues.

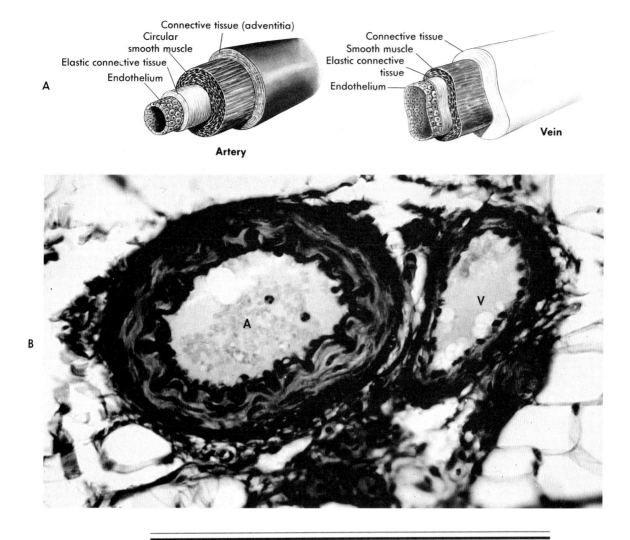

Figure 9-2 Artery and vein; tissues labeled. **A** The tunica media is the dominant layer of the wall of arteries; the tunica adventitia is the dominant layer of the wall of veins. **B** Photomicrograph of an artery (A) and a vein (V).

Large arteries and veins therefore have their own blood vessels, the vasa vasorum, or blood vessels of the blood vessels, that invade the tunica externa and media from the outside.

After the capillaries have passed through tissues, their lumens again enlarge, their walls thicken, and they become veins. Veins are blood reservoirs, holding 70% of the blood supply. They are constructed in much the same way as arteries (Figure 9-2*B*); both have a well-developed tunica interna and externa. However, the tunica media in veins is much thinner and usually has only a scattering of smooth muscle cells mixed with a scattering of collagenous fibers. There are no prominent elastic fibers in the tunica intima or the tunica externa of veins. The veins are moderately elastic, however, and most have some potential for vasoconstriction but they are not designed to withstand the high pressures faced by arteries. Instead, veins have valves spaced periodically through the vessels. These valves permit movement of blood only toward the heart. This movement is improved by physical activity that compresses the veins in various places, squeezing the blood in the only direction it is allowed to move. The largest veins are the superior and inferior vena cavas, both of which enter the heart. The superior vena cava returns blood from the shoulders and head; the inferior vena cava receives blood from lower areas of the body.

The Heart

Embryology. The heart could be described as a thick, valved, highly specialized artery. It forms in the early embryo as a peristaltic tube, moving blood in the same way as the intestines move food. This rhythmic peristalsis becomes a beat as the heart forms a loop and develops valves and septa. The heart completes its development in the fetus, forming four distinct chambers, and it does this without missing a beat.

Composition. The layers of the heart are the same as those of an artery, but because of their high degree of specialization, they are given different names. The tunica intima is known as the endocardium of the heart. It is still composed of a layer of endothelium and fibrous connective tissue. The tunica media becomes an extremely thick layer of interwoven cardiac muscle cells called the myocardium. The tunica externa is known as the visceral pericardium, or just the epicardium. The heart resides in a cavity, the pericardial cavity that contains lubricating pericardial fluid. The fluid is secreted by the outer lining of the heart, the epicardium, and by the outer wall of the pericardial cavity, the parietal pericardium. Sometimes these linings of the heart become infected or inflamed. These conditions are known as endocarditis, or pericarditis, depending on the membranes involved.

Moving the Blood. A longitudinal section through the heart is illustrated in Figure 9-3. The myocardium of the atria is much thinner than the myocardium of the ventricles because the only work the atria do is force blood into the relaxing ventricles, stretching them as they fill. The ventricles, however, must pump blood through all the vascular networks of the body. Therefore they are much stronger. The left ventricle, which must pump blood through the entire body (except for the lungs), is the strongest and thickest.

Blood from the head and shoulders enters the right atrium through the superior vena cava. The inferior vena cava drains the middle and lower portions of the body. When the right atrium constricts, it forces the blood through the tricuspid valve into the right ventricle. This process occurs while the ventricles are relaxing. When the right ventricle contracts it forces the blood through a set of semilunar valves into the pulmonary arteries and in turn to the lungs. This creates pressure in the contracting ventricle and closes the tricuspid valve with an audible snap, preventing the blood from flowing back into the atrium. Keeping the valve from turning inside out and leaking are three sets of strong, thin strands of fibrous tissue, the chordae tendineae. They are attached to several papillary muscles arising from a network of projections of myocardium on the floor of the ventricle. These muscular bundles are called trabeculeae carneae.

When blood returns from the lungs, having lost most of its carbon dioxide and having acquired large amounts of oxygen, it enters the left atrium through four pulmonary veins. When the left atrium contracts, the blood is forced through the bicuspid, or mitral, valve into the left ventricle. This valve too is supported by chordae tendineae attached to papillary muscles. However, although the left ventricle does much more work under even greater pressure than the right, the mitral valve has only two cusps.

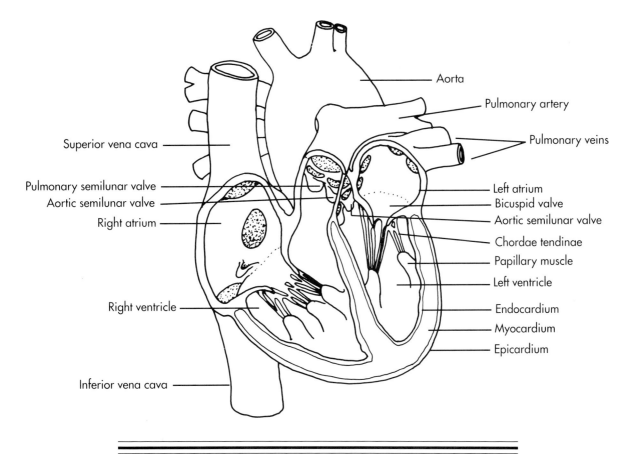

Figure 9-3 labels: Aorta, Pulmonary artery, Pulmonary veins, Superior vena cava, Pulmonary semilunar valve, Aortic semilunar valve, Right atrium, Left atrium, Bicuspid valve, Aortic semilunar valve, Chordae tendinae, Papillary muscle, Left ventricle, Right ventricle, Endocardium, Myocardium, Epicardium, Inferior vena cava

Figure 9-3 The heart shown above is drawn to show the functional components. Note the thicker muscular wall of the left ventricle.

Leaks through this valve occur often during ventricular contraction (systole), and are heard as murmurs with a stethoscope. The murmur sound is due to turbulence in the blood created by valvular narrowing or incompetence. The left ventricle develops murmurs more frequently than the right and while a mild murmur may be harmless, when too much leakage or obstruction occurs due to a malfunctioning valve, the heart may lose critical efficiency.

When the left ventricle contracts, blood is forced through a set of semilunar valves into the aorta. The valves of the aorta (as well as the pulmonary artery) also leak sometimes; this too is heard as a murmur. The aorta supplies blood to the entire body except for the lungs.

The Coronary Circulation. The vasa vasorum of the heart are sufficiently large and important to bear names of their own. They provide what is known as coronary circulation, illustrated in Figure 9-4. Oxygenated blood is provided by two blood vessels arising from the base of the aorta, the right and left coronary arteries. The left coronary artery quickly divides to form two branches. One branch curves behind the heart, later giving rise to the posterior interventricular branch. The other branch from the left coronary artery lies in the interventricular sulcus, a groove lying diagonally across the front of the heart. The right coronary artery is normally smaller and follows the atrioventricular groove behind the heart, with a small marginal branch extending toward the tip of

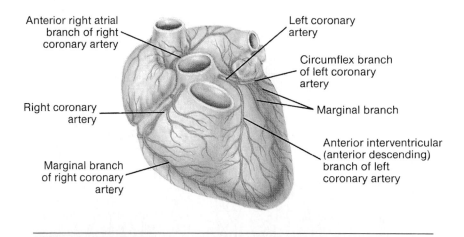

Anterior right atrial branch of right coronary artery

Left coronary artery

Circumflex branch of left coronary artery

Right coronary artery

Marginal branch

Marginal branch of right coronary artery

Anterior interventricular (anterior descending) branch of left coronary artery

Figure 9-4   The large vessels shown nourish the heart. The arteries may become obstructed in cases of coronary artery disease.

the right ventricle. The coronary arteries are subject to debilitating changes that will be discussed later in this chapter.

After passing through the capillaries of the heart, the blood is collected by veins that approximately parallel the coronary arteries. Blood is collected along the interventricular sulcus by the great cardiac vein and along the right atrioventricular groove by the anterior cardiac vein. Both of these empty into the coronary sinus on the posterior side of the heart, as does the dorsally located middle cardiac vein. The coronary sinus empties into the right atrium through a valve.

Generating the Heartbeat. The preceding discussion treated the two sides of the heart as if they worked independently. They of course, do not. To get maximum efficiency, the two atria and then the two ventricles must contract almost simultaneously. The rhythm and "firing order" are determined by a chain of specialized muscle fibers that originate the beat and pass it on. This conduction system of the heart is illustrated in Figure 9-5.

The heartbeat is originated by the sinoatrial (SA) node, also known as the pacemaker. This node is located in the anterior wall of the right atrium. The impulse passes from the SA node to the atrioventricular (AV) node, where it is delayed until the atrial contraction is nearly complete. The impulse is then passed on to the atrioventricular *(AV) bundle* (also known as the bundle of His) where it is delayed further, and passed on to the ventricles by the right and left branches. The AV bundle may originate the heartbeat when impulses originating from the SA node are absent or blocked, but the AV bundle has a slower intrinsic rate. For this reason AV rhythms are slower than normal SA rhythms (approximately 70 times per minute) when transmission between the two is blocked. The electrical impulses originated in the nodes of the heart pass into the tissues around the heart and may be picked up by sensors placed on the skin. Abnormalities in the heart can be indicated by these impulses as they are recorded on an electrocardiograph (ECG).

The heartbeat can originate any place within the conducting system and even outside of the conductive system in the myocardium. When the heartbeat originates any place other than the SA node, the patient is said to have an ectopic pacemaker. This can happen for several reasons and can produce symptoms varying from mild to fatal arrythmias. It is important to note that because of the nature of cardiac muscle, a heartbeat originating anywhere in the atria will spread throughout the atria. If the AV bundle is

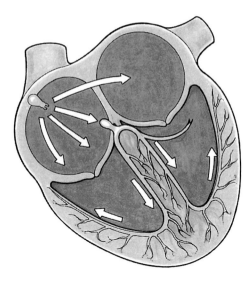

Figure 9-5   The conduction system of the heart. The heartbeat normally originates in the sinoatrial node and is distributed through the lower structures.

functioning, it will "capture" the atrial impulses and pass them into the ventricles. Likewise, a heartbeat originating anywhere in the ventricles will spread through the ventricles and sometimes can pass backward through the AV node into the atria. The conductive system from the AV node through the right and left branches utilizes muscle cells called Purkinje's fibers. If, due to some abnormality, some Purkinje's fibers in the bundle of His develop a rhythm that is more rapid than that of the SA node, they may take over as an ectopic pacemaker.

Coronary Ischemia. Sometimes, because of interruption in the coronary circulation and resulting ischemia, (lack of oxygen delivery to tissues) lactic acid is produced in sufficient quantities to irritate the heart and stimulate spontaneous contraction of cardiac muscle fibers. See Chapter 5 for a discussion of lactic acid. When these contractions are produced in several different places in the myocardium in a very short period, some fibers may still be recovering from contraction and will not be stimulated. Others are just beginning to contract; others are ready to begin contracting. Instead of producing a coordinated contraction, the heart quivers, a condition called fibrillation. Because a fibrillating heart is ineffective and moves no blood, the fibers rapidly exhaust themselves and the heart muscle dies of lack of oxygen. If enough energy remains in the heart, a defibrillator may coordinate the contractions and restore a normal beat. This device passes a strong electric current through the chest and heart. The current causes all the muscle cells to contract simultaneously, allowing all of them to arrive at the excitable stage at the same time. As the contraction moves new oxygenated blood into the heart, the next muscle contractions and subsequent beats hopefully will be coordinated.

Control of the Heart. The rate and force at which the heart beats are controlled by several mechanisms. The sympathetic nervous system can increase both the rate and strength of the heartbeat. It can increase the rate from a normal 72 beats per minute to as much as 250 to 300 beats per minute. The parasympathetic system, operating through the vagus nerve, can suppress both rate and force to produce a heart rate of as little as 20 to 30 beats per minute. The hormone adrenaline, also known as epinephrine, has an effect similar to the sympathetic nervous system on the heart. Both adrenaline and direct sympathetic stimulation of heart muscle result in more rapid and forceful ventricular contractions.

Another factor affecting the strength of heart muscle contraction is the extent to which the heart muscle has been stretched. When larger than normal volumes of blood are returned to the heart by the veins, the atrial muscle fibers are stretched. They respond to this up to a point by contracting with greater strength. This in turn fills the ventricles with more than the usual volume of blood. These too contract with greater force. Consequently, the heart responds automatically to increased or decreased volumes of blood delivered to it by changing the force of its contraction. This relationship between the heartbeat and the changing volumes of blood delivered by the veins is expressed as the Frank-Starling law.

An example of this factor is shown when the brain senses a need for the body to undertake physical action. Stimulation of the sympathetic nervous system constricts large veins in the abdomen, forcing more blood into the venous return to the heart. The heart will respond with a rapid, forceful beat, both as a result of direct sympathetic stimulation and as a result of increased blood flow. The sympathetic system can also stimulate the heart for other reasons. If the brain is deprived of oxygen (ischemia), sympathetic stimulation increases the rate and force of contractions. By contrast, if hemorrhage results in a decrease in blood volume, the heartbeat becomes weaker even though it is beating more rapidly because of sympathetic stimulation.

Feedback to the brain from the cardiovascular system occurs through many mechanisms. Sense organs are located in many large arteries; two particularly important ones, baroreceptors, are located in the carotid sinuses where the external and internal carotid arteries branch in the neck. They respond to high blood pressure by causing the brain to stimulate the parasympathetic nervous system. This system acts in opposition to the sympathetic nervous system to slow and weaken the heartbeat. A sudden decrease in carotid arterial pressure decreases baroreceptor activity, allowing sympathetic stimulation to rapidly increase the rate and force of the heart. You experience this response (positional hypotension) when you suddenly stand or rise from bed. Conversely, extreme pressure on the carotid sinuses in the neck can stimulate the baroreceptors so much that parasympathetic weakening and slowing of the heartbeat (along with vasodilation) can lower the blood pressure to the point that the person faints. The illegal "sleeper hold" of wrestling fame operates by this mechanism. Other arterial receptors called carotid and aortic bodies stimulate the vasomotor centers of the brain and the sympathetic nervous system, increasing vasoconstriction and heart rate. They do this when these bodies are severely oxygen depleted. They stimulate breathing as well.

Various metabolic ions also affect heart function. For example, excess serum potassium interferes with muscle cell polorization which depends on a balance of intracellular and extracellular ions. This causes the heart to become flaccid and dilated, decreasing both the rate and force of the heartbeat. Very large quantities of potassium can block conduction of the impulses through the AV bundle. An excess of calcium has almost the opposite effect, causing spastic contraction.

The Regulation of Blood Pressure

It is obvious that proper functioning of the body requires a uniform delivery of oxygen. The lungs oxygenate the blood and the cardiovascular system moves it. The regulation of cardiac output just discussed is one important mechanism for doing this. Another is the regulation of blood pressure. Maintenance of blood pressure depends on the relationship of cardiac output (the amount of blood moved by the heart in a given unit of time), total blood volume, vascular space for the blood to occupy, and peripheral resistance to blood flow (the resistance created in forcing blood through the bodies arteries, capillaries, and veins.). The kidneys are the key organs in the long-term regulation of blood pressure; vascular resistance, blood volume, and cardiac output control the blood pressure on a moment to moment basis. The part played by the kidneys was discussed in Chapter 8 because of the inseparable relationship between ion balance, urine volume, and blood pressure. The following discussion is provided to help integrate these principles with the circulatory system and for review.

When blood pressure in the kidneys is excessively low, a compound called renin is released. It converts angiotensinogen circulating in the blood to its active form, angiotensin. This conversion will cause constriction of small arteries and veins, increasing peripheral resistance and forcing their blood into the main circulation. This increases the cardiac output, now forcing a larger volume of blood against the increased resistance of the constricted distal blood vessels. The obvious result is an increase in the blood pressure. Finally, angiotensin stimulates release of the hormone aldosterone from the adrenal cortex. The direct effect of aldosterone is to stimulate the renal tubule to absorb salt, resulting in retention of water that would otherwise be excreted by the kidneys. Assuming that some water is consumed, this increases both blood volume and blood pressure.

Another mechanism for the maintenance of blood pressure operates through antidiuretic hormone (ADH), also discussed in Chapter 8. When blood volume and pressure are too low, the posterior pituitary gland releases ADH, increasing the permeability of the distal tubules and collecting ducts to water, thereby increasing water absorption from the urine and increasing blood volume. It should also be mentioned that the atria of the heart have specialized cardiac muscle cells that secrete natriuretic factor (ANF) when they are stretched. This inhibits aldosterone production, therefore lowering blood pressure and increasing urine volume. ANF represents another technique the body can use to keep systems operating within tolerable limits.

Momentary adjustments in blood pressure are regulated by sympathetic "tone," the amount of sympathetic stimulation being given to the body, plus the total blood volume and the stroke volume of the heart. We have alluded to transient hypotension due to position. When suddenly standing after lying down for a period of time, it is common for blood to remain pooled in the dependent (lower) portion of the body. This results in less blood returning to the heart and therefore less blood being pumped. The decreased cardiac output causes a decreased blood pressure, decreased blood delivery to the brain, and a faint or light-headed feeling.

To prevent fainting, compensation begins to occur through baroreceptor activity and central nervous system impulses that all lead to increased sympathetic nerve discharge, resulting in arterial constriction, increased heart rate, and increased force of each beat. These responses cause an immediate increase in blood pressure as a result of more blood being pumped against increased arterial resistance. When the pressure returns to normal, the sympathetic nerves return to their normal output and the status quo has been restored. Sustained excessive sympathetic nerve activity or any factor increasing blood volume, such as excessive salt intake, can lead to sustained hypertension. This illustrates one of the important mechanisms by which salt and stress are associated with hypertension.

THE AGING HEART AND BLOOD VESSELS
Changes in the Arteries

Leonardo da Vinci, in addition to his accomplishments as an artist and maker of anatomic drawings, made one of the first significant observations about the aging process. He wrote: "Veins which by the thickening of their tunics in the old restrict the passage of blood, and by the lack of nourishment destroy the life of the aged without any fever, the old coming to fail little by little in slow death." We know today that the problems of the cardiovascular system are both a partial cause and a result of the deterioration associated with age. Da Vinci's observation, and the later discovery of calcification of the large arteries and increased fibrosis, led to the term arteriosclerosis, or hardening of the arteries.

Arteries undergo changes with age that are not associated with any particular disease. The solubility and flexibility of collagen is altered with age. Age-related precipitation and cross-linking of soluble collagen decreases arterial elasticity. The effect of this loss of elasticity on major vessels such as the aorta is to cause a permanent increase in diameter as well as in length, causing the enlarged aorta to wind tortuously down the trunk of the body. Loss of elasticity also occurs in the peripheral vessels. When combined with atherosclerosis, reduced elasticity results in an increased resistance to each surge of blood, producing an increased systolic pressure, a decreased diastolic pressure, and increased strain on the heart. An increase in the internal volume of the aorta may be of advantage in allowing this vessel to store the surge of systolic blood until it can find its way through the more distal vessels. However, it produces the disadvantage of reduced maintenance of diastolic pressure, causing a wide variation in pulse pressures. The aortic volume increase probably reduces the strain on the heart. While a loss of arterial elasticity results in an increased systolic pressure, the diastolic pressure usually declines or remains the same. Atherosclerosis and other diseases and conditions of the heart and vessels are discussed later in this chapter. See the references at the end of this chapter for texts that discuss the effect of aging on the heart and arteries.

Changes in the Heart

Several changes occur as the heart ages. As with all the organ systems, it is difficult to determine which of these changes result from pure senescence and which represent subclinical disease. Processes believed by most to result from normal aging include lipofucsin accumulation, interstitial fibrosis, collagen cross-linking, benign amyloid accumulation, certain valvular changes (see Chapter 2), and impulse conduction changes. Age-related disuse atrophy combined with lipofucsin accumulation produces the condition described as brown atrophy of the heart.

These cellular and interstitial changes interfere with maximal cardiac muscle function. Collagen cross-linking and increased fibrosis produce subtle losses of elasticity that limit the the ability of the remaining heart muscle to function. When amyloid, a rubbery protein, is deposited in heart muscle tissue, the motion of the heart may be limited. Lipofucsin granules accumulate in the cytoplasm of aged cells. These granules may displace subcellular organelles or interfere with intracellular transportation. These events alone should not significantly compromise cardiac function but they probably contribute to a decrease in cardiac reserve, the extra pumping power a healthy heart has to use in emergencies. Overall, a decrease in cardiac output was reported in early studies, but this has not been confirmed in recent studies.

In addition to the muscle changes just described, senescent changes are seen in the valves and heart chambers. The weight of the aged heart is better correlated with body weight than with age, which suggests that aging by itself does not significantly affect heart weight. However, as body size declines with age, heart size declines proportionately, so it could be said that many elderly are lighthearted. Within the heart, the left ventricle decreases slightly in size with age while the left atrium increases. The valves show a mild fibrosis and calcification that may be related to use, but are present at least to a small degree in all elderly. The cups become slightly fibrotic with focal thickening at the points of closure. This is more prominent in the valves subjected to the highest stresses, the aortic and mitral. The changes in these valves can be confused with valve disease. These changes, if extreme or in combination with other disease, can contribute to valvular stenosis or incompetence.

The conduction system changes at least in part because of the interstitial and intracellular changes noted previously. The SA node shows a decreased number of Purkinje's fibers with fat and fibrous tissue replacement. Similar changes occur in the AV node, the bundle of His, and the right and left ventricular branches. The net result of this is no change in resting pulse, but a greater than normal increase in pulse rate in response to activity. Paradoxically, there is a decline in the maximal possible heart rate. The aged conduction system is susceptible to arrhythmia; one of the common problems seen in the elderly is atrial tachycardia, accelerated atrial contractions that are not coordinated with the ventricular beat. Changes in ECG are associated with heart damage or specific heart diseases.

Diseases and Conditions Associated with Age

Atherosclerosis. Atherosclerosis refers to the formation of fatty, cholesterol-laden, sometimes hemorrhagic deposits, atheromas, in the tunica intima and media of large and medium sized arteries. The term is often poorly separated from arteriosclerosis, which may or may not include atherosclerosis and refers to a fibrous and sometimes calcareous thickening of the arterial wall, making it more rigid. (The linkage of atheromas to high circulating levels of low density lipoproteins and cholesterol is discussed in Chapter 15.) The sequence of events in the formation of an atheroma is illustrated and described in Figure 9-6.

Atheromas develop in all arteries but are especially common in areas of high blood pressure or turbulence, such as the coronary arteries, aorta, iliac arteries, carotids, and cerebral arteries. The thickened vascular wall compromises the vessel lumen, causing a decreased blood flow that may not be significant until maximum flow is needed. In the leg, the condition known as claudication provides an example of how atherosclerosis-restricted blood flow is related to demand. This condition is ischemic pain or weakness in the legs that occurs during exercise and stops when resting. Atherosclerotic aneurysms of the great vessels, the aorta and its tributaries, may rupture and cause sudden death from internal bleeding. In areas such as the brain, damage may result from the pressure of accumulating blood. Atherosclerosis is often the primary element in cerebrovascular accident, myocardial infarction, embolism, and peripheral vascular disease. Atherosclerosis as well as other vascular problems tend to develop more rapidly in diabetics and hypertensive patients than in the general population.

Arteriosclerosis. This is rarely cited today as a particular disease or condition. It refers to loss of elasticity associated with collagen fixation, fibrosis, smooth muscle proliferation and calcification that occurs to a greater or lesser degree as everyone ages. It is sometimes used by physicians as a part of a general explanation for patients who have superimposed, widespread atherosclerosis.

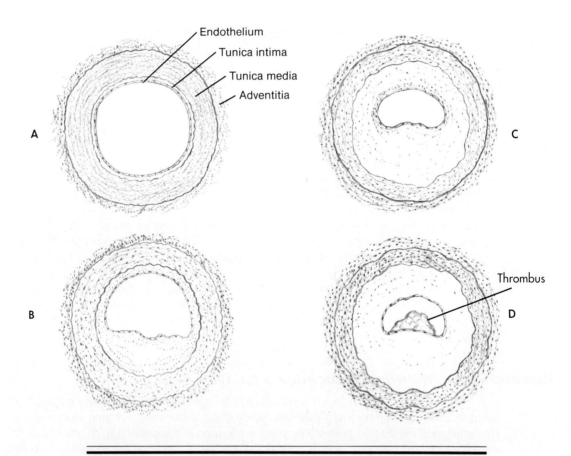

Figure 9-6   **A** through **D** show successive stages in the development of atherosclerotic plaque. The endothelium of an artery is damaged in an area of high pressure and turbulence, such as where arteries branch. Platelets adhere to the damaged area. Platelets stimulate proliferation of smooth muscle cells in the tunica media and intima. The tear is closed off by a platelet-fibrin plug. Lipoproteins and cholesterol accumulate along the endothelium. Macrophages engulf and destroy this material, creating "foam cells." The atheroma enlarges as this process is repeated. The endothelium is stretched and pushed into the lumen of the artery. This obstruction is called atherosclerotic plaque. The endothelium may also rupture, triggering clot (thrombus) formation. Pieces of the atheroma (emboli) may break off and be carried down the bloodstream. The arterial wall, weakened by the replacement of smooth muscle cells with foam cells, may balloon out, forming an aneurysm.

Peripheral Vascular Disease. This condition usually involves the legs and develops when blood delivery to the extremities is inadequate. Most of these patients will have a reduced pulse in the distal leg or foot. Most commonly a patient with this condition will complain of claudication or, in more severe cases when the blood supply is suddenly cut off, the excruciating pain of acute occlusion of an artery. The pain is caused by ischemia of the tissues supplied by the artery. If the blood supply is slowly and progressively interrupted to a part of the body, pain is less prominent as a symptom. It may be aggravated by elevation; some relief may be obtained by keeping the affected limb below the rest of the body.

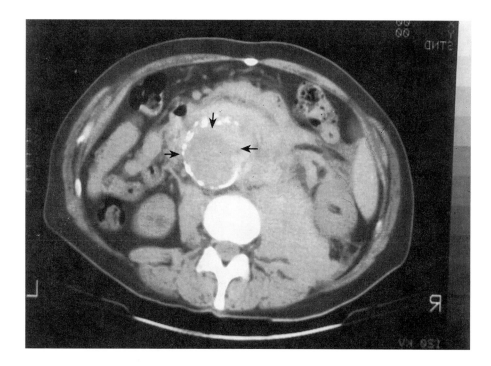

Figure 9-7   CT scan showing aneurysm (arrows). People with an aortic aneurysm this size or larger have a 2-year survival rate of only 50% if left untreated.

Aneurysms. Arterial aneurysms are a common consequence of atherosclerosis. Because they occur in proportion to the duration and severity of the atherosclerosis, most arise after age 60. Aneurysms are ten times more frequent in males than females. In order of decreasing frequency, they are found in the abdominal aorta, popliteal artery, femoral artery, thoracic aorta, and splenic and renal arteries. Abdominal aneurysms are usually discovered when a partially calcified, enlarged aorta is observed during a routine spinal x-ray. An example appears in Figure 9-7. Abdominal pain is usually the signal of imminent rupture or leakage. Untreated, aortic aneurysms over 5 cm in diameter have a 2-year survival rate of only 50%. A frequent complication is distal ischemia due to atheromatous embolism, often to the kidneys or the muscles of the extremities. The most common cause of death from aneurysm is rupture with massive internal bleeding.

Embolism. Emboli are thrombi that have formed in one part of the body, usually in a vein, and have broken loose and lodged some place else. Most commonly, thrombi form in the deep, large veins of the legs or pelvis after a period of immobilization, especially after surgery. They are most likely to occur in patients who have become prone to thrombosis as a result of trauma to a vein, poor blood flow for any reason, excess coagulation factors as in pregnancy or steroidal birth control, or those rare individuals with congenital clotting factor problems.

The thrombi may be associated with a venous inflammation, thrombophlebitis, and can result in pulmonary embolism if the thrombi break free from the veins and lodge in the arteries of the lungs. Emboli occur in the cerebral arteries and elsewhere but these are more likely to arise from a nearby atheroma or from blood clots in the left

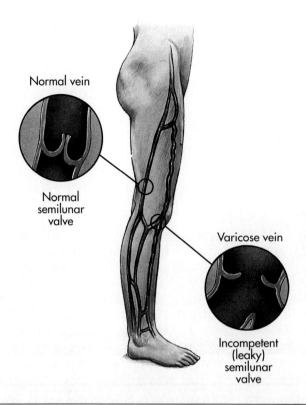

Normal vein

Normal
semilunar
valve

Varicose vein

Incompetent
(leaky)
semilunar
valve

Figure 9-8   Varicose veins result from genetic deficiency of the valves or from constriction of the vessel downstream, causing failure of the venous valves.

side of the heart. Because thrombi arising in any of the veins in the body except the pulmonary veins must pass through the arteriole/capillary filter of the lungs, most venous thrombi produce pulmonary emboli.

Varicose Veins. Varicose veins occur when veins become so expanded that the valves they contain are no longer competent. The valves assist body movement in helping to squeeze the blood toward the heart by not allowing back flow. When the valves in the veins of the legs fail to function, the small veins and capillaries of the legs are filled with blood under continual high pressure. These capillaries and small veins lack the strength to withstand such pressure and they bulge, become inflamed, even rupture, and put irritating pressure on the surrounding tissue. The result can be a painful grotesque array of bulbous veins in the legs and ankles (Figure 9-8). Varicose veins can result from either a genetic deficiency of the valves or from excessive constriction of the veins with a resulting high pressure upstream. Examples of the latter include the pressure on veins caused by pregnancy or by garters and girdles. Another example is hemorrhoids, varicose veins of the anus.

Ischemic Heart Disease. This disease is usually caused by atherosclerosis of coronary arteries with a resultant decreased blood flow. Some instances result from small vessel arteriolosclerosis and are associated with diabetes mellitus. While changes in the coronary arteries may begin early in life, the condition is usually not severe enough to affect heart function until later, especially in men after middle age and women after menopause. The atherosclerotic lesions may be located in any of the coronary arteries.

These arteries may be surgically bypassed, usually using portions of the patient's leg veins. The narrowed portion may also be reopened using balloon angioplasty. (See Figure 9-9). After a period of therapy and healing, the improvement in the elderly patient's cardiac function can often be seen in greatly improved brain function, attitude, and general vigor.

Most episodes of myocardial ischemia and infarction result from narrowing or thrombosis of the coronary arteries mainly due to atherosclerosis. It may cause sudden arrhythmia associated with the tissue damage resulting from the interrupted blood supply. Contributing to or precipitating the blockage may be vasospasm near the site of the atheroma or thrombosis. A thrombus is present in 85% of patients who survive a heart attack more than 48 hours. It may be the cause of the problem but some argue that the thrombosis may be secondary to the vasospasm and decreased blood flow. The end result of a complete interruption in blood supply to a piece of heart tissue is infarction, death of a portion of the heart muscle. Alternately, ischemic heart disease can be caused by small emboli or other blockages occurring in a series of small arterioles, killing small groups of muscle fibers over a long period in separate incidents. This progressively weakens the heart as muscle fibers are lost and replaced by nonfunctional fibrous tissue. Of course sudden blockage of a major vessel can produce massive damage, killing so much of the myocardium that sudden death is the only possible outcome.

Immediately following blockage of a coronary artery, tissue necrosis begins, although this is generally not evident until 12 hours or more have passed. The first microscopically observable changes are cellular disintegration with loss of cellular detail and nuclear degeneration. Within 24 to 48 hours, neutrophils and macrophages invade the infarct and begin removing necrotic tissue. Finally, fibroblasts invade and begin replacing the dead muscle removed by the macrophages with collagen that will form a scar. At about 10 to 14 days, the muscle is most weakened by necrosis with only a scattering of repairing fibroblasts and the heart at this stage is subject to rupture.

Throughout this process, blood supply to the damaged area of myocardium will be slowly improving due to the development of collateral circulation. This involves the invasion and expansion of capillaries and arterioles into the ischemic heart muscle from unblocked blood vessels nearby. Each time infarctions occur, the remaining myocardium hypertrophies as it responds to the increased work load. This increased cardiac muscle size combined with the scar tissue results in a weakened but somewhat enlarged heart.

The most common location for myocardial infarction (MI) is the thickest muscle mass, the left ventricle. If the infarction is large enough to cause dysfunction of this ventricle, blood will back up in the pulmonary tree and produce symptoms of congestive heart failure (See Chapter 6). Myocardial infarction can also be associated with decreased blood pressure leading to decreased kidney output and consequent water retention. This, along with the increased venous pressures, may produce edema, especially in the "dependent" portions of the body, the ankles and feet.

The patient who has just suffered an MI must avoid excessive stress. During the early stages of infarction the myocardium is irritable and in the first 24 to 48 hours can develop fatal arrhythmias. For this reason, patients with MI are kept quiet, free of pain, and at rest and their heart rhythms are carefully monitored for the first few days. They are often given supplemental oxygen in the hope of saving heart muscle that has been deprived of oxygen but is not yet dead. Once the patient has been stabilized and there is no evidence of ongoing heart damage, a progressive activity program can be begun.

Heart Block and Arrhythmias. Cardiac conduction defects are common in the elderly with coronary artery disease. Any part of the conduction system of the heart may be damaged by infarction or ischemia, resulting in arrythmias. Complete heart

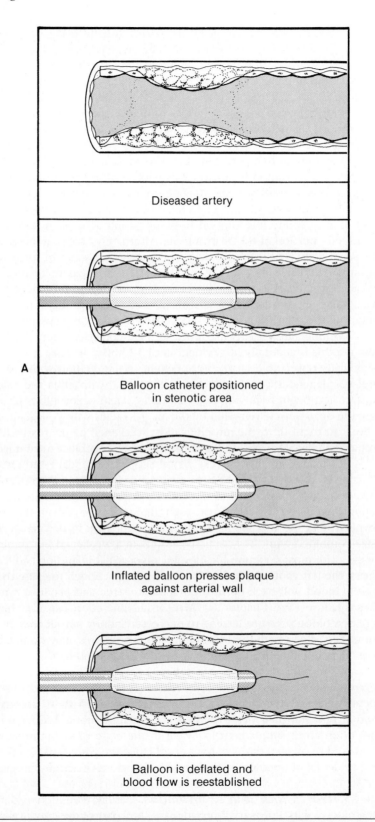

Diseased artery

A

Balloon catheter positioned
in stenotic area

Inflated balloon presses plaque
against arterial wall

Balloon is deflated and
blood flow is reestablished

Figure 9-9   For legend see opposite page.

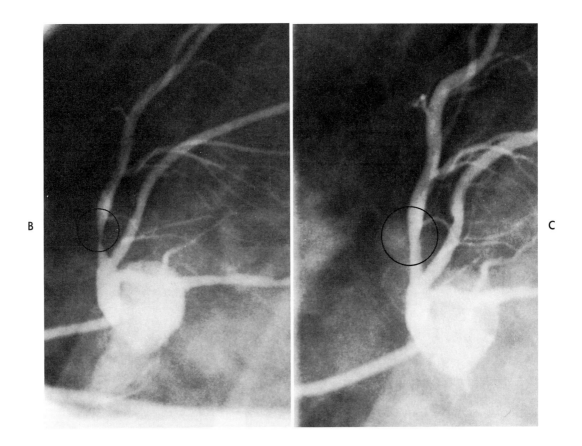

Figure 9-9 **(A)** Balloon angioplasty procedure (opposite page). Above, coronary arteriograms before **(B)** and after **(C)** angioplasty.

block occurs when the conduction system between the atria and ventricles is interrupted so that these two parts of the heart operate under their own pacemakers. This generally produces ventricular bradycardia with atrial tachycardia, (slowed ventricular beat, rapid atrial beat) and inefficient ventricular filling, sometimes with a murmur. Partial heart blocks commonly produce complex rhythms and may lead to fibrillation or to periods of up to 30 seconds without a beat. Since the brain can withstand only up to 5 seconds without oxygen, this produces fainting spells. This condition is often treated with implantation of an artificial pacemaker under the skin of the chest. This battery-operated device emits a dependable charge, creating a regular beat (Figure 9-10). One problem with such devices is that the battery must be changed about every 5 years. Because arrhythmias are often the result of ischemic heart disease, many people who would have received pacemakers 15 years ago, now have bypass surgery. The increased blood supply often restores the normal rhythm.

Valvular Heart Disease. As mentioned earlier, the atrioventricular and semilunar aortic valves are subject to thickening and calcification. With age and use, the valves experience small insults that result in reparative changes. If wear and tear are excessive due to factors such as hypertension, atherosclerosis, or congenital abnormalities, the normal valvular changes will be accelerated, resulting in deformity of the lobes of the valves and dysfunction (see Figure 9-11).

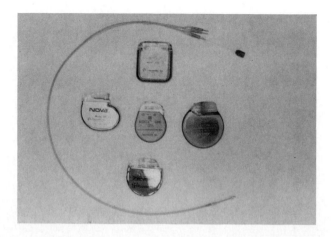

Figure 9-10 Pacemakers like this are inserted under the skin of the chest. They correct irregular heartbeat by emitting regular electrical impulses.

The muscular tissue around the mitral valve normally contracts like a sphincter during systole, helping the valve to close. This contractility is lost when the mitral annulus is heavily calcified. Over 90% of women over 90 years old show this level of calcification. This calcification, combined with damage to the papillary muscles or chordae tendineae, can produce mitral valve failure. Damage to the tricuspid valve is less common and usually less severe.

One of the most common valvular problems in the elderly is aortic valve calcification and fibrosis. As this occurs, the normally pliable valves become fixed. This results in increased resistance to blood flow through the aortic valve. In some cases the valve cannot close and allows blood to flow back into the ventricle. Both conditions produce murmurs and increase the work load on the heart. Figure 9-12 shows prosthetic valves used to replace diseased heart valves.

Congestive Heart Failure. Many of the conditions just discussed can result in the heart's inability to pump adequate amounts of blood. The result is decreased ability of the heart to meet increased physical demands, and failure to keep the blood flowing evenly through its chambers. The failure creates a backup of blood behind the weaker side of the heart. Left-sided failure results in pulmonary edema; right-sided failure produces signs such as swollen ankles, distended neck veins, and liver congestion. Drugs such as digitalis (Digoxin) are used in these circumstances to strengthen the heart's contractile power.

Hypertension. When the diastolic blood pressure exceeds 95 mm Hg, hypertension is said to exist. In the elderly, the loss of elasticity in the aorta and peripheral arteries does not allow them to expand during systole and contract during diastole. For this reason, the elderly often have an elevated systolic pressure, or systolic "hypertension." This form of hypertension is of less consequence unless the systolic pressure exceeds 190 mm Hg. Diastolic hypertension is associated with increased incidence of many diseases including heart failure, myocardial infarction, atherosclerosis, strokes, and renal failure. It can be classified under the two broad headings of essential hypertension (unknown cause) and renovascular hypertension. These conditions are discussed in Chapter 8.

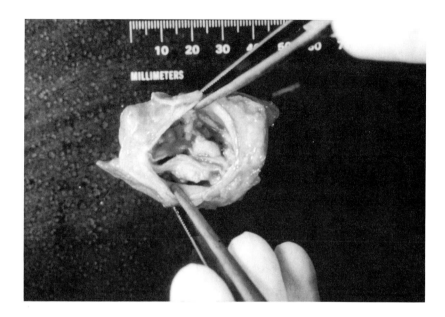

Figure 9-11   This aortic semilunar valve shows the fibrotic thickening and calcification typical of severe aortic stenosis.

© COURTESY OF ST. JUDE MEDICAL, INC.

Figure 9-12   The valve shown above is used to replace diseased heart valves.

CASE HISTORIES

❖ Aneurysm

A.L. is a 64-year-old executive who collapsed while at his desk after complaining of the sudden onset of a severe backache. There was no known precipitating activity and he apparently felt fine until the onset of his pain. He was quickly brought to a local hospital by ambulance. The transport record revealed his initial blood pressure to be 54/? with a thin rapid pulse. Despite vigorous administration of fluid during the ambulance trip, he was still in shock and showed no improvement in his blood pressure at the time of arrival at the hospital emergency room (ER).

In the ER the patient was observed to be a tall, thin man in profound shock. He was nonresponsive to any stimuli and was pale, with a cold, clammy skin, a nonpalpable pulse, and no recordable blood pressure. His eyes were fixed, dilated, and did not respond to light. The deep tendon reflexes were hyperactive with positive Babinsky's sign (big toe moving upward when bottom of foot is stroked). All these symptoms indicate hypotension with brain damage due to hypoxia. During the rapid assessment phase of his admission, he went into respiratory arrest that could not be reversed.

At autopsy A.L. was found to be 6 feet 5 inches tall with abnormally long arms, hands, and fingers, a condition known as arachnodactyly. The cause of death was a large left hemothorax, an accumulation of blood in the left lung cavity that was due to a ruptured dissecting (tissue-separating) aneurysm of the aorta. The aortic intima showed only mild atherosclerosis without plaque formation. The only other significant finding was the presence of an intact aneurysm of the anterior communicating artery of the brain. Microscopic examination of the normal and dissected aorta showed degenerative changes of the muscular tunica media that preceded the intimal tear and dissection. The final diagnosis was aortic aneurysm and massive hemorrhage into the left thorax.

Comment. Most ruptured aortic aneurysms are due to atherosclerosis. Only a few are due to other causes such as Marfan's syndrome. Most of the time when the wall of such a large vessel is breached, profound shock and death quickly follow. Exceptions occur usually when the hemorrhage occurs in an enclosed area where only a limited amount of blood can accumulate before the resulting tissue pressure prevents further bleeding.

In Marfan's syndrome there is a tissue defect leading to degeneration of vascular walls (cystic degeneration of the tunica media) and a characteristic habitus: tall, thin, with abnormally long fingers and arms ("Abe Lincoln" appearance). In addition to the aorta, aneurysms may occur within the cerebral vessels. Patients may have valvular heart disease and detached lens problems.

❖ Pulmonary Embolism

P.S. is a retired secretary who fell and fractured her left hip. She was hospitalized and the hip was pinned. She was feeling progressively better until the fifth hospital day when she experienced acute dyspnea (difficulty in getting enough air). The problem developed suddenly while she was learning to use a walker.

When she was examined, she was very anxious and in severe respiratory distress. Despite a respiratory rate (RR) of 38/min, her lips were cyanotic, indicating lack of oxygen. Her other vital signs included mild tachycardia but a normal blood pressure. No abnormalities were heard in her heart, and her lungs were clear to percussion and auscultation (stethoscope examination). Her left lower leg was slightly edematous with a midcalf diameter of 18.3 cm while the right measured 16.5 cm. Palpation revealed the left calf to be tender and dorsiflexion of the left toes caused pain.

Laboratory examination produced a normal chest x-ray. Her arterial blood gases on normal room air were: pH, 7.50; O_2, 60 mm Hg; and CO_2, 18 mm Hg, indicating decreased amounts of both oxygen and carbon dioxide. The ECG, urine analysis (U/A), CBC, and chemistry profile were normal. A ventilation/perfusion lung scan revealed a segmental defect in the right midlung field with adequate ventilation and decreased perfusion compatible with a large pulmonary embolism. No blood was getting into a portion of the lung (decreased perfusion) while normal numbers of alveolae were being filled, (adequate ventilation).

She was treated with nasal oxygen, bed rest, a heat tent to the left leg, and heparin (an anticoagulant). During the next week she completely resolved her pulmonary problems and lost her leg pain. The left calf remained slightly larger than the right.

Comment. When a blood clot (thrombus), formed in a large vein, fragments or breaks loose, the loose thrombus (embolism) often passes through the right atrium and ventricle and lodges in one of the pulmonary arteries. The resulting vascular blockage does not prevent that portion of the lung from being aerated but it prevents blood from entering that portion of the lung and being exposed to the air. This prevents adequate oxygenation of the red cells and results in apparent air hunger (dyspnea) when the actual defect is inadequate blood flow to a portion of the lung. Most of these emboli begin in the large veins of the legs and pelvis. If the blockage to blood flow were sufficiently large, death could be instantaneous. Most cases result from trauma to the legs, enforced bed rest such as hospitalization, or hypercoagulable states such as those induced by estrogens, smoking, or deficiency of antithrombin factors. Some of the early birth-control pills contained large dosages of estrogens and were associated with pulmonary emboli.

❖ Peripheral Vascular Disease

RD, is a 68-year-old night watchman who developed a nonhealing ulcer on the heel of his right foot. The ulcer appeared 6 months ago and had been gradually enlarging despite soaks, topical antibiotics, and dressings. It began as a small, slightly painful, reddened area on the weight-bearing portion of the heel. It quickly broke down and began progressively enlarging with a reddened, wet rim of surrounding skin. Systemic antibiotics, debridement, or the surgical removal of dead tissue, elevation of the foot, and clean dressings during a hospitalization arrested the process and the lesion receded to 0.5 cm. However, after a month of home ambulation, the ulcer again began to enlarge.

His past medical history included 20 years of insulin dependent diabetes mellitus. Five years prior to the ulcer RD had limited his daily walks because of two-block claudication (pain in the legs due to insufficient circulation after walking two blocks), most severe in the right leg. Three years prior to admission, congestive heart failure was diagnosed on the basis of an S3 heart sound, distended jugular veins, and pedal edema (swelling of the feet). Heart sounds are recorded as S1 (closure of the tricuspid and mitral valves), S2 (closure of the atrial and pulmonary semilunar valves), and sometimes an abnormal third heart sound, S3. Current medications include 30 units of NPH (long-acting) and regular insulin divided into a BID (twice daily) injection, digitalis to strengthen the heart, and a thiazide diuretic. He smoked three packages of cigarettes a day until 3 years ago.

On physical examination he appeared older than his stated age, in no acute distress. His vital signs were normal with the exception of mild systolic hypertension. His HEENT (head, eyes, ears, nose, and throat) were unremarkable with the exception of arcus senilis (a fatty ring of degenerative tissue around the cornea,), and a loud systolic bruit or blood sound arising from the carotid bifurcation indicating unusual turbulence in the flow of blood. The heart, chest, abdomen, and neurologic examination revealed no abnormality. The peripheral pulses were normal in the upper extremities but the femoral pulses were both absent, as were the popliteal (back of knee) and anterior tibial (lower leg) pulses. Both legs were well muscled. The calf skin was relatively hairless. The right heel was distorted with a wet, foul-smelling ulcer that measured 3.4 cm in diameter. In addition, the base of the first toe and the medial aspect of the third were unnaturally darkened and firm.

Laboratory examination (Doppler scans) revealed no detectable blood flow in the right pedal and popliteal vessels with a marked decrease in the femoral. The flow in the left system was decreased. An aortogram showed both common iliacs severely occluded with good collateral circulation in the left leg but few collateral vessels in the right. The ECG demonstrated an old anterior MI plus nonspecific ST-T wave changes compatible with ischemia or digitalis effect. The urine analysis showed 2+ urinary glucose and trace ketones, indicating poor diabetes control. The chemistry profile showed 230 mg/dl of glucose with 350 mg/dl of triglycerides (normal to 200) and a cholesterol of 330 mgm/dl (ideal values are less than 200).

After a trial therapy of 7 days of soaks, topical and systemic antibiotics, and hyperbaric oxygen with no improvement, a right common iliac endarterectomy was attempted but without success. Subsequently, a right, above-knee amputation was carried out.

Comment. Peripheral vascular disease can occur on its own, but in our society it is often associated with diabetes mellitus, excess lipids and cholesterol, or smoking. It often first manifests as pain on exercise (claudication) that is proportional to the degree of arterial blockage. Due to the slow nature of the vascular disease, there is time for the development of alternate blood flow pathways, or collateral circulation. If the alternate blood flow is not sufficient to maintain the needs of the tissues, the tissues die. This produces ulcers and, if extensive, gangrene. During development of the disease the vessels lose their elasticity and for this reason can not respond well to vasodilator medications. If the blocked vessels can not be reopened mechanically, the compromised tissues must be removed. If left in situ, secondary infections will occur, spread, and lead to sepsis and death.

❖ Myocardial Infarction

J.P. a 58-year-old man, was brought to the emergency room with an apparent heart attack. The pain began 1 hour earlier while JP was climbing the stairs. It was accompanied by dizziness and a profuse cold sweat and did not respond to rest or two sublingual nitroglycerine tablets. The pain was intense and felt like his chest was being crushed. It was precordial (anterior) and radiated to his left arm. His medical history included treated hypertension of 7 years duration and 2 years of angina pectoris manifested as dull precordial pain with exercise that abated with rest or nitroglycerine. On average he had needed two nitroglycerine tablets per month.

When admitted to the emergency room, he appeared anxious, in pain, and acutely ill. His skin was bathed in a cold, clammy sweat. His vital signs included mild hypotension (BP 108/76) and a rapid pulse rate of 96 beats per minute. The heart sounds included a faint extra beat, an S3, sometimes present in heart attack or heart failure. The lungs contained increased fluids indicated by fine rales (crackles heard with a stethoscope). An emergency electrocardiogram (ECG) showed features of a myocardial infarction with Q waves and large S-T segment elevation. Subsequent ECGs revealed a large evolving anterior myocardial infarction.

Admission cardiac enzymes were normal. CK (creatine kinase) rapidly became elevated with a peak level of 2140 IU/L (normal < 200) at 36 hours; 24% of the total was due to the MB heart fraction. The LDH peaked at 6320 IU/L (normal <200) on the third hospital day. The LDH 1 fraction was the most elevated. On the third day a friction murmur suggestive of pericarditis was noted over the anterior chest. This can occur if the infarction extends through the epicardium and causes epicarditis.

JP was treated with intramuscular pain medication, nasal oxygen, TPA (tissue plasminogen activator to dissolve thrombus), and digitalis. On the first hospital day a coronary angiogram demonstrated focal areas of slight narrowing in the interventricular branch of the left and of the main channel of the right coronary arteries, and evidence of slight obstruction. The impression was one of large anterior MI due to coronary thrombosis that had partially spontaneously lysed. He was discharged on the seventh but died suddenly on the tenth. An autopsy revealed rupture of the myocardium through a healing, large anterior MI with cardiac tamponade (accumulation of fluid in the pericardial cavity, resulting in compression of the heart).

Comment. Today 80% or more of all MIs are believed to result from coronary thrombosis superimposed on atherosclerosis. For this reason, modern therapy is directed toward clot lysis, opening the blocked arteries, and other techniques to overcome the chronic vascular problem. TPA is a clot-dissolving agent that has the advantage of being an activator of a naturally occurring substance and of only affecting preformed clots. Other forms of therapy are directed at increasing flow by enlarging the area around the atherosclerotic obstruction or by diverting flow around the blocked vessel. The former is accomplished by inflating a balloon on the end of a catheter inserted through the narrowed vessel (angioplasty technique). The latter is accomplished by bypass surgery where veins or other arteries are used to supply blood past the obstructed vessel.

The heart musculature has large amounts of enzymes needed to store and release energy; these enzymes are somewhat specific to heart muscle, the CK-MB and LDH-1 fractions. By determining the amounts of these subtypes of the LDH and CK in serum, one can determine if the enzyme elevations are due to damage to the heart or to other muscle damage. The heart wall is weakest from 10 to 14 days after the MI, when the localized myocardial degeneration is nearly complete and the necrotic area has not been adequately replaced with scar tissue. It is characteristically during this time that myocardial rupture may occur.

❖ Aortic Stenosis

J.G. is a 53-year-old physician who has had a long history of valvular heart disease. Although he had been aware of a heart murmur since his teens, it first became serious at the age of 39 when he applied for additional life insurance. Evaluation with sonography, chest x-ray, ECG, and phonocardiography were inconclusive but life insurance was not issued. Repeat studies conducted at the age of 43 documented a degree of aortic stenosis with a gradient of 18 mm Hg across the valve (normal less than 5). He was completely asymptomatic but had a mild hypertension (140/90) that disappeared without therapy. Periodic follow-up revealed a slow but steady disease progression and lately he has noted faintness on exertion. On his last evaluation, the appearance of valvular insufficiency suggested a need for valve replacement.

On evaluation he was healthy with the exception of a loud systolic murmur that obscured the second heart sound and radiated into the carotid arteries. It was so loud that he could hear it any time it was quiet and it kept him awake at night. A chest PA (posterior-anterior) and lateral x-ray showed LVH (left ventricular hypertrophy) plus calcification in the area of the aortic valve. The ECG confirmed the LVH and showed an old MI. Cardiac catheterization revealed a thickened left ventricular wall with good motion; poorly moveable cusps in the aortic valve with high-grade stenosis and moderate insufficiency. The coronary arterial system was unremarkable.

The diseased aortic valve was removed and replaced with a prosthesis. The removed valve was extensively calcified and rigid. It appeared to have only two cusps but the amount of nodularity and calcification made it difficult to be sure of this. In 6 months J.G. had returned to normal activity.

Comment. With the advent of cardiac valve replacement and with greater access to health care, few people in the United States die of untreated valvular disease. On average, a mechanical valve can last 10 to 30 years before requiring replacement. Due to the aging of our population and the stress in the aortic valve, it is one of the most common valve replacements. If the valve has a congenital abnormality such as a bicuspid rather than tricuspid construction, the stress will result in acceleration of the degenerative changes. The resulting fibrosis and calcification prevents the valve from opening normally, creates a stenosis or narrowing of the channel, or it prevents the valve from closing all the way. This insufficiency results in blood flowing back into the ventricle from the aorta when the ventricle is supposed to be filling with blood from the atrium. This unusual flow of blood produced the loud murmur that kept J.G. awake.

BIBLIOGRAPHY

Belt E: Leonardo da Vinci's study of the aging process, *Geriatrics* 7:205-210, 1952.

Blumenthal HT: *Athero-arteriosclerosis as an aging phenomenon.* In Goldman R, Rockstein M, ed: *The physiology and pathology of human aging,* San Diego, 1975, Academic Press.

Kitzman DW, and others: Age-related changes in normal human hearts during the first 10 decades of life. Part II (maturity): A quantitative anatomical study of 765 specimens from subjects 20 to 99 years old, *Mayo Clinic Proc* 63:137-146, 1988.

Klima M, Burns TR, Chapra A: Myocardial fibrosis in the elderly, *Arch Pathol Lab Med* 114:935-942, 1990.

Kottle BA: *Disorders of the blood vessels.* In Pathy MJS, ed: *Principles and practice of geriatric medicine,* New York, 1985, John Wiley & Sons.

Lie JT, Hammond PL: Pathology of the senescent heart: anatomic observations on 237 autopsy studies of patients 90 to 105 years old, *Mayo Clinic Proc* 63:552-564, 1988.

Stamler J, and others: Hypertension screening of 1 million Americans. Community hypertension evaluation clinic (CHEC) program of 1973 through 1974, *JAMA* 235:2299-2306, 1976.

Williams BO: *The cardiovascular system.* In Pathy MJS, ed: *Principles and practice of geriatric medicine,* New York, 1985, John Wiley & Sons.

Wright JR: *Cardiovascular and pulmonary pathology of the aged.* In Rcichel W, ed: *Clinical aspects of aging,* Baltimore, 1983, Waverly Press.

10 *The Lymphatics, Blood, and Immunity*

❖ *Relevant Anatomy and Physiology*

THE LYMPHATIC SYSTEM
Structures and Functions

Blood, the lymphatic system, and the immune process are highly interdependent. The immune process involves specialized white blood cells, often centered in lymphatic structures such as the lymph nodes and the spleen. Many of these specialized cells are transported in blood and lymphatic vessels. The thymus, a small gland located behind the sternum, plays a very important part in the development of some lymphocytes. Because of these interdependencies, these three topics are presented here in a single chapter. The primary blood transportation mechanism, the cardiovascular system, has been described in Chapter 9. We will begin this chapter with discussion of the transportation of the nonvascular body fluids through the lymphatic system.

The lymphatic system consists of a network of interconnected tubules distributed throughout the body (Figure 10-1). Like blood vessels, the lymphatic vessels are composed of endothelial cells. Large lymphatic vessels have a smooth muscle layer but their walls are thin, even when compared with veins. Lymph channels drain fluid that has leaked from capillaries into the body's interstitial (the areas between the cells) spaces and transport it back to the bloodstream. On the way, the fluid is forced to circulate through lymph nodes where bacteria and other undesirable agents may be rendered harmless (Figures 10-2 and 10-9, *A*, and Color Plate 13, *A*). Large concentrations of lymph nodes occur in the inguinal, abdominal, axillary, and cervical areas of the body. Most lymphatic fluid enters the bloodstream through the thoracic duct at the junction of the left internal jugular and left subclavian veins. Much less fluid is returned through the right lymph duct at the junction of the right subclavian and right internal jugular.

The abdominal lymph vessels also absorb and transport nutrients, especially fat. Lymphatic vessels called lacteals extend into the villi of the intestines, where they absorb fat and certain other large molecules. These vessels pass through a series of abdominal channels, the cisterna chyli and encounter many lymph nodes before passing on to the thoracic duct and bloodstream. These nodes remove foreign material such as bacteria, and are a center for the formation of antibodies against foreign substances.

There is no pumping counterpart to the heart in the lymphatic system. Even so, the lymph must be "pumped" against considerable resistance and often uphill so that the

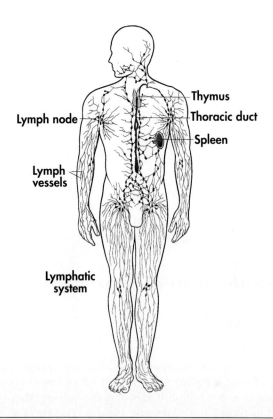

Lymph node

Thymus

Thoracic duct

Spleen

Lymph vessels

Lymphatic system

Figure 10-1   The lymphatic system collects fluid from throughout the body and returns it to the blood vascular system. On the way the fluid is filtered through lymph nodes. Concentrations of lymph nodes are labeled above.

interstitial spaces can drain. This is accomplished primarily by two mechanisms. First, when a person moves, the muscles contract and organs shift, compressing parts of lymph vessels. This squeezes the fluid out of the compressed portion of the vessel. Valves, spaced periodically along each vessel, permit the fluid to move only in one direction. When the pressure is released, the vessel can again be filled. By this means, normal daily movements continually drive the lymphatic fluid toward the access points into the vascular system. In addition, smooth muscle fibers in the large lymph vessels can contract in response to being stretched. When a large lymph vessel is stretched because it is filled with lymphatic fluid, it contracts, pushing the fluid in the direction the valves permit. The total lymphatic return to the blood is about 120 ml per hour for a resting person. This may increase fivefold during activity.

Because small amounts of fluid are always leaking from capillaries into the interstitial spaces, the body depends on the lymphatic system to return that fluid to the bloodstream. If this fails to take place rapidly enough, either due to blockage of the lymphatic channels or to excessively rapid loss of fluid from the blood, edema develops. Sur-

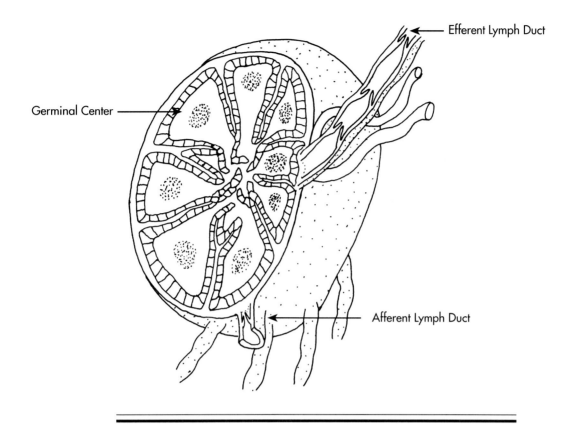

Efferent Lymph Duct

Germinal Center

Afferent Lymph Duct

Figure 10-2 The nodules of lymph nodes contain colonies of B and T lymphocytes. The lymphocytes reproduce, causing swelling when they are fighting an infection.

gical removal of lymph vessels or nodes can cause transient edema until new lymph channels form. If proteins accumulate in the interstitial spaces and are not removed, they can cause fluid to leave the blood and enter these spaces by osmosis as well as by normal leakage. This too, obviously contributes to edema.

The Thymus

The thymus functions as a staging area for the T lymphocytes during prenatal and early postnatal development. In the thymus these cells "learn" to identify the body's tissues so that they do not launch an immune attack against them. The thymus also produces thymosin, a hormone that stimulates the maturation of T lymphocytes. Thymosin also normalizes excessively high or low levels of supressor T lymphocytes. Rarely, children are born without a thymus. These children are extremely suceptible to infection and must live in protected, sterile environments such as that shown in Figure 10-3.

Blood Composition and Function

Blood is divided into two components: the formed elements and the plasma. All the formed elements in blood, including the red and white blood cells and platelets, arise

© AP/WORLD WIDE PHOTOS, INC.

Figure 10-3 This boy was born without a thymus and therefore lacked the T cell organization of the immune response to infection. Such individuals must live in a sterile environment until researchers find a way to replace thymic function.

from a single progenitor cell, the hemopoietic stem cell (see Figure 10-4 and Color Plate 12). The plasma is the liquid portion of blood and everything dissolved in it. The term *hematocrit* is sometimes used for all the formed elements of blood (the packed cell volume) but is also often used as an approximate measure of the percentage of blood volume that is red blood cells, erythrocytes. The average healthy male has a hematocrit of 42 to 52 and the average female 37 to 47. This means that when a sample of blood is centrifuged, 37% to 52% should be erythrocytes, about 1% leukocytes and platelets, and the remaining portion should be plasma (Figure 10-5). The hematocrit of both sexes varies with the geographic elevation, activity level, and general health of the individual. Today most medical reports use hemoglobin instead of hematocrit as a measure of oxygen-carrying capacity. There are conditions however, where knowledge of the hematocrit is more useful than the hemoglobin value. There are other situations where knowledge of both is essential. The hematocrit is about three times the hemoglobin value. Therefore a hematocrit of 37 to 42 will usually approximate a hemoglobin of 12.3 to 14.0.

The hemopoietic, or blood forming, tissue occurs mostly in the marrow cavities of the axial skeleton in adults. It is responsible for production of all the blood cells. One of the earliest evidences of aging in the body occurs before birth, when most blood formation shifts from the spleen and liver to the marrow of the long bones. Usually by age 20 it has shifted again, this time to the membranous bones, such as the ribs and sternum.

Figure 10-4, opposite page. Hemopoiesis is the process of blood formation. The drawing shows the lines of descent of each type of blood cell in the bone marrow.

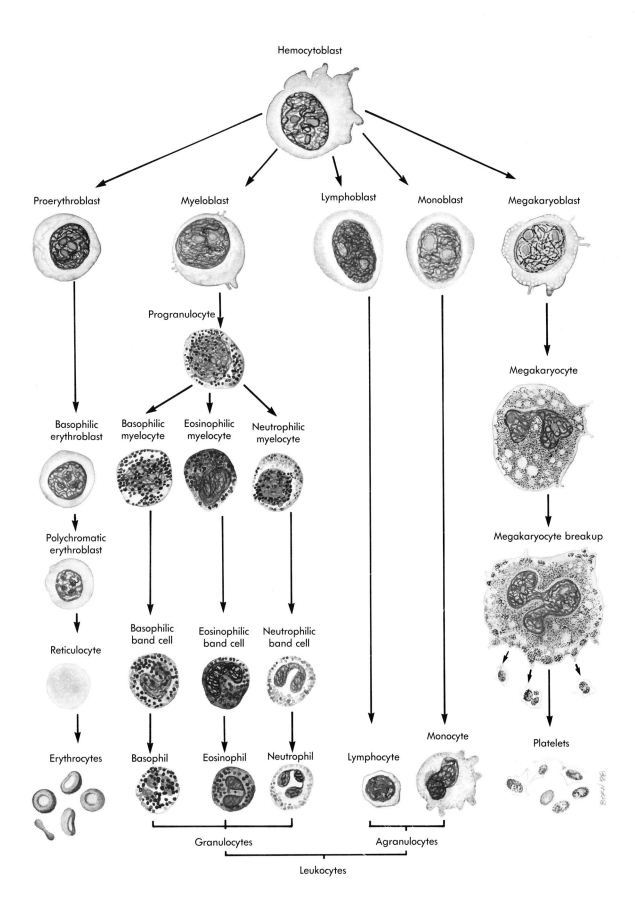

Hemocytoblast

Proerythroblast   Myeloblast   Lymphoblast   Monoblast   Megakaryoblast

Progranulocyte

Megakaryocyte

Basophilic
erythroblast

Basophilic
myelocyte

Eosinophilic
myelocyte

Neutrophilic
myelocyte

Polychromatic
erythroblast

Megakaryocyte breakup

Reticulocyte

Basophilic
band cell

Eosinophilic
band cell

Neutrophilic
band cell

Erythrocytes   Basophil   Eosinophil   Neutrophil   Lymphocyte   Monocyte   Platelets

Granulocytes   Agranulocytes

Leukocytes

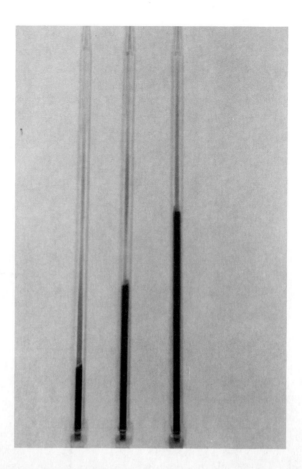

Figure 10-5 The hematocrit tube on the left shows anemia; the one in the center is normal; and the one on the right shows polycythemia. Hemoglobin values are about one third of the hematocrit percentage.

The red blood cells carry oxygen and are produced in response to the body's oxygen needs. For this reason, very slightly lower than normal hematocrits are found in healthy people who live at sea level where the oxygen concentration in the air is high, and in people who are healthy but normally inactive and do not require much oxygen. Similarly, those who live at high elevations and exercise vigorously are on the high end of the normal hematocrit scale. Logically, if people are deprived of adequate oxygen for other reasons such as partial heart or lung failure, their hematocrits will eventually rise as well.

Anemia and Polycythemia

A hemoglobin of 13.7 (hematocrit of 41) or less for men is considered anemia, as is a hemoglobin of 12 (hematocrit of 36) or less for women. It can be hemorrhagic, caused by hemorrhage, aplastic, caused by bone marrow inactivity, hemolytic, caused by rupture of erythrocytes, or nutritional. Nutritional anemia is dietary, usually a deficiency in iron, folic acid, or vitamin B_{12}. In the latter case, the anemia is called pernicious and is caused by failure to absorb the vitamin. Anemic patients usually appear pale, lack en-

energy, have difficulty keeping warm, and can be more susceptible to other diseases.

A hematocrit above 53 for men and 48 for women is generally termed *polycythemia* (an increase in RBC numbers) and can result from many conditions. These hematocrits usually, but not always, correspond to similar increases in hemoglobin values. The common causes of polycythemia include chronic respiratory disease with hypoxia, and living at very high altitudes. False (spurious) polycythemia is due to contraction of plasma volume by dehydration. Polycythemia vera is an overproduction of all the bone marrow elements; red cells, white cells, and platelets. It has many complications and often terminates in leukemia. Any excessive increase in formed elements produces an increase in the viscosity of the blood and therefore creates an increased load on the heart. It also increases the likelihood of vascular thrombosis.

The Formed Elements

The formed elements of blood are the platelets and the red and white blood cells. These are illustrated in Figure 10-4 and described in Table 10-1.

Lymphocytes and the Immune Response

One category of formed elements in the blood has important functions in managing the immune response.

Lymphocytes. They arise in the bone marrow from stem cells called lymphoblasts. They are active in the immune response against foreign materials called antigens. Lymphocytes generate either a cellular inflammatory response or form antibodies against antigens. They differentiate in the early embryo to form three morphologically similar but functionally different kinds of cells. Some of these are further subdivided in function. Figure 10-6 on page 199 illustrates the origin and functions of the three types of T lymphocytes.

T Lymphocytes. Early in life these cells migrate from the bone marrow to the thymus, where they undergo changes that permit them to direct the immune process and mount a cellular response to foreign antigens. This migration is why these cells are called *T* lymphocytes. T lymphocyte activity begins when one or more of the cells encounters a foreign material (antigen) on the surface of a cell such as a monocyte or a B lymphocyte. The T lymphocyte then begins to divide, producing a population of cells specifically targeting that antigen which is often a virus, a TB organism or a foreign tissue. This reproduction occurs primarily in the lymph nodes and the spleen; consequently chronic infections can cause enlargement of these organs.

T lymphocyte populations are specialized for many different jobs and accordingly, bear different names. Some of the cells produced are *killer T lymphocytes.* These attach to and attack antigens. They accumulate around the antigen and ultimately destroy it (cellular immunity). Others are *suppressor T lymphocytes.* Suppressor T cells can shut down the immune response. It has been suggested that a decline in the suppressor population may contribute to increase in allergic reactions in the elderly or to some autoimmune diseases. The programming and response of a T lymphocyte is shown in Figure 10-7, page 200.

A third form of T lymphocyte is the *helper T lymphocyte.* Helper T lymphocytes release various substances that aid the immune response. One of these is *interleukin-2* (IL-2), a compound that is released from sensitized T cells and stimulates other sensitized T lymphocytes to reproduce. Another substance, *macrophage chemotaxic factor,*

Table 10-1 Formed Elements of the Blood

CELL TYPE	DESCRIPTION	FUNCTION
ERYTHROCYTE	No nucleus; normal concentration 5.4 million/mm^3 (men) 4.8 (women)	Transports oxygen and carbon dioxide
LEUKOCYTE		
Neutrophil	Spherical cell; nucleus with two to four lobes connected by thin filaments; immature forms with a single band of nuclear material ("bands") seen in infection or inflammation; granular cytoplasm	Phagocytizes microorganisms
Basophil	Nucleus with two indistinct lobes; stain blue-purple; granular cytoplasm	Releases histamine, which promotes inflammation, and heparin, which prevents clot formation
Eosinophil	Nucleus often with two lobes; stain orange-red or bright red; granular cytoplasm	Releases chemicals that reduce inflammation
Lymphocyte	Round nucleus; cytoplasm forms a thin ring around the nucleus	Produces antibodies and other chemicals responsible for destroying microorganisms; responsible for allergic reactions, graft rejection, tumor control, and regulation of the immune system
Monocyte	Nucleus round, kidney, or horse-shoe shaped; contains more cytoplasm than does lymphocyte	Phagocytic cell in the blood; leaves the blood and becomes a macrophage, which phagocytizes bacteria, dead cells, cell fragments, and debris within tissues can also become stationary (tissue macrophages) and reside in various parts of the body. Enables organization of the immune response
PLATELET	Cell fragments	Necessary for blood clotting

Modified from Seeley, Stevens, Tate: *Anatomy and Physiology,* ed 2, St Louis, 1992, Mosby.

attracts macrophages to the location of the antigen. The contribution of macrophage chemotaxic factor can be seen when an organ or tissue is transplanted from one body to another. Unless they are perfectly matched (an unlikely situation unless the donor is an identical twin), a population of T cells is produced that releases macrophage chemotaxic factor, causing macrophages to accumulate around the transplanted tissue and destroy it. A similar response may result in destruction of many cancers originating in the

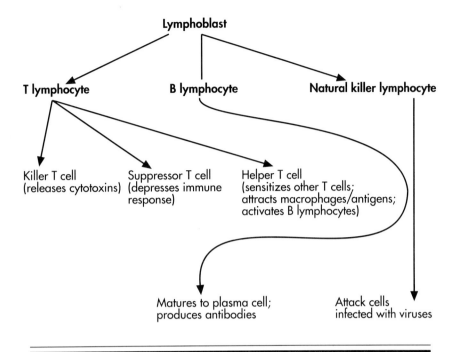

Figure 10-6   Diagram shows the origin and functions of the different lymphocytes involved in the immune process.

body. Other substances that helper T cells release help activate the B lymphocyte's response to the antigen. A delicate balance of helper T cells (that begin immune reactions) and suppressor T cells (that stop the immune process) orchestrate both the humoral (fluid carried) and cellular aspects of immunity.

Natural Killer Lymphocytes (NK Lymphocytes). Natural killer lymphocytes are particularly effective against some viral diseases. When bound to a virus-invaded cell or another antigen, killer T cells release cytotoxins, destroying the invaded cell. With some minor upper respiratory infections, a cold for example, the virus does little damage when it invades the cells of the throat. However, when the NK cells attack the invaded cells, the resulting inflammation may produce a very sore throat. The remains of the dead cells may also create an environment for bacteria, sometimes leading to a *Streptococcus* infection (strep throat).

B Lymphocytes. B lymphocytes learn to recognize the body's tissues during fetal life in the bone marrow. However, the "B" does not signify *b*one marrow. It signifies the *b*ursa of Fabricius, where the function of B cells was discovered in birds. The B lymphocytes are continuously produced in the bone marrow and shed into the blood and lymphatic system. In the lymphatic system they are trapped in the lymph nodes that filter and concentrate both these and T lymphocytes. When activated by an antigen and stim-

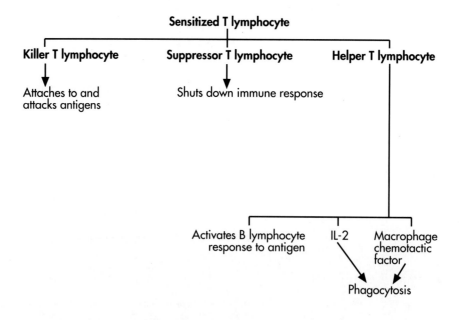

Figure 10-7   Development of specialized T lymphocytes. Some T lymphocytes sensitized to a specific antigen remain as "memory cells", enabling a rapid response years later if the same infective organism reappears.

ulated by helper T lymphocytes, the B lymphocytes transform into plasma cells and produce antibodies. These antibodies are specific to the original antigen and attach to it, aiding in its destruction. They are important in controlling bacterial infections, allergies and auto immune processes. Like T lymphocytes, the sensitized B lymphocytes reproduce clones that produce a large amount of the desired antibody. If a different antibody is needed, the process must be repeated with sensitization to the new antigen and subsequent activation and reproduction of a different clone of B lymphocytes. When the infection is defeated, a few "memory cells" remain so that the process can restart rapidly when the infective organism appears again. B lymphocytes also are important in presenting antigens to T lymphocytes. The action of B lymphocytes is shown in Figure 10-8.

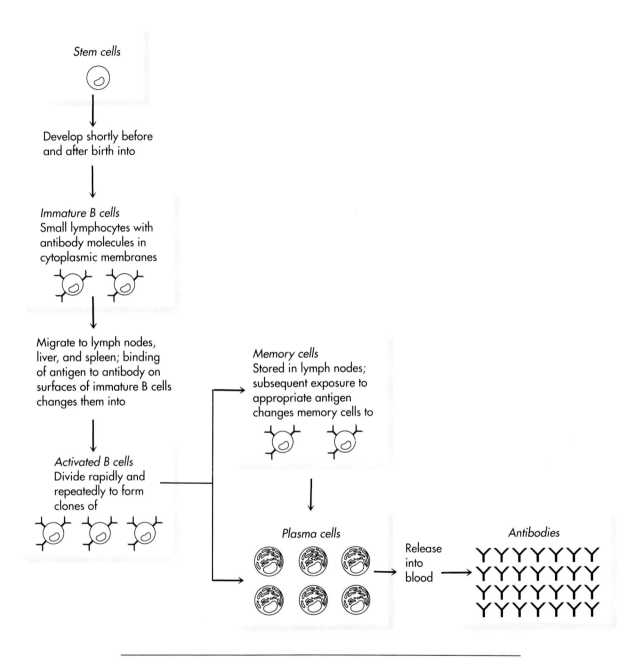

Figure 10-8 When a B lymphocyte (B cell) encounters an antigen, it reproduces, creating a clone of cells. Some of them will become memory cells while others will mature into plasma cells, producing antibodies that inhibit or, through a number of mechanisms, destroy the infecting organism.

Table 10-2   Protein Components of Plasma

NAME	%	FUNCTION
Albumin	4.5	Lubricates and maintains osmotic pressure
Globulins	2.3	
Alpha		Antiinflammatory substances
Beta		Involved in iron metabolism, lipid transport, antigen/ antibody activity
Gamma		Antibodies
Fibrinogen	0.3	Precipitated by clotting factors to form thrombus (clot)

The Plasma

Plasma is the portion of blood that is not composed of formed elements. The volume of plasma is about 3 liters in an adult. Plasma carries and lubricates the cellular elements. It also dissolves and transports gases, nutrients, waste products, and hormones throughout the body. Plasma filtrate (plasma that has filtered out of the blood vessels) bathes the tissues in appropriate concentrations of ions and nutrients. Plasma is a clear liquid that is mostly water containing varying amounts of dissolved compounds. Approximately 7% of plasma consists of proteins. Table 10-2 summarizes the protein components of plasma and their functions.

Following a meal, plasma contains an increased load of nutrients, fats, amino acids, and sugars. These will be cleared from the blood and incorporated into the tissues. Minerals such as sodium, potassium, and calcium will be maintained within narrow limits; trace amounts of many other metals will be present. Waste products, including urea, uric acid, and ammonia are carried in the plasma. Bicarbonate and other buffers act to keep the pH (acidity or alkalinity) within narrow limits. Nitrogen is the primary dissolved gas in plasma, with lesser amounts of oxygen and carbon dioxide.

Clotting and Control of Bleeding

Leaks appear in blood vessels continually as a result of normal activity, especially in the capillaries. Most of the leaks are small, but if they were not immediately repaired it would not be possible to maintain normal blood volume or pressure. Bleeding is controlled by three mechanisms:

Vascular Spasm. When a blood vessel containing muscle is severed, especially with considerable trauma such as when the vessel is crushed, the smooth muscle fibers contract, decreasing the size of the vascular lumen. This action decreases blood flow and loss.

Platelet Plugging. When platelets encounter tears in blood vessels, especially those exposing collagen fibers, they rapidly swell and produce radiating cytoplasmic extensions. These adhere to the collagen as well as to other platelets. Activated platelets are capable of activating other platelets and can release a factor (thromboplastin) that starts blood coagulation. Small leaks in vessels are often plugged by platelets alone. Large leaks may require all three mechanisms.

Thrombus formation. When a blood vessel is injured, thromboplastin (or prothrombin activator) is generated. This compound works as a catalyst in the conversion of prothrombin, an alpha globulin formed in the liver, to thrombin. Thrombin is an enzyme that catalyzes the conversion of the dissolved fibrinogen to fibrin threads. These threads adhere to each other, to platelets, to the damaged vessel wall, and to erythrocytes to form the thrombus, or clot. A few minutes after the thrombus has blocked the opening in the blood vessel, it begins to contract, pulling the torn edges of the vessel together. Further clotting is prevented because the blood flowing past the thrombus rapidly dilutes the clotting factors and carries them away to be eventually removed by the liver. Anticoagulants present in the blood further suppress clotting.

Thromboplastin formation, a complex process, involves several clotting factors and two different sequences, depending on whether the process is extrinsic (stimulated by tissue fluid) or intrinsic (stimulated by changes in the vessel wall). At least 14 clotting factors including thromboplastin, platelets, prothrombin, and fibrinogen have been described. The clot will eventually be removed as the healing blood vessel releases tissue plasminogen activator. This occurs as fibrous tissue invades and replaces the blood clot. Stoppage of bleeding results followed by restoration of blood vessel continuity.

Immunity

Immunity is the resistance of an organism to disease. The granulocytes and monocytes are important components of immunity because they capture and break down foreign proteins into antigenic units. Equally important are the T and B lymphocytes and the thymus that sets up the T lymphocyte system. The part that these structures play in the immune response should now be understood. Immunity is traditionally described as active, passive, or natural.

Active Immunity. This type of immunity is acquired after lymphocytes are exposed to an antigen. Normally the antigen is a disease-causing organism. In most instances, lymphocyte sensitization, cloning, subsequent production of antibodies, and the sensitization and cloning of T cells will occur before the disease overwhelms the body. In many diseases such as smallpox and polio, the body's immune response is often too slow to prevent damage or death from the disease. To combat this, vaccines are used. These are antigens that have been killed or weakened by heat or a chemical process, or genetically altered by growing them in the tissues of another animal. As long as they remain similar enough to the original antigen to stimulate an appropriate antibody response, they can be injected before exposure to the disease to give the immune system a head start. Ideally, vaccines produce an immunity as long lasting and effective as having survived the disease. Some vaccines produce a localized symptom of the disease, such as the localized "small pock" from the smallpox vaccine.

Passive Immunity. A temporary immunity, passive immunity results from antibodies that have come from a source outside the body. A fetus acquires passive immunity when antibodies pass through the placenta into its bloodstream. Some additional passive immunity may be produced by antibodies passed on to the newborn through the mother's milk. These antibodies will stay in the bloodstream about 6 months, giving the baby resistance to the diseases that the mother has had. Passive immunity is also created when a person is injected with a serum. A serum is a collection of antibodies produced in other humans or nonhuman animals. When a person has a disease that typically develops more rapidly than the immune system can respond, these antibodies can be used as a first line of defense. It is important to remember that these antibodies are themselves a foreign protein. The patient's B lymphocytes will be manufacturing antibodies not only against the disease antigen but often against the serum antibodies as well. This creates the risk of producing a serum sickness that can be fatal. In any case, passive immunity contributed by the serum will normally disappear as the antibodies are eliminated from the blood, either by natural attrition or by the immune response.

Natural Immunity. Protection provided by natural immunity results from antibodies that are produced automatically by B lymphocytes without an obvious exposure to an antigen. Examples are the anti-A and anti-B blood factors present in people with type O blood. We have many circulating antibodies whose function is not known. It has been speculated that many naturally occuring antibodies are actually produced against common antigens. The anti-A and anti-B factors may be cross-reacting antibodies whose reactions to A and B antigens are coincidental.

❖ *The Aging Lymphatics, Blood and Immunity*

Aging Lymphatics and Thymus

The lymphatic system extends throughout the body. Yet it contains few grossly recognizable tissues other than the thymus, large lymph vessels, nodes, and parts of the spleen. Probably for this reason, aging of this system has been poorly studied. Few age-related changes have been described. It has been suggested that a decline in the immune response may be related to a decline in the production of lymphocytes.

The thymus undergoes more obvious aging than any other lymphatic organ. It is very large at birth in proportion to the rest of the body and reaches maximum size at about age two. At the time of sexual maturity the thymus begins involution, a slow process of of cell loss and replacement of cells with fatty tissue. Production of thymosins begins to decline at around age 20. By age 60, production of these hormones has nearly stopped. Since the thymus is the place where T lymphocytes mature, the decrease in number of functioning thymic cells appears to be associated with an observed increase in the number of immature lymphocytes in the gland. While this is going on, it should be noted that some age-related autoimmune diseases such as myasthenia gravis are associated with increased abnormal thymic activity. Removal of the thymus may result in remission of the disease. Removal of the thymus in an elderly person seems to have little effect on health because large populations of differentiated T cells exist throughout the body. However, one cannot help but suspect that the observed decline in the responsiveness of the immune system with age is related to loss of thymic function.

Aging of Blood

Age-related changes in most components of blood seem to be slight. In most cases, abnormal blood values are a result of disease. There may be a slight decrease in total red cell count and a slight increase in the volume of each cell. The ability to produce blood cells seems to decrease only slightly if at all. There may be a decline in some subpopulations of lymphocytes and possibly in monocytes. Erythrocyte sedimentation rate (the rate at which erythrocytes fall in plasma) is subject to many factors. It increases with allergies, certain inflammatory diseases, anemia, and cancer but may not increase, as has been suggested, purely as a result of age. The World Health Organization (1969) set the lower values for hemoglobin at 13 and 12 g/dl for men and women respectively, *regardless of age.* Erythrocyte count was set at 5.4 and 4.8 $\times$ 10^{12}/liter respectively for men and women *regardless of age.* Anemia does occur with greater frequency in the elderly but this is usually due to other underlying conditions.

The Aging Immune Response

Several T cell-mediated immunologic functions have been shown to decrease with age. A decrease in suppressor T cell activity in mice has been reported. A similar decline in suppressor T cells in humans could permit the development of autoimmune disease. There appears to be a decline in the T cells' ability to assist B cells in providing a primary response to an antigen. There also is a decline in the primary antibody response. There has been no observable decline in macrophages with age. In general there is a decline in the ability to fight infections and an increase in the autoimmune and suspected autoimmune-related diseases with age.

Claman points out: "It is a paradox that, while immunologic capacities seem to decline somewhat, the incidence of autoantibodies increases with age." Rheumatoid factor, an antibody to human immunoglobulin, is observed in 2.1% of healthy people 20 to 35 years old, and in 33% of all 90-year-olds. Increases are also seen in antinucleoprotein and antithyroglobulin antibodies, sometimes resulting in disease. The immune system can be said to change with age, seemingly to our disadvantage.

Autoimmune Disease

Autoimmune disease occurs when the immune system fails to recognize some tissue in the body as "self." This is usually because an exposed protein of the tissue has been modified. This modification can result from temporary or permanent attachment of a chemically active substance (a free radical or even a prescribed drug, for example). The substance can alter the affected protein and be released, damaging others. The altered proteins are hard to detect but many antibodies specifically targeting affected tissues have been found.

This text discusses most of the diseases believed to be of autoimmune origin. Each is addressed in the chapter discussing the system in which the disease is manifest. The summary in Table 10-3 illustrates that all the systems of the body are vulnerable.

The following diseases and conditions of the immune system and blood are generally associated with aging. However, the relationship of many of these to advancing age is poorly understood.

Table 10-3   Some Diseases Suspected of Being Autoimmune in Origin, and the Systems They Affect

DISEASE	SYSTEM(S) AFFECTED
Systemic lupus erythematosus	Tissues throughout body
Psoriasis	Integumentary (Chapter 3)
Rheumatoid arthritis	Skeletal (Chapter 4)
Myasthenia gravis	Muscular (Chapter 5)
Idiopathic pulmonary fibrosis	Respiratory (Chapter 6)
Asthma (some forms)	Respiratory (Chapter 6)
Atrophic gastritis	Digestive (Chapter 7)
Glomerulonephritis	Urinary (Chapter 8)
Grave's disease	Endocrine (Chapter 12)
Addison's disease	Endocrine (Chapter 12)
Multiple sclerosis	Nervous (Chapter 13)
Alzheimer's disease	Nervous (Chapter 13)

Anemias. Elderly patients with anemia are typically pale, complain of weakness, and are easily exhausted by their normal activity. Since most anemias develop gradually, these patients may not even be aware of a change in their health. Their symptoms may become more severe, however, with anginal pain, intermittent claudication, irritability, forgetfulness, and dizziness. Anemia in the elderly can be caused by numerous mechanisms resulting in either decreased production or increased destruction of blood cells.

The most common cause of anemia in both the elderly and the young is iron deficiency. In the elderly, iron deficiency anemia is as likely to be caused by chronic, often unseen, bleeding, as it is to result from dietary problems. After menopause in women, or at any time in men, the cause of iron deficiency should be investigated. Often it is found to be a colon cancer, ulcer disease, or uterine (endometrial) cancer. These conditions result in blood loss in excess of iron replacement.

Many elderly develop iron deficiency anemia as a result of dietary factors. For example, iron is found in meat, and poor dentition can make eating meat too difficult. Other individuals become apathetic, or movement becomes painful and they neglect to prepare adequate meals for themselves, resulting in a "tea and toast" diet.

Atrophic gastritis, discussed in Chapter 7, can lead to poor iron absorption, but this is unusual. Gastritis is more often associated with decreased production of intrinsic factor, a protein needed to absorb vitamin B_{12}. Decreased levels of B_{12} will lead to pernicious anemia, so named because before the discovery of vitamin B_{12}, it was a fatal disease. With pernicious anemia, the blood has large, red cells, or megaloblasts, also seen in cases of folic acid deficiency.

Another important cause of anemia in the elderly is acute blood loss. Just as gradual occult (unseen) bleeding from a tumor or ulcer can result in iron deficiency, acute bleeding can cause severe symptoms or death. Cancers, gastrointestinal ulcers, colon diverticulae or arteriovenous malformation are common causes of sudden blood loss in the elderly.

Chronic disease is a common cause of anemia in the elderly. For poorly understood reasons, iron metabolism is altered by long-standing disease processes. The result is an anemia that often resembles iron deficiency but will not respond to iron therapy. The diseases commonly associated with this form of anemia are frequently found in the elderly. They include arthritis, both rheumatoid and osteoarthritis, chronic renal failure, cirrhosis or chronic hepatitis, and hypothyroidism.

Drug-induced hematologic problems occur as frequently as "natural" blood diseases in the elderly population. This is because people take increased numbers of medications as they age. Many elderly have arthritis; most appropriate medications for this condition can be associated with low white cell or platelet counts and anemia.

Malignancies of the bone marrow are more common in the elderly and are often associated with anemia. Some cancers such as leukemia or multiple myeloma originate in the marrow; other cancers metastasize there. Breast and prostate cancers are examples. Bone marrow malignancies (dysplastic states) can result in too few or too many of the marrow elements: red cells, white cells, or platelets. Dysplastic states often evolve into leukemia.

Diseases and Conditions Associated with Age

Lymphoma. Lymphoma is one of two broad categories of lymphocytic cancer. The other, lymphatic leukemia, will be discussed later. Lymphomas are composed of solid masses of lymphocytes or monocytes (histocytes) usually arising in the lymph nodes or spleen (Plate 13, *B*, and Fig. 10-9). These cancers are more common in the

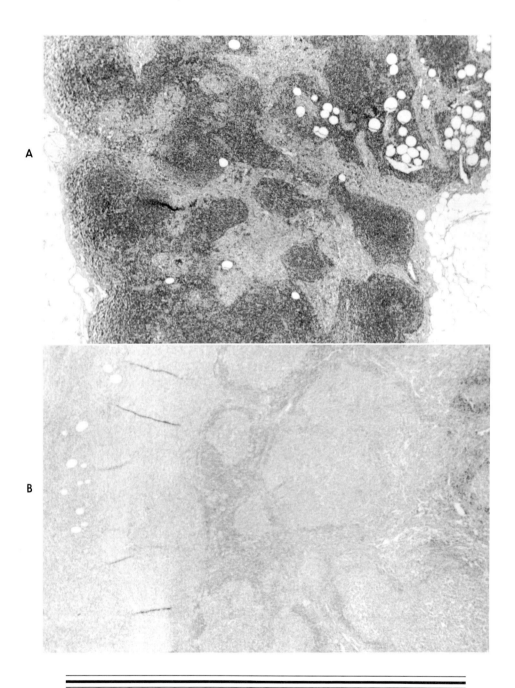

Figure 10-9 A normal lymph node **(A)**, and one swollen with chronic lymphatic leukemia **(B).** The malignant cells are T lymphocytes.

elderly than in the young. They can be slow growing, relatively benign malignancies, or may quickly lead to death.

Multiple Myeloma. A distinct form of lymphoma, multiple myeloma is a cancer that first appears at age 40 to 50. Its presence is often first noted as back pain or pathologic fracture. It is a malignant disease of plasma cells that normally function to produce antibodies against infection. The excessive growth of these cells may destroy and replace bone, thereby weakening the bone and producing the pain that is often the presenting symptom.

Multiple myeloma characteristically produces multiple areas of bone destruction. The bones most frequently affected are the vertebrae, ribs, bones of the cranium, pelvis, femur, clavicle, and scapula. The process is often insidious, with tumor cells present 10 to 20 years before diagnosis. The malignant plasma cells normally produce abnormal gamma globulins, or para-proteins; these are often present in the blood and urine. Because of the frequent use of blood tests, many instances of abnormal para-proteins are found without any evidence of myeloma. In time a significant portion of these patients will develop myeloma. The usual cause of death from this disease is severe anemia resulting from the replacement of the blood-forming bone marrow by malignant cells. Sometimes death results when the tumor spreads to other organs. Another disease is frequently the terminal event because of the inability of the immune system to launch an adequate defense.

Leukemias. Typically, lymphatic leukemia has been a disease of both the young and the old, sparing middle-aged people. The young tend to have acute lymphatic leukemia, an aggressive but potentially curable disease; the elderly tend to have an incurable, more slowly progressing type, chronic lymphocytic leukemia (CLL). There are several forms of leukemia. The following types occur in the elderly with some degree of frequency.

Chronic Lymphocytic Leukemia. CLL is a malignant proliferation of lymphocytes. The cells appear in the blood in various stages of maturation, although with many more mature than immature cells. Often the disease is diagnosed when a routine examination of an apparently healthy individual reveals a high white cell count. Most of the white cells are mature lymphocytes but they may be fragile and rupture during laboratory preparation of blood smears. When broken on a peripheral smear, these cells are called smudge or basket cells. The bone marrow, lymph nodes, liver, and spleen can all be infiltrated by these cells. Figure 10-10 shows gross and microscopic views of a spleen affected by malignant lymphocytes.) Chronic lymphocytic leukemia often progresses very slowly. Patients may survive for years without needing treatment.

As the disease progresses, the lymphocytes crowd out the other blood-forming elements in the bone marrow. At this time patients notice fatigue from the anemia, or unusual bleeding as a result of the low platelet count. Increased susceptibility to infection results from the combination of decreased granulocytes and the altered immune response. Death often results when chronic leukemia is transformed into acute lymphatic leukemia.

Myeloid Leukemia. This condition results from the malignant production of granulocytes and their precursors. In the chronic form of this disease, the marrow and circulation become flooded with nearly mature neutrophils. These cells eventually infiltrate the spleen, liver, and various other organs. While the mature cells look relatively normal, they have multiple defects that make it difficult for them to fight infections. One of the characteristics of these defective cells is a lack of an enzyme common in most granulocytes. Most elderly patients with chronic myeloid leukemia survive about 4 years

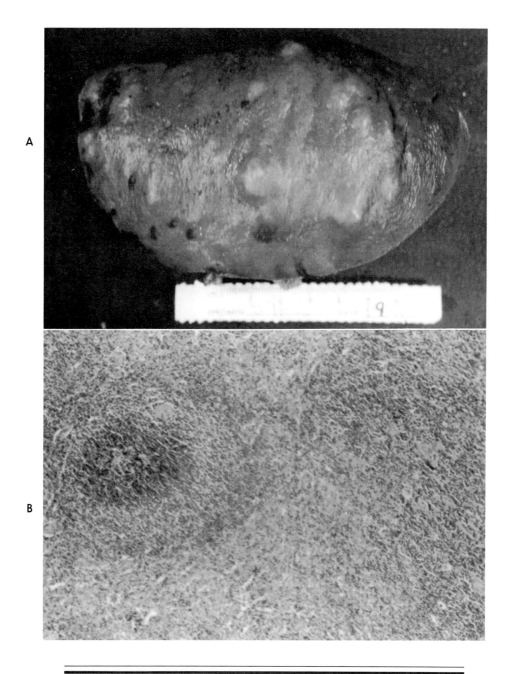

Figure 10-10   Gross **(A)** and microscopic **(B)** views of spleen, showing infiltration by malignant lymphocytes.

and die when the disease evolves into the acute form. The acute disease generally kills in 6 months. Both the acute and chronic form of myelogenic leukemia occur in the elderly, but the chronic form is more common. The median age for chronic myelogenous leukemia is 52 years.

Myeloproliferative Disorders. These include malignant or premalignant conditions arising in the bone marrow. They include the chronic leukemias just discussed, polycythemia rubra vera, and several difficult-to-classify processes seen in the aged. Among these are refractory anemia with blasts (preleukemia), essential thrombocytosis, myelofibrosis, and various cytopenias. In general, all of these culminate in acute leukemia. However, there can be a 10 to 15-year hiatus between the diagnosis of the myeloproliferative process and the manifestation of acute leukemia.

Myelofibrosis is an unusual process that occurs in people over 50 years old. In cases involving this disease, the bone marrow is slowly converted into fibrous tissue, forcing out the blood-forming elements. The result is a decrease in all cells made in the marrow. The cause of the disease is unknown but paradoxically, the first symptom can be a markedly elevated platelet count (over 1,000,000 platelets per mm). A compensatory mechanism may occur in which the bone marrow elements start producing blood in the spleen and liver, causing enlargement of those organs. Death is usually related to pancytopenia or may be an acute leukemia.

Immune Deficiency. Immune deficiency probably develops to a degree in all aging people. As discussed in nearly every system in this text, many diseases appear in the elderly partially as a result of the body's decreased ability to combat them. Common examples of these are the reappearance of the chicken pox virus (Herpes zoster) as shingles later in life and the decreased ability to suppress cancer as age advances. That the immune system is closely tied to cancer resistance is illustrated by the recent emergence of acquired immune deficiency syndrome (AIDS). In this disease, the master of the immune system, the helper T cell, is destroyed by a virus. The patient generally dies of opportunistic infections unusual in the healthy and often with cancer. A rare cancer, historically found only in the aged, Kaposi's sarcoma, has become a likely event in the relatively young patient whose immune system has been damaged by AIDS virus.

Acquired Immunodeficiency Syndrome (AIDS). AIDS is a disease most often acquired during ages 15 to 40. The disease has a tendency to remain dormant for as much as 15 years and possibly even longer in some individuals. Because it may take 5 years or more for the disease to kill after symptoms appear, it can be expected to become a disease of the elderly. The final stages of the disease are extremely debilitating and often require placement in a nursing home or hospital.

AIDS is acquired by exposing the blood of an uninfected person to body fluids containing the human immunodeficiency virus (HIV). Infection usually occurs through sexual contact, reuse of needles (common among intravenous drug users), or transfusion with contaminated blood. The virus enters T helper lymphocytes and causes the nuclei of these cells to produce more of the virus, eventually killing the cells. Infected cells are unable to fight disease, and as the disease progresses, the helper T cell population declines. The infected individual becomes acutely ill from infections that would normally be relatively harmless or would not cause an infection at all. Kaposi's sarcoma, a cancer appearing as purplish lesions on the skin, is common among AIDS patients and virtually unknown among people with normal immune systems. While some progress is being made in slowing the rate at which new T cells become infected, and in prolonging the life of infected people, the disease is still universally fatal.

CASE HISTORIES

❖ Multiple Myeloma

A.R. was a 63-year-old man who returned to his physician because of recurring back pain. He had been treated and even hospitalized for back pain many times in the past 20 years. In the past the pain had always responded to bed rest and antiinflammatory agents. The last episode occurred 2 months prior to this visit to his doctor and was diagnosed as flank pain secondary to a urinary tract infection (UTI). The pain responded to antibiotic therapy. His medical history also included an appendectomy at age 16, gastric surgery for ulcer disease at age 43, and a partial thyroidectomy for removal of a benign thyroid tumor 1 year ago. He is concerned that his flank pain has returned. It changes with body movement but does not worsen with laughing, coughing, or sneezing. He has no urinary problems.

Physical examination found nothing abnormal. The x-ray examination revealed compression fracture and lucency (lack of normal density) of the eleventh thoracic vertebra, and osteoporosis with mild degenerative changes in the spine. The kidneys were large and faintly visible with an intravenous pyelogram (IVP). Laboratory tests showed a hemoglobin of 8.9 (normal > 14) with a normal mean corpuscular volume (MCV). Serum iron was normal, as were calcium and alkaline phosphatase. BUN was elevated at 68 (normal to 22) as were creatinine (9.1; normal to 1.5) and uric acid (12.0; normal to 6.0). A urine sample contained 4.8 g of protein (normal is 0), which electrophoresis showed to be Bence Jones protein (light chain portion of immunoglobulin). The Bence Jones proteins were only faintly present in the blood but massively present in the urine. Bone marrow examination showed 50% of the cells to be immature plasma cells. A diagnosis of multiple myeloma was made. Kidney biopsy revealed the renal tubules to be blocked by large amounts of protein, suggesting myeloma nephropathy.

A.R. was treated for the myeloma and put on renal dialysis. In addition, plasmapheresis was started in hopes of reversing the renal tubule blockages. The tumor was brought under control but the renal function continued to deteriorate. A.R. died after 4 years of renal dialysis with the myeloma still under control.

Comment. This case is interesting in many respects. Multiple myeloma shows an abnormal immunoglobulin protein in the serum in 90% of the cases. This tumor was producing only the light chains of the immunoglobulin and these readily passed through the kidneys into the urine because of the protein's small size. Because large amounts of these proteins were present in the urine at the time of the dehydrating IVP, they precipitated and blocked the nephrons. This test is now done only when the patient is overhydrated. Since the malignancy was controlled, the complicating renal failure became the major problem for A.R.

❖ Autoimmune Hemolytic Anemia

R.A. is a healthy 58-year-old man retired from the U.S. Air Force who noticed that the whites of his eyes were discolored. Fearing liver problems, he went to his physician even though he felt fine. His medical history included moderate hypertension for which he was taking α-methyl dopa.

When examined he was found to be in excellent physical condition. Mild jaundice was noted in the sclera of the eye but it was not visible in his skin. Jaundice could not be detected even when the skin was blanched with a finger or a glass slide. His spleen was barely palpable (very slightly enlarged). Evaluation of his blood showed anemia with numerous red cell fragments and spherocytes (abnormal spherical erythrocytes instead of the normal biconcave shape). The LDH was elevated, as was the bilirubin at 4.8 mg/dl (normal < 1.5). Most of the bilirubin was indirect, the non water-soluble type which indicated these molecules had not yet passed through the liver. A direct and indirect Coombs test conducted to find antibodies attached to the red cells or free in the serum was positive for red cell protein attachment. The remaining laboratory tests were normal.

A diagnosis of drug-induced, autoimmune, hemolytic anemia was made. The α-methyl dopa was discontinued. Within 3 months all of the laboratory findings reverted to normal and the jaundice had disappeared.

Comment. The elderly take many medications and run the risk of drug-induced diseases. One of the common drug-induced problems is alpha methyl dopa hemolytic anemia. In 5% to 10% of all patients taking the drug, an IgG class antibody is produced that binds to the red cells. In some cases the red cells are destroyed by the antibody, resulting in anemia and its consequences. Fortunately, the antibody almost always disappears after the drug is discontinued.

❖ Iron Deficiency Anemia

E.K. is an 81-year-old widow who has been "all tired out" for the last few months. The onset of her illness was poorly defined and aside from a lack of energy, she has had no other complaints. She reported being easily winded when climbing the stairs or walking for any prolonged distance. She has been on replacement thyroid therapy for the last 15 years without any change in dosage. She told her doctor she had experienced no recent weight gain or loss, but she has been eating poorly since she lost her husband 8 years ago. Lately her diet has deteriorated to tea and toast.

E.K.'s physical examination revealed her to be a thin, frail-appearing female who looked to be her stated age. Her skin was pale, as were the mucosal membranes of her eyelids and fingernail beds. (A pale appearance is hard to see on the sun-exposed portion of the skin but is easily seen in these mucosal areas because they lack natural pigmentation.) No jaundice was seen; this suggested that any anemia would not be due to rapid destruction of red blood cells. Bilirubin, the breakdown product of hemoglobin, would probably produce a yellowish green coloration if rapid hemolysis of blood had been occurring. Her thyroid was irregularly nodular and enlarged, consistent with autoimmune thyroid damage.

Laboratory evaluation produced evidence of severe anemia. Her hemoglobin was 4.2 (normal 12 to 16 and her mean corpuscular volume (MCV) was 62 (normal 82 to 96). Anemias with decreased MCVs are those with small erythrocytes such as are produced with iron deficiency, congenital decreased hemoglobin production (thalassemias), and fragmentation anemia. The platelets were elevated at 650,000 and the reticulocyte count was 0.9% (normal 1.5%). Reticulocytes are slightly immature erythrocytes. Three checks of stools for occult blood were negative. Tests for folate and B_{12} were normal but the serum iron was 10 (normal >50) and the TIBC was 524 (normal < 490). A diagnosis of iron deficiency anemia was made. She was treated with oral iron and intramuscular iron injection. Within 10 days her reticulocyte count was up to 22% and within 4 months her hemoglobin was 11.2 g/dl.

Comment. This is a classical case of iron deficiency anemia. The most common nutritional defect among the elderly is lack of iron. Due to a decreased expenditure of energy, some of the elderly decrease their food intake. They lack the interest to vary their diet. When this occurs, a tea and toast diet may result. In all anemias, the hemoglobin value is low but in situations such as this, the presenting hemoglobin may be dramatically low. This is because the patients are inactive and the loss of hemoglobin is slow and steady. They are still able to function despite the decreased oxygen-carrying capacity of their blood.

❖ Refractory Anemia

R.V. is a 78-year-old retired farmer who visited his physician complaining of tiredness. He had noticed a lack of energy about 2 months earlier but believed that it was caused by lack of sleep. Despite having returned to a normal sleeping pattern, he remained tired. His examination revealed anemia. He was referred to a hematologist. The hematologist found that R.V. had been bothered by bad pruritis after bathing. His diet was adequate and balanced, and he had not noticed bloody stools that would have indicated bleeding into the intestinal tract. The physical examination was unremarkable and showed no lymphadenopathy (enlarged lymph nodes). The CBC revealed mild anemia with 10.8 g/dl of hemoglobin, an elevated MCV of 108, and normal platelets and white cells. Examination of the peripheral blood smear confirmed macrocytic anemia without any other special features. Serum iron, B_{12} and folic acid were all within normal limits (WNL). R.V.'s chemistry profile and urinalysis were similarly normal. A bone marrow biopsy and aspiration showed no evidence of tumor, but revealed a megaloblastic change and slightly increased cellularity. R.V. was treated with multivitamins and his progress was followed; no specific disease could be diagnosed.

During the next 5 years R.V. had numerous reexaminations and therapies including stomach and small bowel x-rays, a Schilling test for vitamin B_{12} malabsorption, upper GI endoscopy, repeat bone marrows, and trials of pyridoxine, B_{12} injections, oral folate, and iron. All the examinations were normal and there was no response to any of the therapies. During the same time period, the anemia worsened and he required periodic transfusions. At the age of 84 R.V. began to notice bruising and the CBC revealed decreased platelets and increased WBCs with immature forms. The bone marrow aspirate now was compatible with acute myelomonocytic leukemia. R.V. did not respond to therapy and died after a few months.

Comment. This case provides a good example of what used to be called preleukemia and is now known as one of the myeloproliferative (or myelodysplastic) disorders, or refractory anemia without excessive blasts. These conditions are characterized by a lack of some bone marrow element, often red cells, that is not a typical anemia and does not respond to medication. The anemia is usually megaloblastic (large red cells) and the marrow is megaloblastoid (producing large red cells). A high proportion of these cases evolve into malignancies. No current therapy has slowed or altered this progression, but because of the slow disease evolution, elderly patients often die of other causes before developing the terminal bone marrow pattern.

❖ Pernicious Anemia

See the pernicious anemia case history presented with the digestive system on p. 134.

❖ Chronic Lymphatic Leukemia

D.G. is a 68-year-old man who was found to have an elevated white blood cell count. Aside from this abnormality and a mild hypertension controlled by diuretics, he has been healthy. There was no evidence of infection, but the WBC was 17,300 with 68% lymphocytes, some of which were atypical. The remaining laboratory studies showed no abnormalities. The physical examination was normal, with specific attention being given to a lack of lymph node and spleen enlargement (splenomegaly). Enlargement of these organs could indicate lymphoma. A review of a peripheral blood smear by a pathologist revealed increased numbers of mature-appearing lymphocytes with many smudge cells and an occasional prolymphocyte. The suggested diagnosis was chronic lymphatic leukemia (CLL). A bone marrow examination revealed mild hypercellularity with a diffuse increase in lymphocytes suggestive of CLL.

D.G. was followed medically without symptoms or therapy for 4 years. During this time his WBC remained elevated with no significant change in the total cell count or distribution. At the age of 72, D.G. entered a residential care facility because of failing memory and difficulty in caring for himself. At age 73, his WBC began to increase, with increased numbers of immature cells. When hospitalized for an episode of pneumonia he was found to have many small but enlarged lymph nodes and slight splenomegaly. His bone marrow was hypercellular with many immature lymphocytes. A chemistry profile revealed increased LDH and uric acid, waste products found in many rapidly dividing cells. D.G. was treated for acute lymphatic leukemia and went into remission. After 7 months of good health, D.G.'s acute leukemic process reappeared. It did not respond to therapy, and he died.

Comment. Chronic lymphatic leukemia and the lymph node counterpart, well-differentiated lymphocytic lymphoma, are indolent diseases that are often left untreated. Given enough time they almost always evolve into acute leukemia. In such cases there is inadequate production of red or white blood cells or platelets. In any of the bone marrow malignancies there is a large turnover of cells and an increase in cellular waste products. These can be measured in the blood, as in this case were LDH and uric acid. The latter is a product of nuclear metabolism.

❖ Acute Leukemia in an Adult

R.G. was a 71-year-old baker still practicing his trade when he observed that he was bruising easily. Minor bumps would result in large ecchymoses (diffused accumulation of blood outside the vessels in the skin). His wife was concerned that he appeared pale and tired easily. She convinced him to see his doctor.

When examined, he was found to be pale with numerous old and recent hemorrhages in his skin. His lymph nodes and spleen were diffusely enlarged. Laboratory evaluation revealed a markedly low platelet count of 12,000 (normal over 100,000), mild anemia with a hemoglobin of 9.8 g (normal over 14) and a markedly increased WBC 97,000 (normal to 11,000). In addition, his urine contained protein and blood. His stool was positive for occult blood. R.G.'s chemistry profile was abnormal for high uric acid, LDH, and serum glutamate oxalic transaminase (SGOT). Because platelet counts below 20,000/mm³ can result in fatal hemorrhages following minor trauma, he was hospitalized for platelet transfusion, further tests, and therapy.

A bone marrow examination found hypercellular marrow with 80% blasts. These cells and the increased cells in the peripheral blood were too immature to identify using the usual Wright-Giemsa stain, but a special stain revealed them to be myeloblasts, granulocyte precursors. A diagnosis of acute myelogenous leukemia was made. R.G. was started on chemotherapy.

On his third hospital day, the day of his first chemotherapy treatment, he developed a fever and cough. Chest x-rays showed fluffy infiltrates in both lungs, suggesting pneumonia; blood cultures showed *Pseudomonas* organisms. Despite intensive antibiotic therapy, R.G. died on his seventh hospital day.

Comment. This is one of the two common leukemias found in adults. The other is the more slowly progressing, chronic lymphatic leukemia. Acute myelocytic leukemia has been resistant to therapy, with only 20% of patients surviving for 30 months. With younger, healthier patients, complete cures can be effected in some by complete destruction of all the hemopoietic tissue and replacement by bone marrow transplant.

The major complications result from lack of blood elements. Decreased platelet counts can result in hemorrhages, especially in the brain. The abnormal white cells are nonfunctional; many patients die from infection because of lack of antimicrobial action. As a rapidly proliferating cell line, the leukemic cells liberate lots of nucleic acids, elevating the uric acid level in the blood. The cytoplasm from the dying cells releases large numbers of other enzymes such as SGOT and LDH.

❖ Acquired Immunodeficiency

W.W. was a 78-year-old father of a nurse. He was referred to a physician by his daughter when he developed a persistent low-grade fever, cough, and shortness of breath (SOB). His symptoms had developed gradually over a 2-week period and did not respond to oral antibiotics. Outpatient posterior-anterior and lateral x-rays (PA and L) revealed bilateral scattered fluffy infiltrates in all lung fields, diagnostic of pneumonia. His past medical history included a large-cell, diffuse lymphoma requiring extensive chemotherapy with many blood transfusions 3 years earlier. In addition W.W. complained of a sore mouth and slight difficulty in swallowing.

On examination he was noted to be febrile, with a rapid pulse and respiratory rate. The head, eyes, ears, nose, and throat (HEENT) were positive for a patchy, thick, red and white membrane covering the throat and tongue. His heart sounds were within normal limits (WNL) without murmur. His chest was clear to percussion with normal movements of the diaphragm, but scattered fine rales could be heard bilaterally. His abdomen, extremities, and neurologic examinations were negative with the exception of mild confusion. His skin was unremarkable. Palpable lymph nodes were noted in the axilla and groin. The nodes were twice normal size, soft, and freely moveable. This is unlike the enlarged nodes of lymphoma, which are often hard and fixed to the surrounding tissue.

W.W. was started on nasal oxygen and IV antibiotics (tobramycin and a second-generation-cephalosporin). A Gram's stain of the throat membrane demonstrated numerous fungi, and a culture grew *Candida albicans.* A biopsy of the largest lymph node from the right axilla was performed. It showed no evidence of lymphoma but did show benign hyperplasia with prominent plasma cells, a situation compatible with HIV infection. Bronchoscopy produced smears positive for *Pneumocystis carinii.* An HIV antibody screen was positive, as was the confirmatory test, the Western blot. The diagnosis of acquired immunodeficiency syndrome was made. The patient was treated with appropriate antibiotics and recovered fully from pneumonitis. Unfortunately, his confusion worsened and he became delusional. These neurologic problems were believed to be due to AIDS encephalopathy. He was placed in a long-term care facility.

Further history revealed no evidence of IV drug abuse, homosexuality, extramarital affairs, or sexual intercourse with anyone except his wife, who was negative when screened for HIV. Numerous units of blood and blood products were used during the treatment for lymphoma in 1985. All of these units were HIV negative and none of the donors had manifested AIDS or were HIV positive. However, one of the donors had died in a car accident within months of his blood donation. Further checks of this donor found that he was in a high risk group for HIV infection.

Comment. Human immunodeficiency virus (HIV) is the virus associated with acquired immunodeficiency. Current (1988) estimates suggest that one million people (1% of the U.S. population) are infected with the virus and that many others will become infected. Probably all of those infected will develop the clinical syndrome (AIDS) and many of these will have neurologic symptoms. At this writing AIDS is 100% fatal. The increasing numbers of AIDS victims requiring acute and long-term care threaten to overwhelm our medical system. It is a particularly severe burden in high incidence areas like San Francisco, New York, and Washington D.C.

The disease is manifested when the virus destroys the helper T lymphocytes. This cell is now considered the coordinator of our entire immune system. When it is no longer available, a complete failure of immunity occurs, leading to life-threatening opportunistic infections and the development of malignancies. Two infections common in AIDS patients and rare in immunologically competent people are *Pneumocystis* pneumonia and chronic cryptosporidiosis diarrhea. A common malignant tumor in AIDS patients is Kaposi's sarcoma; it too is rare in the non-AIDS population.

BIBLIOGRAPHY

Banerjie AK: *The hemopoietic system.* In Pathy MJS, ed: *Principles and practice of geriatric medicine,* New York, 1985, John Wiley & Sons.

Bloom ET and others: *Immunity and aging.* In Pathy MJS, ed: *Principles and practice of geriatric medicine,* New York, 1985, John Wiley & Sons.

Claman HN: *Immunologic changes in aging.* In Schrier WS, ed: *Clinical internal medicine in the aged,* Philadelphia, 1985, WB Saunders.

Haynes BF, Fauci AS: *Clinical immunology.* In Braunweld, and others, eds: *Harrison's principles of internal medicine,* ed 11, New York, 1987, McGraw-Hill.

Hyams DE: *The blood.* In Brocklehurst JC, ed: *Textbook of geriatric medicine and gerontology,* New York, 1985, Churchill Livingstone.

Lipschitz DA: *Anemia.* In Exton-Smith AN, Weksler ME, eds: *Practical geriatric medicine,* New York, 1985, Churchill Livingstone.

Robbins SL, Angell M: *Basic pathology,* Philadelphia, 1971, WB Saunders.

Stites DP and others: *Basic and clinical immunology,* ed 5, New York, 1984, Lang.

World Health Organization: Anemias, 1922-1976 mortality statistics, *WHO Statistical Rep* 22:409-427, 1969.

11 *The Reproductive System*

RELEVANT ANATOMY AND PHYSIOLOGY
General Functions

The primary function of the reproductive system is the production of eggs and sperm. Because procreation requires extensive behavioral, physiologic, and anatomic adaptations, the needs of the reproductive system are reflected in sexual differences in all of the other systems of the body, from integument and bone to the endocrine system. There are many examples of the close relationship between the the central nervous system and the endocrine system regarding reproduction. Most of us are aware that the normal menstrual cycle can be interrupted by worry, stress, or disease, or that visualizing certain sexual phenomena can be stimulating. The aging process affects both the physical and behavioral aspects of reproduction, but both women and men can continue to enjoy both their sexual differences and sexual stimuli throughout their lives.

Male Reproductive Anatomy and The Male Reproductive Cycle

Testicular Function. The male reproductive organs consist of paired testes that usually descend into the scrotum before birth. When mature, they have two functions: the production and maturation of sperm and the production of androgens, the hormones responsible for male secondary sexual characteristics. The path of the sperm can be followed from each testis to the outside of the body (Figure 11-1). On the surface of each testis is an epididymis that stores sperm and transfers them to the vas deferentia. Each vas deferens (also called ductus deferens), conducts sperm from the epididymis through the inguinal canal in the axils (the angles formed by the abdomen and the abducted femur) of each leg, into the abdomen to the single prostate gland (Figure 11-1). The duct of a seminal vesicle empties into each vas deferens shortly before the vas deferens enters the prostate. Both the two seminal vesicles and the prostate gland add alkaline fluid to the sperm, giving the semen most of its volume. The seminal vesicles contribute about 60% of the fluid of the semen; the prostate contributes only about 30%. The vas deferens joins the urethra within the prostate. The two ducts of the Cowper's or bulbourethral glands enter farther down the urethra. These glands provide lubrication during intercourse and may contribute up to 10% of the fluid of semen. The urethra conducts the semen and Cowper's gland secretions to the end of the penis and outside of the body. Although the quantity of sperm produced declines with advanced age (more abnormal sperm are produced), all of these organs usually retain their function throughout life.

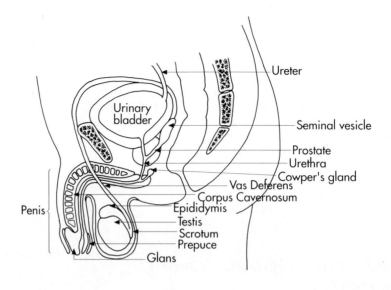

Figure 11-1 The male reproductive system.

The Penis and Erection. The penis must become erect and rigid to gain passage into the vagina. This is accomplished through sexual stimulation, causing increased activity of the parasympathetic nervous system. The parasympathetic and sympathetic nervous systems are discussed in Chapter 13. Arteries leading into the spongy vascular tissue of the penis dilate, filling the spongy tissue with blood. The increased pressure of this expanded spongy tissue compresses veins that would normally drain the tissue, thus slowing the loss of blood from the penis. The resulting further engorgement of the spongy tissue causes full erection. Urination is impossible during ejaculation because of compression of the urethral sphincters. Further, the bladder neck sphincter is closed by the sympathetic nervous stimulation that causes ejaculation. Impotence, or failure to achieve full erection, can occur in mature men of any age as a result of a variety of psychologic and physical factors. It is possible to achieve ejaculation without full erection, however, and tactile stimulation of the genital region is pleasurable whether or not ejaculation occurs. Therefore impotence does not necessarily imply that a man is unable to experience sexual pleasure.

The Male Hormonal System. Puberty begins as the pituitary gland becomes less sensitive to the negative feedback of the sex hormones, (In men, androgens such as testosterone; in women, estrogens and progesterone circulating in the blood stream, "feed back" to suppress pituitary gonadotropic hormones). This decreased sensitivity allows the anterior pituitary gland to produce more gonadotropins (luteinizing hormone [LH] and follicle-stimulating hormone [FSH]). As adolescent males mature, the influence of rising levels of LH on the interstitial cells of Leydig in the testes causes testosterone levels to increase (Figure 11-2 and Color Plate 14, *B*). This process continues until testosterone levels become high enough to reduce production of LH-releasing factor from the hypothalamus. This interaction between LH-releasing factor and testosterone exists throughout life and regulates the blood testosterone levels.

Testosterone and other androgens are responsible for the secondary sexual char-

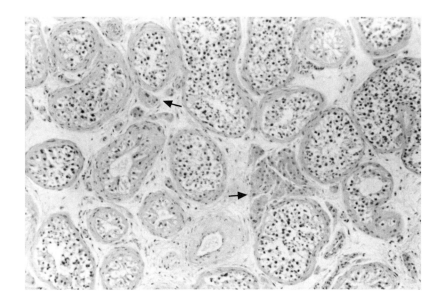

Figure 11-2 Sperm are produced in the seminiferous tubules by mitosis of spermatogonia on the outside and two meiotic divisions starting with primary spermatocytes closer to the middle. Male hormones are produced by the interstitial cells of Leydig between the seminiferous tubules (arrows).

acteristics seen in males and the maturation of sperm. They cause deepening of the voice, increased rates of muscle formation, abdominal distribution of fat, growth of facial and other body hair, growth of the penis, libido, and baldness in genetically susceptible men. Androgens also cause increased aggressiveness, such as is seen in roosters and bulls. The amount of aggressiveness or the way it may be expressed is determined in part by the subject's behavioral environment and perhaps in part by his genetics. Androgens and their estrogen counterpart in females cause increased activity of certain cells by anchoring on receptors on the cells. The resulting increased growth of the vocal cords for example, causes deepening of the voice. This will happen with people of either sex when exposed to large dosages of androgens.

Sperm Production. Spermatogenesis takes place under the influence of FSH and testosterone on the seminiferous tubules. This process follows mitotic division of spermatogonia into primary spermatocytes. The primary spermatocytes undergo the first meiotic division to form secondary spermatocytes; these produce spermatids as a result of the second meiotic division (Figure 11-3, *A*). Meiosis involves two consecutive cellular divisions in which the four cells produced have one-half the number of chromosomes as the original parent cell. (Mitosis occurs when the two cells produced by a cell division are genetically identical to the parent.) The spermatids therefore mature into sperm with half the number of chromosomes present in a normal body cell. The most mature spermatogenic elements are located toward the center of each seminiferous tubule, near its lumen; the larger spermatogonia and primary spermatocytes are located toward the outside of the tubule.

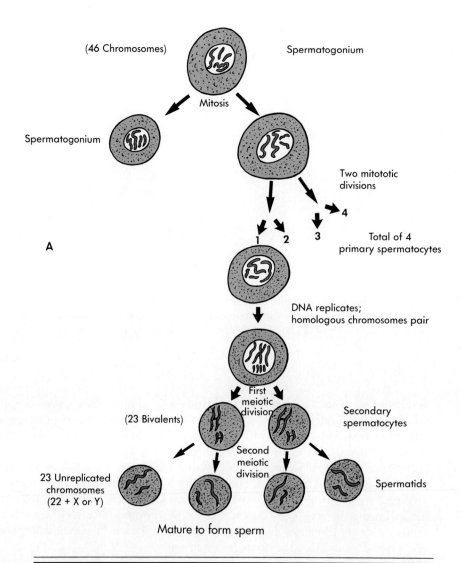

(46 Chromosomes)

Spermatogonium

Mitosis

Spermatogonium

Two mitototic
divisions

4

1   2   3   Total of 4
primary spermatocytes

A

DNA replicates;
homologous chromosomes pair

First
meiotic
division

(23 Bivalents)

Secondary
spermatocytes

Second
meiotic
division

23 Unreplicated
chromosomes
(22 + X or Y)

Spermatids

Mature to form sperm

Figure 11-3   Spermatogonia and oogonia divide by mitosis to form primary spermatocytes or primary oocytes. These divide by meiosis I to form secondary spermatocytes or secondary oocytes. These divide by meiosis II to form spermatids or ootids, which mature to form sperm or ova. Cytoplasm is conserved in a single secondary oocyte, the other becoming a polar body that is eventually destroyed. Likewise, only a single ootid from each division receives significant cytoplasm, the other becoming a polar body. Thus a single primary spermatocyte produces four sperm but a single oocyte produces only one ovum. **A** Shows spermatogenesis; **B** shows oogenesis.

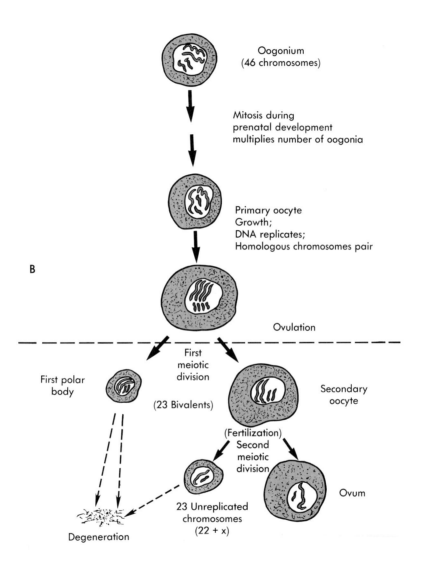

Oogonium
(46 chromosomes)

Mitosis during
prenatal development
multiplies number of oogonia

Primary oocyte
Growth;
DNA replicates;
Homologous chromosomes pair

B

Ovulation

First
meiotic
division

First polar
body

(23 Bivalents)

Secondary
oocyte

(Fertilization)
Second
meiotic
division

23 Unreplicated
chromosomes
(22 + x)

Ovum

Degeneration

Female Reproductive Anatomy and Reproductive Cycles

Functional Anatomy. The female reproductive organs consist of a pair of ovaries that rest in the abdominal cavity, each at the end of an oviduct (also called a Fallopian tube or uterine tube; see Figure 11-4, *A, B*). The ova, or eggs, are produced by the ovaries and are contained in fluid-filled follicles (primordial follicles). Ova are produced by meiotic division just as occurs with sperm production in the testes (see Figure 11-3, *B*). A major difference is that in ovum formation, or oogenesis, one of the two products of each meiotic division degenerates as a polar body; the other retains all the cytoplasm of the dividing cell. The two meiotic divisions produce only a single ovum in oogenesis, but four spermatozoa in spermatogenesis. A second difference is that all the primary oocytes, the cells that divide by meiosis to ultimately form ova, are formed during the intrauterine development of female babies. They do not complete meiosis until the child reaches sexual maturity, many years later. They then complete the process as the cells are ovulated at the rate of about one each month until the process ceases at menopause. In males, production of primary spermatocytes, the cells that divide by meiosis to ultimately produce sperm, continues throughout life, producing millions of sperm per month, although sperm production, like the production of ova, does not begin until puberty.

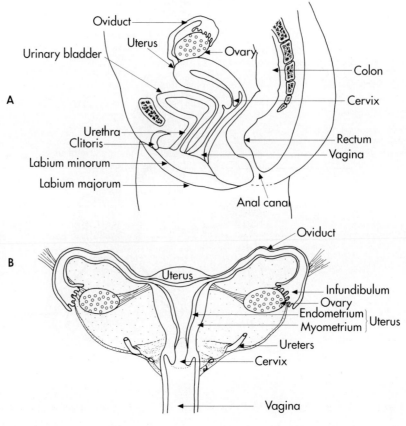

Figure 11-4   The female reproductive system **(A)**. The path of ovulation **(B)**. Ova are ovulated from the ovary, pass through the infundibulum into the fallopian tube and on into the uterus. During menstruation, the secretory endometrium of the uterus is shed through the cervix into the vagina and out of the body.

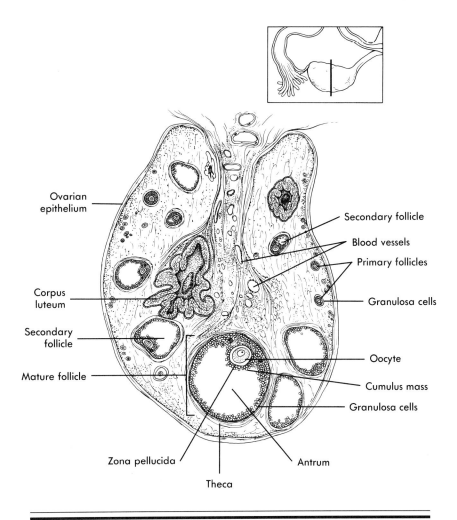

Ovarian epithelium

Secondary follicle

Blood vessels

Primary follicles

Corpus luteum

Granulosa cells

Secondary follicle

Oocyte

Mature follicle

Cumulus mass

Granulosa cells

Zona pellucida

Antrum

Theca

Figure 11-5   Estrogens are produced by the follicular epithelium surrounding the ovarian follicle. After ovulation, the follicular epithelial cells organize to form a corpus luteum and produce progesterone as well as estrogens.

Estrogenic hormones are produced by the cells around the egg, or follicular epithelium, as the follicles grow (Figure 11-5). After an ovum is expelled from the ovary (ovulated), the remaining follicular epithelium forms the corpus luteum, a temporary gland that secretes both estrogens and progesterone. The path of an ovum can be traced through the female reproductive tract in Figure 11-4, *B*. The ovum passes from the ovary into the fimbriated end of one of the fallopian tubes. The two fallopian tubes are convoluted structures extending from each ovary to the uterus. The uterus is a centrally located, muscular structure that develops a vascular, secretory, nutritive lining in preparation for the implantation of a fertilized ovum. It functions as a direct source of nutrition for the embryo and later becomes a part of the placenta. Entrance to the uterus from the vagina is restricted by the cervix, or neck, at the base of the uterus.

If an ovum is not fertilized, menstrual fluid consisting of the shed uterine lining and associated blood passes through the cervix into the vagina (the tube between the

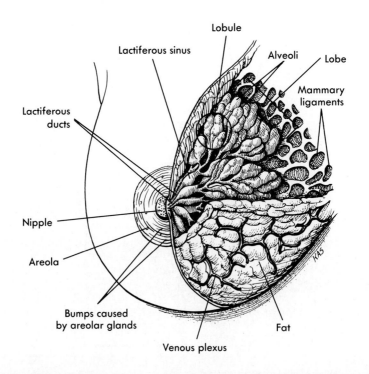

Lobule

Lactiferous sinus

Alveoli

Lobe

Mammary ligaments

Lactiferous ducts

Nipple

Areola

Bumps caused by areolar glands

Venous plexus

Fat

Figure 11-6 The mature female breast consists of glandular alveoli that drain into branches of ducts. Before exiting through the nipple, these ducts enlarge to allow for storage of milk. Most of the volume of the breast is fat. The stretching and loss of supporting ligaments and other tissue cause the breasts to lose their firmness with age.

cervix of the uterus and the vestibule) it passes to the outside of the body through the vestibule (Figure 11-4, *B*). The vestibule is a shallow cavity surrounded by the labia minora; it also receives the urethra, carrying urine from the urinary bladder. Surrounding the labia minora are two large fleshy folds, the labia majora.

The linings of the uterus, cervix, and vagina are composed of cells that divide rapidly under the influence of estrogenic hormones. Tissues with rapidly dividing cells tend to be more prone to cancer than nonproliferating or slowly proliferating cells.

The mammary glands are a product of embryonic skin, actually modified sweat glands, but are included here rather than with skin because of their response to female reproductive hormones. They provide nutrition for children for a varying period after birth. The breasts attain full development in mature women. Milk is produced from the glandular alveoli deep within the breast (Figure 11-6). The milk passes through ductules which join to form lactiferous ducts. These carry the milk to small storage structures just under the darkly pigmented areola. The milk leaves the breast through the nipple. The areoli of developed breasts have numerous small bumps that are less-developed mammary glands called areolar glands. They may have a lubricating function during lactation (milk production). The breast is supported by ligaments and is protected by large deposits of fat.

The Female Hormonal Cycle. The female hormonal cycle actually consists of two related cycles, one involving the ovaries and another involving the uterus. The ovaries reach maturity under the influence of rising FSH levels during the early part of the

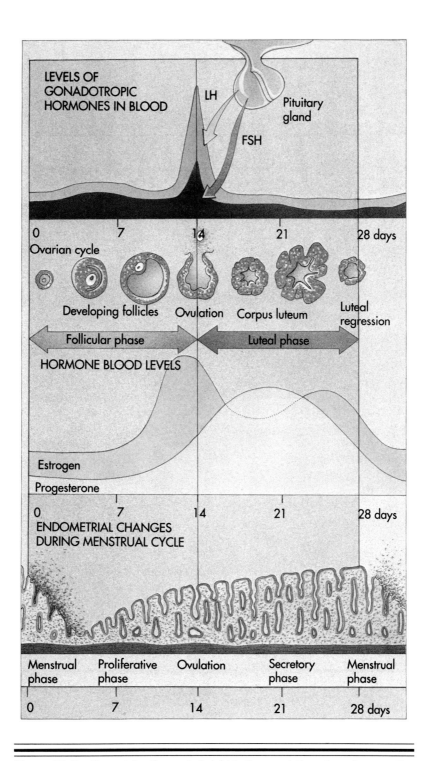

Figure 11-7   Except for the period right before ovulation, there is a negative feedback relationship between follicle stimulating hormone (FSH) and estrogens, and also between luteinizing hormone (LH) and progesterone.

second decade of life. The reproductive cycle begins when estrogens produced by the developing follicular epithelium reach a high enough level to feed back and reduce the levels of FSH-releasing factor produced by the hypothalamus. Figure 11-7 illustrates the hormonal changes associated with a single month of the female reproductive cycle. During part of this time LH levels are rising, and when a proper balance of hormones is attained, ovulation occurs; the ovum is released, and a scar remains on the ovary. A corpus luteum is formed from the ovarian scar under the influence of LH. Progesterone levels rise rapidly as the corpus luteum begins to function. After a few days the progesterone levels begin to feed back against the hypothalamus, reducing the levels of LH-releasing factor. Ultimately the LH production from the pituitary gland declines, the corpus luteum degenerates, and the levels of progesterone and estrogens it produced begin to drop. At the end of the monthly cycle, the levels of progesterone become so low that the lining of the uterus cannot be sustained. It degenerates and hemorrhages, producing the vaginal flow known as menstruation.

The estrogens, primarily estradiol, produce the female secondary sexual characteristics. These include breast growth, deposition of extra subcutaneous fat, the feminine shape of the pelvis, the pubic hair pattern, and the deposition of fat in the upper legs and pelvic region. They are also responsible for some female behavioral characteristics and growth of the clitoris, the structure anterior to the exit of the urethra that responds to tactile stimulation and contributes to sexual excitement.

The male and female breast are identical except for the influence of estrogens and progesterone. It has been demonstrated that the male breast, with appropriate hormone support, can develop to full female size and produce milk. Estrogens cause increased growth of supporting tissue, development of the duct system, and deposition of fat. Progesterone causes growth of the milk-producing tissue of the alveoli of the breast but actual milk production is stimulated by lactogenic hormone from the pituitary gland. Some monthly growth and decline occurs in the breasts of adult premenopausal women in response to monthly hormonal fluctuations, but much of this is due to edema associated with progesterone.

The uterine cycle involves both estrogens and progesterone. These hormones are both low at the beginning of the cycle and during the first 5 days of menstruation. Under the influence of estrogens during the next 11 days of the cycle, the stromal and epithelial cells remaining after the last menstruation rapidly proliferate. This results in substantial thickening of the inner lining, the endometrium. During this time the blood vessels invade the growing tissue and endometrial glands begin to form. On about day 12 of the cycle, progesterone levels increase. The influence of progesterone remains high for the next 12 days. Progesterone is produced first from the preluteal follicular epithelium under the influence of increasing LH, and later from the corpus luteum, formed in part from these cells. Proliferation of the endometrium continues through most of the second half of the cycle. Under the influence of progesterone, the glandular tissue of the endometrium becomes secretory and the endometrium again doubles in thickness. At the end of the cycle both the secretory and most of the nonsecretory endometrium deteriorate and menstruation begins.

THE AGING REPRODUCTIVE SYSTEM
General Comments

Female reproductive systems age to the point of nonfunction before any of the other systems. Such nonfunction, or menopause often occurs during a period of decline over several years, usually between the ages of 45 and 60. Even though some men may remain capable of complete reproductive function until their 80s or later, male reproductive systems also show unmistakable signs of aging. In both males and females, the aging of the reproductive system involves declines in the production of sexual steroid hormones.

The Aging Man

Men do not normally experience a sudden decline in reproductive function comparable to the menopause of women. After age 50 men experience a gradual decrease in testicular mass. There also is an increase in size of the prostate gland, often reducing the ability to empty the urinary bladder (see Chapter 8). Often associated with the decline in testicular mass is a decline in the number of interstitial cells of Leydig. It is also possible that the Leydig cells decline in their ability to respond to luteinizing hormone. Possibly both because of their decline in number and reduced responsiveness, the Leydig cells' production of testosterone seems to decline more or less steadily in men age 45 and older. However, the testosterone decline may be less than expected because of decline in renal clearance of the hormone. Reduced testosterone produces a decline in many of the secondary sexual characteristics, probably contributes to muscle loss, and possibly also contributes to the decrease in aggressiveness that is usually referred to as the "mellowness of age." While there is a decline in rate of growth of facial hair, many hairs become more bristly and shaving continues to be necessary. The vocal cords are reduced in thickness and the voice acquires a thin, raspy quality. The penis decreases in size with age, and the erectile tissue becomes compromised because of loss of elasticity and contractility of the blood vessels.

Sperm production declines after age 50; the motility and viability of the sperm produced decreases. The total number of sperm produced by an average 65-year-old is about 30% less than what is typical for a man of 25. Even so, many men retain enough sperm-producing ability to fertilize an egg even until their 80th year and beyond. A decrease in the diameter of many of the seminiferous tubules is associated with the decrease in sperm production, as is an increase in the fibrous connective tissue of the tubular walls and the interstitial spaces.

The secretions that constitute part of semen also decline with age. Even though the prostate gland increases in size, prostate secretions may decline significantly, along with the contractility of the prostate muscles that force the secretions into the urethra. The seminal vesicles decrease in weight after age 60; their contribution to semen declines as well.

The Aging Woman

Menopause. The aging of the female reproductive system can be separated into menopause (the completion of a comparatively sudden aging) and gradual postmenopausal aging. Menopause is defined as the date 1 year past the last menstrual period. During the reproductive life of a typical woman, about 450 primordial ovarian follicles out of over 500,000 oocytes grow to ovulatory size and are ovulated. The remaining follicles never develop into ova but degenerate. By the age of 45 or 50 only a few remain to be stimulated by FSH secreted by the pituitary gland. This results in progressively less estrogen being produced. The normal fluctuating female cycle eventually ceases. With this, ovulation ceases, no corpora lutea are formed, and FSH and LH levels continue to climb because of lack of estrogen and progesterone feedback. Surges in LH produce responses by the adrenals, releasing adrenal androgens and cortisol. These, combined with the unaccustomed low estrogen levels, may be associated with depression and fatigue, hot flushes with extreme flushing of the skin, sensations of suffocating and dyspnea, irritability, anxiety, and psychic disturbances that sometimes develop into psychoses. Approximately 15% of women require treatment for symptoms associated with menopause. This usually involves estrogen replacement therapy, often in combination with progesterone.

Postmenopause. Postmenopausal age-related changes are associated primarily with near-deprivation levels of estrogen over the remainder of a woman's life. The ovaries atrophy and undergo a fibrotic thickening. Without the stimulation of estrogens and progesterone, the ligaments and ducts of the breasts degenerate and the absence of progesterone causes loss of replacement of secretory cells. This causes the breasts to atrophy and become flaccid and pendulous. The oviducts, uterus, vagina, and external genitalia decrease in size. The lining of the vagina becomes thin and loses much of its elasticity. The vestibular glands and their secretions decline. Lack of lubrication of the vestibule in addition to decline in vaginal secretions may cause intercourse to be painful. The vaginal secretions that do occur do not seem to inhibit bacterial growth as they did in earlier years; vaginal infections become more common. These infections commonly involve yeasts in younger women. In the elderly, bacteria that are normally nonpathogenic in a hormonally supported vagina are usually involved in vaginal infections.

Effects of estrogens on nonreproductive organs become evident after menopause, particularly regarding the effect on bone metabolism. Beginning at this time and lasting for several years, the body undergoes significant calcium loss caused by increased activity of osteoclasts normally suppressed by estrogens. Up to 2.5% of the body's calcium can be lost each year for the first few years after menopause. The rate of loss normally declines after this time. Estrogens also appear to have antiatherosclerosis properties. Before menopause, atherosclerotic heart disease is unusual in women, but after menopause the female rate of myocardial infarction approaches that of the male. The loss of estrogens contributes to the thinning of skin and the depletion of subcutaneous fat.

Estrogens present in postmenopausal women are produced primarily by the adrenals. Produced with them are adrenal androgens that cause development of some male secondary sexual characteristics. The voice may drop in pitch, and facial hair may appear. Pubic and cranial hair becomes more sparse; some hairs become more coarse. More fat may be deposited in the abdomen than would normally occur. Many women complain that fat is more easily acquired in the lower body and upper legs. This may be caused by the overall decline in metabolism resulting from the low level of estrogens.

It is difficult to separate the changes that occur from direct aging from those that are associated exclusively with decline in hormone levels. Most of those presented in the following discussion involve both, but it should be obvious which are primarily hormonal.

Treatment with Estrogen Replacement Therapy. We have mentioned that estrogen replacement therapy (ERT) is often used in treatment of severe symptoms of menopause. Also mentioned is the fact that many postmenopausal women suffer many other symptoms. These include osteoporosis, increased cardiovascular disease, vaginal thinning and irritation (often with loss of sexual pleasure), premature skin wrinkling, sagging of breasts, and even loss of vigor. To combat these, especially osteoporosis, ERT has been prescribed to as many as 24% of the women over 70 in the United States. Because this is not done without risk, the physician must weight the benefits against the hazards. Although ERT is believed to reduce the development of cardiovascular disease, large dosages have been associated with increased blood clotting and embolism. Withdrawal vaginal bleeding may occur and endometrial hyperplasia (excessive growth) is likely. The greatest concern involves the possible increased risk of uterine cancer. For this reason progesterone is often given in the latter part of the cycle. This combination has not been implicated in increasing uterine cancer. Additional concern exists regarding acceleration of undiagnosed breast cancer.

Diseases and Conditions Associated with Advancing Age

Impotence. The inability to produce or sustain erection of the penis for intercourse is one of the most common complaints of aging men. While the condition may be psychologic in origin or have psychologic overtones, it is often caused by loss of neurologic or vascular function. Either the nervous control or the erectile vessels themselves may be impaired. This loss of function may be caused by neuropathy, diabetes mellitus, thyroid disease, trauma, or drugs such as alcohol, sedatives, or antihypertensive agents. This may be corrected by artificial devices that reduce the flow of blood leaving the penis or by surgical implants. The implants are moderately flexible or inflatable devices that provide rigidity.

All men over 70 and most men at an earlier age suffer some loss of sexual function short of impotence. It requires a longer period of stimulation to produce an erection. The penis tends to be less firm when erect. More time is required to build to the climax that causes ejaculation. The amount of semen produced is less than in the earlier years; the intensity of the pleasure generated also is usually reduced. The recovery period following ejaculation increases. All this may contribute to decreased sexual activity. The decrease in sexual activity may, in turn, contribute to further loss of function.

Gynecomastia. Some men develop breast enlargement due to breast gland proliferation. Because this process is similar to female breast enlargement it is known as gynecomastia, or female breast. It is most common in puberty but also can be seen in the aged. In the latter case it is also associated with medication, especially digitalis and antihypertensive agents. It may also occur with estrogen therapy, testicular tumors, cirrhosis, or thyroid disease. Treatment is done only for cosmetic reasons and involves surgical removal of the proliferating tissue.

Adenocarcinoma of the Prostate. This malignancy becomes more common with advancing age. It is the third most common non-skin cancer of men (after lung and colon cancer) and is twice as common in black men as it is in whites. The incidental prostate cancer rate in autopsy studies varies from 15% to 50%, with the higher rates found in studies of older patients.

Prostate cancer usually develops on the dorsal side of the gland and is often palpable from inside the rectum. A rectal check of the prostate gland is a normal part of the physical examination of men. In young men prostate cancer can metastasize rapidly; it is a dangerous disease. In the elderly the malignancy tends to grow more slowly. Because prostate cancer is often somewhat testosterone-dependent, removal of the testes, antiandrogen medication, and estrogen therapy may greatly retard the progress of the disease. It has been said that all men will get prostate cancer if they live long enough.

Benign hypertrophy of the prostate is unrelated to prostate cancer. This condition, extremely common in elderly men, causes difficulty in emptying the urinary bladder. See the discussion of benign hypertrophy in Chapter 8.

Testicular Cancer. Testicular cancer is relatively rare, accounting for only 1% or 2% of all malignant tumors in men. It is a heterogeneous disease with one form prevalent in the 25 to 50 age group and the other type having the highest incidence in 50 to 70-year-olds. The latter tumors (seminomas) are sensitive to both radiation and chemotherapy, making them a treatable disease. The 5-year survival rate for seminomas is 90%. Most testicular tumors present either as a palpable mass or as testicular pain. The tumors often have metastasized at the time of diagnosis, making early discovery especially important.

Atrophic Vaginitis. This condition results from a lack of vaginal support by estrogens. Without estrogens, the vaginal epithelium becomes thinned, the supporting tissues lose blood vessels, and the elastic tissue is replaced by inelastic fibers. The net result is a thin, atrophic epithelium that lacks the normal elasticity and secretions. Such tissue is susceptible to infections and responds poorly to trauma such as inadequately lubricated intercourse. Senile vaginitis can be due to specific organisms, but in many cases no recognized pathogen can be found. The most common symptom is intense vaginal irritation. There may or may not be bleeding due to superficial ulceration. Scar tissue resulting from either infection or trauma may restrict or, rarely, close the vaginal canal. Topical or systemic estrogen therapy is the most effective treatment.

Contributing to vaginitis associated with intercourse is the decline in production of lubricating fluid, or vaginal sweating by the walls of the vagina. Gynecologists argue that the large glands of the vulva do little to lubricate the vagina. For whatever contribution they do make, it takes a much longer period of stimulation to cause their fluid to be released, and less total fluid is produced.

Prolapse of the Uterus, Bladder, and Rectum. Prolapse of the uterus occurs when the uterus descends into the vagina, and on occasion descends through the vaginal opening. Uterine prolapse is associated with weakening of the supporting ligaments of the uterus. Two similar conditions occur in elderly women; herniation of the urinary bladder into the vagina, or cystocele, and herniation of the large intestine into the vagina, or rectocele. All of these conditions are referred to as "pelvic relaxation."

Reasons for development of these conditions include genetic predisposition, damage incurred during childbirth, and estrogen deprivation. Development of the condition in the elderly almost always involves the latter often in combination with a history of the former two. The most common symptoms relating to pelvic relaxation are (1) a protruding mass from within the vagina; (2) urinary incontinence with stress such as walking, running, jumping, or laughing; or (3) difficulty in evacuating the rectum, frequently requiring manual or mechanical assistance to accomplish defecation.

Uterine prolapse is most commonly detected as a result of a "full" feeling in the vagina, incontinence, or infection. Sometimes it is found only after the vaginal walls are felt projecting between the labia. In severe cases there may be ulceration and bleeding. The condition is treated with surgery to tighten the pelvic floor and resuspend the uterus. Sometimes a supportive ring (pessary) is installed at the top of the vagina to hold the uterus in place.

Cervical Cancer. The incidence of cervical cancer declines after menopause. However, it remains a threat well into old age as the loss of the immune defenses becomes a greater factor than the decreased estrogen levels. Cervical smears should be conducted along with mammograms and thorough breast examinations periodically throughout life, beginning as early as the first sexual activity.

Younger, sexually active women have experienced an epidemic of a new form of cervical cancer associated with human papilloma virus (HPV). This tumor often develops and evolves into an invasive cancer in as few as 2 years. In the older population, cervical cancer is a more indolent disease and often looks basaloid (like the basal cells of skin). These tumors progress much more slowly from their origin to "carcinoma in situ" (have not invaded beyond the tissue of origin) and then to invasive and metastatic cancer. It is the latter disease that kills due to local complications or distant metastasis. The

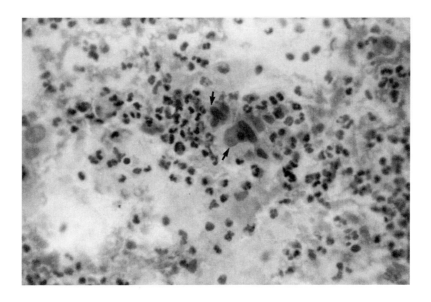

Figure 11-8   This positive Pap smear indicates a developing cervical cancer (arrows).

5-year survival rate for cervical carcinoma in situ is nearly 100%. (Figure 11-8 and Color Plate 14, *C* show a Pap smear with evidence of developing cervical cancer.)

Endometrial Cancer. The most common malignancy of the female reproductive system endometrial cancers; 80% of these occur after menopause. Increased disease risk is associated with obesity, diabetes, hypertension, and infertility. The most common symptom is postmenopausal vaginal bleeding, but some cases are discovered as a result of uterine enlargement. This tumor is only rarely detected by Pap smear; periodic pelvic examinations are required for early diagnosis. Because this tumor is typically slow to metastasize, early diagnosis can be associated with good survival. The normal treatment for uterine cancer is removal of the uterus (hysterectomy) followed by whatever additional cancer therapy is appropriate. The overall 5-year survival rate for endometrial cancer is 85%.

Ovarian Cancer. These cancers can occur any time in the post-puberty life of women but most occur after menopause (Figure 11-9 and Color Plate 14, *D*). The main difficulty with this tumor is its occult nature. Typical ovarian cancers do not cause symptoms until they are far advanced. This is because the ovaries are small and difficult to feel or examine. In most cases the cancer has spread in the peritoneal cavity to the peritoneal surface, the other ovary, or to the surface of the uterus. Luckily, many of these tumors remain localized, affecting only the peritoneal lining, and do not invade the other tissues. Although this establishes a longer survival rate, many patients develop massive ascites (a-SĪT-ēz),or fluid in the abdominal cavity.

Breast Cancer. Until very recently, this was the most common form of internal cancer among women. It is now second to lung cancer. Breast, or mammary, cancer arises from the lactogenic tissue of the breast. Because of its origin in glandular tissue, it is classified as an adenocarcinoma (Figure 11-10). It may invade the pectoral muscles

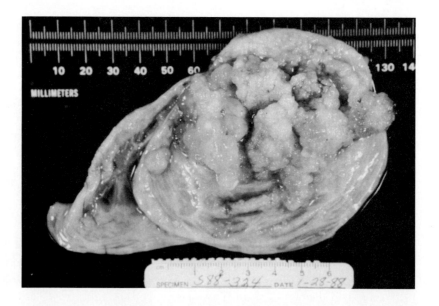

Figure 11-9   Benign ovarian cysts are very common, but large cysts also can occur with ovarian cancer, as is shown with this bisected ovary.

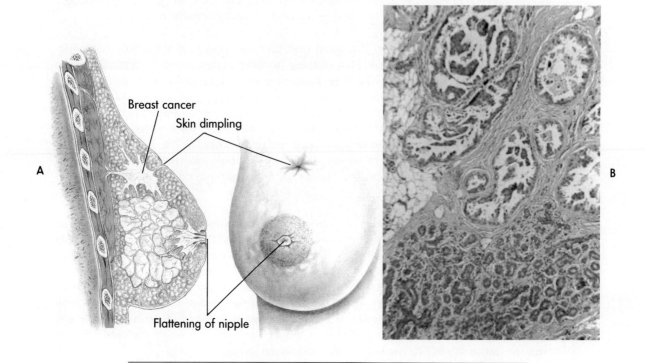

Figure 11-10   Adenocarcinoma of the breast usually retains its glandular nature even when metastatic to other parts of the body. **A,** Diagram of internal and external effects of adenocarcinoma; **B,** photomicrograph.

beneath the cancerous breast as well as the other breast. Metastasis occurs through the lymphatic channels and blood vessels. When passing through the lymph channels, metastatic cells are sometimes trapped in the lymph nodes. A check of these nodes is a reasonable measure of whether metastasis has occurred. Metastasis through the vascular pathways frequently involves spread to the lungs, liver, or spine.

These cancers may arise at any age after maturity but most commonly arise after age 40. Advancing age does not preclude breast cancer; in fact the incidence rises with age. In postmenopausal women, normal atrophy of the mammary glands minimizes the amount of palpable tissue in the breast, leaving only fat and supporting tissue. Therefore any palpable lesion that is not fat may be a malignant tumor. Benign lumps are even more common than malignant ones; the most common cause of these is fibrocystic change. Because breast cancers tend to metastasize while they are small, monthly self-examination, annual examination by a physician, and periodic mammograms will help to detect them before they have spread.

Sexual Activity Among Elderly People

We have presented many of the mechanical problems involving the sexual organs of elderly people. We now need to examine sexuality as an act of affection and pleasure. Elderly couples sometimes drift away from sexual intercourse because the woman experiences vaginal discomfort, or because the man requires more stimulation to produce erection and a longer period of penile-vaginal contact to produce ejaculation. Pleasure from orgasm may also lessen. Two misguided beliefs may also interfere with continuing that sexual activity; (1) the belief that such activity is "abnormal" in the elderly, and (2) that intercourse that does not result in ejaculation is "incomplete." Estrogen creams can greatly reduce most vaginal discomfort. Patience and affection can make a prolonged period of stimulation pleasant. In addition, both men and women can enjoy close physical contact and tactile stimulation even if orgasm does not always result.

Difficult situations can occur when an elderly person has lost a spouse and is placed in a nursing home. It is normal for many people to still desire sexual contact. This desire may be compounded by the need to try to fill the emotional gap left by the loss of the wife or husband. Most nursing home environments do not allow the privacy necessary for courtship or sexual activity. Additionally, there is likely to be pressure on the home management from the children to prevent such "ridiculous" activity, especially when it may divert a parent's attention (or resources) to someone outside the family.

CASE HISTORIES

❖ Impotence

R.C. is a 56-year-old contractor who has experienced decreased libido culminating in impotence. The problem has developed gradually over the last 10 years. Although he has had problems maintaining an erection for years, it has become worse since his coronary bypass surgery 4 years ago. R.C. believes that a contributing factor is the failing health of his wife, who is an obese diabetic with poor blood flow and is developing foot ulcers. His only medications are aspirin (an antiplatelet agent) and dipyridamole. He has fathered two children.

His physical examination showed R.C. to be obese, short, and with normal vital signs. The head, eyes, ears, nose, and throat (HEENT) results were normal, as were examinations of his chest, heart, and abdomen. A sternotomy scar indicative of past heart surgery was noted. The neurologic

234 Biology of Aging

system was intact. Genital examination showed a small penis and bilaterally small testes in the scrotum. Laboratory results were normal for CBC (complete blood count), U/A (urine analysis), chemistry profile, and thyroid function tests. Two serum testosterone tests were low.

R.C. and his wife underwent sex therapy with a slight improvement in performance. After 6 months of therapy, R.C. was placed on weekly testosterone injections and noted marked improvement in his libido and sexual satisfaction. After 2 years of weekly injections, he elected to have surgical implantation of a permanent inflatable penile prosthesis.

Comment. Impotence, or waning sexual ability, is a fact of life for the aging male. Peak sexual performance occurs in the male in the early 20s; both desire and performance decline with age. In spite of this, all men with a patient and willing partner are capable of satisfactory sexual intercourse, regardless of their age. However, this is only true for a normal male. R.C., despite fathering two children, has probably always had low testosterone production and low sex drive. This condition has worsened with age, producing impotence. Without a correctable medical cause for his disorder such as medication effect, acute illness, hormonal imbalance, or obvious psychiatric condition, he was tried on multiple, progressive therapies. Each one helped but R.C. was not satisfied until he received an implantable and controllable prosthesis.

❖ Gynecomastia

D.M. is a 61-year-old carpenter who noticed painless asymmetric enlargement of his left breast. He first noticed the change 1 month ago and has not observed any change in breast size since then. He has not seen any blood or other visible drainage from the nipple. His medical history is significant for an appendectomy as a child and a coronary bypass 3 years ago. His only medication is digoxin used to strengthen the heart. He smokes an occasional cigar and is a social drinker.

His physical examination revealed a small inguinal hernia on the left and the left breast mass. The mass was firm and rubbery and located just beneath the areola. Its margins were poorly defined and it was not tender or fixed to the underlying tissues. Laboratory examination showed normal results for a chemistry profile, CBC, and U/A. A mammogram revealed a benign-appearing soft tissue density with no calcification and no scirrhous (scarring) features.

The patient was assured that the breast lump was probably benign and probably related to his medication. Even so, he wanted the lesion removed if there were any chance of cancer. Since certainty is impossible with an unexamined lesion, an excisional biopsy was performed. The tissue showed development of new ducts and acini in young fibrous tissue, or gynecomastia.

Comment. Gynecomastia is a common problem most often seen in puberty. In the elderly male it is often related to medication. Digoxin for example, can stimulate the estrogen receptors in certain male breasts. Unlike this case, many instances are related to excess estrogen. Estrogens accumulate in patients with cirrhosis, in certain testicular tumors, and are sometimes given to treat prostate cancer.

❖ Prostate Cancer

H.V. is an intelligent, active 64-year-old man who came to our hospital for monitoring of his prostate tumor. His present illness began 5 years earlier when his local family physician felt a small, 0.5 cm hard nodule on his prostate during a rectal exam. He was referred to a urologist who conducted a biopsy of the area through the rectum and found a moderately differentiated adenocarcinoma of the prostate that appeared to be limited to the gland. Acid and alkaline phosphatase levels were within normal limits. With such limited disease, H.V. was offered a chance of cure by a radical prostatectomy. When it was explained that this could cause impotence, he declined and sought "serum" therapy in Mexico where he was "cured."

Three years later he returned with severe, nonexertional back pain not related to position. Spinal x-ray revealed multiple osteoblastic bone lesions believed to be metastatic cancer. His prostate was now enlarged and rock hard on the right side with extension of the tumor into the adjacent tissues and into the right seminal vesicle. Serum studies showed prostate cancer markers to be elevated with acid phosphatase (AP) of 10.8 IU/L (normal <1.5), prostate specific antigen (PSA) of 45.2 µg/ml (normal < 4.0), and an alkaline phosphatase of 424 IU/L (normal <150). These values indicated widespread tumor growth and involvement of bone. Symptomatic treatment using radiation therapy to the spine was performed. Further symptomatic control was offered consisting of bilateral orchiectomy (testis removal) and estrogen therapy. The patient again declined and sought a new chemical orchiectomy developed in Canada.

He returned to us having received an experimental antitestosterone drug, an experimental LH - FSH suppressor, and large dosages of corticosteroids. He requested and received periodic serum monitoring of testosterone, LH, FSH, cortisol, and tumor markers. The therapy was intended to keep his testosterone, pituitary gonadotropins, and adrenal steroids near zero. Hopefully, this chemical suppression of androgens would slow the growth of his tumor and decrease or stabilize the serum level of the tumor markers. His back pain had returned and he was using an orthopedic back brace for stabilization. As long as his back was held in place he was free of pain. The therapy had resulted in impotency, gynecomastia, and loss of libido. The tests results indicated success with the monitored substances but showed clear disease progression. Testosterone, LH, and FSH were all low and cortisol was high. The tumor markers AP and PSA continued high. After 2 years of further therapy and clinical deterioration, H.V. entered the hospice program and died.

Comment. Palpable early prostate cancers limited to the gland (stage B) are not all curable using radical prostectomy, but the 10-year survival ranges from 50% to 79%. H.V. chose a completely untested form of therapy and his tumor progressed, becoming metastatic in a few years. When he developed back pain, the new Canadian therapy he chose was probably as effective as the more conventional therapy of estrogens and orchiectomy. Many cancers leak protein into the serum that can be used as tumor markers. AP and PSA are very specific markers for prostate cancer because almost no other tumor makes them. Unfortunately, early curable prostate cancers often do not leak detectable amounts of these substances into the serum and PSA may be elevated with benign prostate hypertrophy.

❖ Female Feminization

C.A. is a 78-year-old grandmother of eight who began to bleed vaginally. She is gravida III, para III, (3 pregnancies; 3 live births) and had normal spontaneous menopause at age 52. She had not had any vaginal bleeding until 5 months before seeing her physician, when she noticed spotting. During the last 4 months she has had two episodes of menstrual-type bleeding requiring multiple pads over 4 or 5 days. She was not taking any medication and had no other symptoms.

Her physical examination revealed normal vital signs. She appeared younger than her stated age, with smooth, supple skin. Her breasts were pendulous and atrophic but there was no vaginal atrophy. Pelvic examination showed a large symmetric uterus without masses. Adnexa (ovaries and fallopian tubes) were not palpable. The remaining portions of the physical were normal for her age.

Laboratory evaluation revealed a normal but bloody Pap smear with increased estrogen effects for her stated age. The CBC, U/A, and chemistry profile were within normal limits. Serum hormonal evaluation revealed normal cortisols. A 24-hour urine collection for 17 keto and ketogenic steroids (sexual steroids and cortisol precursors made in the adrenal gland also was normal). The serum testosterone was normal.

A dilation of the cervix and curetting of the endometrium (D and C) revealed abundant fragments of a benign but slightly atypical endometrium. A hysterectomy with bilateral salpingo-oophorectomy (removal of fallopian tubes and ovaries) was performed. The uterus was 180 g, enlarged even for child-bearing years. There were no tumors but the endometrium showed simple benign hyperplasia. Both ovaries were atrophic, but the right contained a 1 cm sharply defined, deep brown nodule which represented a benign hilar cell tumor. Postoperatively, C.A. did well but desired estrogen replacement.

Comment. This is a very interesting and unusual case of a hormonally active tumor. The usual sources of sex hormones are the ovaries, testes, or adrenal glands. All of these are capable of producing either androgens or estrogens. In the usual case, symptoms are those of *inappropriate* hormone production, such as masculinization of a female. The opposite sometimes occurs, producing a morphologically perfect but reproductively nonfunctional female as a result of congenital testicular dysfunction, resulting in feminization of a genetic male. With C.A., the only reason symptoms were noted was because of the post-menopausal nature of the patient. Excessive estrogen production in a woman of childbearing age can easily be overlooked. Hilar cells normally produce testosterone but like all germ cells, they have total potential. They were producing large amounts of estrogen in this patient.

❖ Atrophic Vaginitis

N.R. was a 48-year-old married insurance executive who had no children. She went to her physician because intercourse had become painful and she had begun to experience postcoital spotting, vaginal burning, and discharge. She indicated that what had been a very pleasurable and compatible sexual relationship with her husband had now become a frightening and uncomfortable ordeal.

N.R.'s medical history included an abdominal hysterectomy with removal of both ovaries and fallopian tubes at age 38 because of endometriosis. This was followed with hormone replacement therapy (HRT). After 1 month of HRT, N.R. developed pelvic thrombophlebitis and all hormones were discontinued. For the ensuing 8 months she was troubled with "hot flushes" but these gradually disappeared.

Pelvic examination revealed marked thinning of the vaginal mucosa, a thin yellow discharge, and loss of vaginal elasticity. Microscopic examination of vaginal fluid revealed numerous white blood cells and bacteria. These findings were consistent with atrophic vaginitis secondary to estrogen deprivation.

N.R. was seen by an internist who reevaluated her diagnosis of thrombophlebitis and believed that this condition was probably a complication of her abdominal surgery. The internist advised her that she could probably safely restart HRT. N.R. was started on estrogen topical vaginal cream as well as oral estrogen. After 3 weeks, her vaginal discharge disappeared and the pain from intercourse subsided. Six weeks after initiating therapy, the topical cream was discontinued and she was advised to continue the oral estrogen medication indefinitely unless problems related to the drug appeared.

Comment. Atrophic vaginitis is a common problem following the natural decline in estrogens associated with menopause. It can be even more of an interruption of married life when it occurs earlier because of removal of the ovaries, the normal source of estrogens. In this case replacement estrogens were discontinued because of their known tendency to cause clotting and the appearance of pelvic thrombophlebitis (inflammation of the walls of the veins of the pelvic region and the formation of clots (thrombi) in the affected veins). When it was concluded that the thrombophlebitis was probably a result of the surgery and the patient was not predisposed to the disease, the ERT was reinstated and N.R. quickly recovered.

❖ Difficult Menopause

P.W. is a 54-year-old housewife who has been married for 31 years and has four children. She went to her physician because of intense "hot flushes" that have become so severe that they interrupt her sleep. To avoid disturbing her husband and the rest of her family, she has been arising and engaging in some activity to pass the night. She is concerned that her libido has declined, and unlike her usual demeanor, she has become withdrawn and sullen. She attributes these changes in behavior to lethargy and insomnia as well as generalized skeletal discomfort.

Her medical history shows that she experienced menopause at age 48 and that her problems began about 1 year before that. She has recently lost 15 pounds. Her physical examination revealed a blood pressure of 140/80. All of her other findings were within normal limits except for diminished estrogen effect and mild osteoporosis involving thoracic and lumbar vertebrae.

P.W. was started on estrogen daily for the first 25 days of each month. Progesterone was prescribed for days 16 through 25 of each month. Two weeks after initiation of the hormones, her hot flushes disappeared. She reported a general feeling of well-being and began to sleep better. Following the second month of her therapy she redeveloped menstrual periods, which she considered something of a nuisance, but this inconvenience was diminished by the other benefits of the therapy. A baseline duel photon study for osteoporosis was done. She was advised to supplement her diet with 2 grams of calcium daily and to return in 1 year to reevaluate the osteoporotic changes.

Comment. This case is unusual in that undesirable behavioral changes seemingly resulted from loss of sleep that in turn was caused by the intense vasomotor responses associated with menopause. In most patients who experience severe emotional problems at the time of menopause, estrogen deprivation is not considered the sole cause. Preexisting emotional problems or other circumstances occurring at that time are often more important. Because increased uterine malignancy is reported in patients who have received systemic estrogen of this dosage alone, concomitant use of progesterone is required.

❖ Cervical Cancer

M.R. is a 73-year-old clerical worker who developed postcoital spotting 18 months ago. After intercourse she would see a few small flecks of blood on the toilet paper when she went to the bathroom. In addition she had a thin, watery vaginal discharge at about the same time which she assumed to be a mild infection. Recently she had developed persistent lower back pain that was not related to position and seemed worse at night. She visited her physician because of the backache and mentioned spotting.

M.R.'s pelvic examination revealed that the cervix, right vaginal apex, and soft tissue of the right pelvic floor were unusually hard. Visual inspection showed the cervix to be covered with a necrotic, friable (easily disintegrates) neoplasm that bled easily. A biopsy and Pap smear were obtained and documented invasive squamous cell carcinoma. Her last Pap smear had been at the age of 63, just before she had retired.

Gynecologic consultation believed that the tumor extended to the bony pelvis. A CT scan confirmed destruction of part of the adjacent pelvis, lymph node enlargement, and obstruction of the right ureter. A scan of the liver revealed defects believed to represent metastatic cancer. M.R. was believed to have incurable disease and was treated palliatively with pain medication and radiotherapy to the bony pelvis. Despite these treatments, the back pain became unbearable and required a cordotomy, or interruption of the pain-carrying fibers in the spinal cord. She died of widespread tumor 13 months after diagnosis.

Comment. Cervical cancer will occur in one of every 50 women over the age of 40. When detected early, while still within the cervical epithelium (in situ), it is completely curable by simple hysterectomy (removal of the uterus and cervix). Detection at this early stage requires periodic Pap smear evaluations. Many older women think these screening tests are unnecessary and stop having them. For this reason, cervical cancer deaths increase in the older age groups.

❖ Adenocarcinoma of the Ovary

M.L. is a 78-year-old retired nurse who noticed increasing abdominal girth. This change began insidiously, but she had become aware of increasing abdominal size 3 months ago. She also noticed a recent loss of appetite. There appeared to be no other problems. Specifically, she reported no weight change in spite of her larger belly and no ankle swelling, respiratory problems, or change in bowel habits. She was under treatment for mild hypertension and traumatic arthritis of the knees. The latter condition had required her to enter a communal care center.

Her physical examination was for the most part normal. The chest was clear to percussion and auscultation but the diaphragm was high and did not move well. The heart sounds were regular and without abnormality, and the breasts were free of masses. The abdomen was nontender and without masses or organomegaly but was enlarged. Percussion showed shifting dullness, a sign of ascites (abdominal fluid). The rectal examination revealed a firm, nodular cul de sac (the area between the rectum and uterus). The vaginal exam showed a normal-size involuted uterus and bilateral masses.

Laboratory evaluation showed a hemoglobin of 9.8 (normal over 12) with normal-size cells. The WBCs and platelets were normal. The chemistry profile showed a "tumor pattern" with an elevated LDH of 358 IU/L (normal below 180), uric acid of 11.2 μg/dl (normal below 8.0, and alkaline phosphatase of 312 IU/L (normal below 150). Paracentesis (removal of ascitic fluid) showed a malignant effusion with total protein of 4.8 gms, LDH of 510, and adenocarcinoma cells present on cytologic examination.

An exploratory laporotomy showed bilateral tumor masses replacing the ovaries plus numerous serosal implants (tumors on the outer tissues of other organs). A hysterectomy with bilateral salpingo-oophorectomy (removal of the uterus with the fallopian tubes and ovaries) and omentectomy (removal of the fatty, apron-like extension of the peritonium) were performed. Histologic examination revealed that both ovaries had been replaced by a poorly differentiated papillary serous adenocarcinoma. Tumor cells studded the serosa of the peritoneal cavity and the uterus and created large masses on the omentum. After M.L. recovered from surgery, chemotherapy was given intraperitoneally. She continued to have ascites and continued tumor growth. She died of ovarian cancer 3 years after surgery.

Comment. Ovarian cancer is very difficult to detect. Unlike cases of breast or cervical cancer, the tumor is difficult to screen for. There are often no symptoms until the tumor has progressed too far to be treated successfully. Adenocarcinoma of the ovary usually involves the surface epithelium of the ovary. This involvement allows cells to drift off into the abdominal cavity. This allows the tumor to spread to the surface of other abdominal organs and to the peritoneal lining. Recently new blood-borne tumor markers have been developed that may be helpful in screening for ovarian cancer. At present, however, they are not reliable enough for use.

❖ Mammary Carcinoma

T.H. was an 81-year-old great grandmother who came to her doctor because of shortness of breath. She had had the problem for several months and it became worse when she caught a cold. She noticed it especially when she was lying down. Her medical history revealed no recent surgeries or other medical problems. She was taking no medications.

Her physical examination showed her to be be in mild respiratory distress with 24 respirations per minute (normal 18). She showed dullness to percussion in the lower left lung fields, with muffled breath sounds. The right lung showed increased breath sounds as well. Her left breast was distorted by a large tumor mass that was fixed to the chest wall and had invaded the skin. The nipple was inverted and the skin overlying the mass was adherent. Multiple enlarged lymph nodes were palpable in her left axilla (armpit).

Despite the evidence of extensive tumor, she had a normal chemistry profile, CBC, and urinalysis. Radiologic examination confirmed a left pleural effusion (fluid in the lung cavity), numerous pleural nodules, enlarged internal mammary lymph nodes, and a breast mass invading the left chest wall. Bone, liver, and spleen scans revealed no evidence of tumor. Thoracentesis (removal of fluid from the chest wall) revealed cells compatible with primary breast cancer.

A simple mastectomy was performed, removing the breast and the primary tumor, a scirrhous adenocarcinoma. The breast tumor was $7 \times 4.5 \times 2.5$ cm. Estrogen and progesterone assays revealed 301 femtomoles (a unit of estrogen responsiveness) and 4.5 femtomoles respectively, indicating that the tumor was stimulated by estrogen. T.H. was irradiated, placed on antiestrogen medication (Tamoxifen), and discharged.

Comment. Visual inspection of the breast suggested the correct diagnosis. It is apparent that in spite of obvious breast disease, T.H. ignored the cancer. Because of this denial, she waited to see her doctor until respiratory symptoms secondary to invasion of the thoracic cavity had appeared. This is unfortunate because the tumor seemed to have remained localized in the breast for a long period and would probably have been curable with surgery. Metastasis in younger people usually occurs much more extensively and earlier than in this case. Because of the extensive invasion and lymph node involvement at the time of her examination, it was determined that radical surgery would not have extended her life and only the breast was removed. Because this tumor was responsive to estrogen, removal of the estrogens produced by her adrenals and her body's metabolism could be expected to slow the progress of the disease.

BIBLIOGRAPHY

Andrews J: *Gynaecology of the elderly.* In Pathy MJS, ed: *Principles and practice of geriatric medicine,* New York, 1985, John Wiley & Sons.

Blacklock NF: *The genitourinary system—the prostate.* In Brocklehurst JC, ed: *Textbook of geriatric medicine and gerontology,* ed 3, New York, 1985, Churchill Livingstone.

Breschi L: *Common lower urinary tract problems in the elderly.* In Reichel W, ed: *Clinical aspects of aging,* ed 3, Baltimore, 1989, Williams & Wilkins.

Brown ADG: *The genitourinary system—gynaecological disorders in the elderly.* In Brocklehurst JC, ed: *Textbook of geriatric medicine and gerontology,* ed 3, New York, 1985, Churchill Livingstone.

Danforth DN, Scott JR, eds: *Obstetrics and gynecology,* ed 5, Philadelphia, 1986, JB Lippincott.

Freed SZ: *Genitourinary disease in the elderly.* In Rossman I, ed: *Clinical geriatrics,* Philadelphia, 1986, JP Lippincott.

Glowacki GA: *Geriatric gynecology.* In Reichel W, ed: *Clinical aspects of aging,* ed 3, Baltimore, 1989, Williams & Wilkins.

CHAPTER

12 *The Endocrine System*

❖ *Relevant Anatomy and Physiology*

GENERAL INFORMATION AND PRINCIPLES

The endocrine system is a complex arrangement of glands that secrete hormones controlling the metabolism and activity of the body. Hormones are compounds produced by one tissue or organ that regulate the activity of other tissues or organs. Many hormones control only isolated organs and often affect only specific cells within these organs. In men for example, interstitial cell stimulating hormone (luteinizing hormone) affects only the interstitial cells of Leydig in the testis. Other hormones, such as thyroxin, affect virtually every cell in the body. A cell must contain receptors for a particular hormone to be affected by it. Cells containing receptors for a particular hormone are called target cells for the hormone.

Hormones usually operate through a system of negative feedback mechanisms (Figure 12-1). This means that when there is too little of a regulated substance, the appropriate endocrine gland produces a hormone causing another tissue or gland to increase production (or release) of that needed compound. The increased level of that compound then feeds back negatively on the endocrine gland, causing production of the hormone to decline. Many examples of this process have been presented in the preceding chapters of this text and will be reviewed here. When one studies the endocrine system, it becomes evident that "homeostasis," the steady state, is not a steady state at all. Almost everything is allowed to fluctuate within tolerance limits. This provides adaptability, a big advantage that enables the body to adjust to changing conditions.

Endocrine glands lack ducts. Their hormonal products diffuse through interstitial fluid directly into the bloodstream. Because some hormones are absorbed by only a few target cells, they may be present in the bloodstream in extremely small amounts. Some endocrine glands are scattered cells with very little organization. The gastrin-secreting cells of the gastric mucosa or the secretin producing cells of the duodenum are examples of the diffuse neuroendocrine system. Others, such as the islets of Langerhans located in the pancreas, are organized tissues within organs that have another purpose. Still others are separate, identifiable endocrine organs, such as the pituitary or thyroid glands.

The location of the endocrine organs is shown in Figure 12-2. Many of these organs have already been discussed in conjunction with the body system they affect. Those that have not will be discussed in greater detail in this chapter.

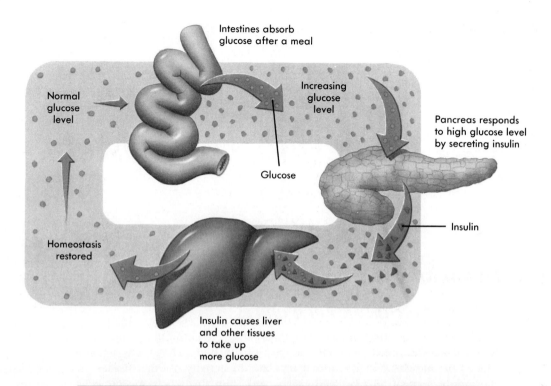

Intestines absorb
glucose after a meal

Normal
glucose
level

Increasing
glucose
level

Pancreas responds
to high glucose level
by secreting insulin

Glucose

Insulin

Homeostasis
restored

Insulin causes liver
and other tissues
to take up
more glucose

Figure 12-1   A negative feedback mechanism regulates secretion of most hormones, allowing fluctuation within tolerable limits. Here an increase in blood glucose triggers secretion of insulin. Insulin promotes uptake of glucose by cells, restoring blood glucose to its normal (lower) level.

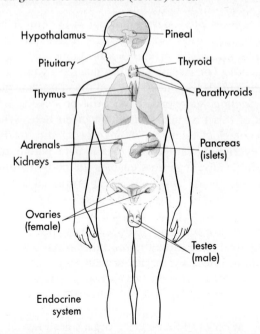

Hypothalamus — Pineal

Pituitary — Thyroid

Thymus — Parathyroids

Adrenals — Pancreas (islets)

Kidneys

Ovaries
(female)

Testes
(male)

Endocrine
system

Figure 12-2   The endocrine organs. Some endocrine tissues are scattered in cells in other organs. For example, gastrin is secreted by cells in the gastric mucosa.

THE PITUITARY

The pituitary gland, also known as the hypophysis, has been called the master gland because in addition to controlling many nonglandular body tissues, it, in close association with the hypothalamus, controls many of the other endocrine glands. Figure 12-3 shows the pituitary and target organs of the principal pituitary hormones. The pituitary is formed from two different embryonic ectodermal tissues. The anterior pituitary, or adenohypophysis, is formed from outpouching of the skin of the roof of the mouth, skin ectoderm. The posterior pituitary, or neurohypophysis, is formed from a downward extension of the brain, neural ectoderm. These two parts of the pituitary release very different hormones with very different functions.

The anterior pituitary is associated with the hypothalamus. The hypothalamus contains neurosecretory cells whose products release hormones (therefore, their name "releasing factor") from specific cells in the pituitary. Other hypothalamic neurosecretory cells stimulate organs elsewhere in the body. Opinions differ on whether the hypothalamus is an endocrine gland and on whether its products should be called hormones. Some define a hormone as a compound produced by an endocrine gland, and an endocrine gland as "glandular tissue producing a hormone." Because neurosecretory cells are not typical glandular tissue, some do not accept the term *hormone* applied to their products. For practical purposes, the authors define a hormone as any compound produced by one organ or tissue that affects the behavior of another organ or tissue. Accordingly, any tissue or organ that produces a hormone is functionally an endocrine gland. The posterior pituitary gland is virtually an extension of the hypothalamic portion of the brain. The anterior pituitary gland is only connected to the hypothalamus by a vascular network. This vascular network carries releasing hormones from the hypothalamus to the anterior pituitary. Figure 12-4 shows the relationship between the two pituitary glands and the hypothalamus.

Although the pituitary gland is often called the master gland, we have to give credit to the hypothalamus for controlling the pituitary. Hypothalamic secretions are influenced by the brain and by circulating blood levels of many substances that are being regulated. The hypothalamus provides one of several links between the nervous system and the endocrine system. Therefore the endocrine system can work alone to cause changes in the body, or its action can be initiated or modified by the nervous system. Of course, the nervous system can work alone to cause rapid changes in the body. Changes initiated by the endocrine system are usually slower and longer-acting than changes initiated by the nervous system.

Adenohypophyseal Hormones

Growth Hormone (GH). Also known as somatotrophic hormone (STH), growth hormone is released throughout life in response to growth hormone releasing factor (GHRF) issued by the hypothalamus. GHRF stimulates the pituitary to release GH in response to hypoglycemia (or inadequate blood sugar), increases in circulating amino acids, or exercise. GH promotes the formation of glucose from the body's fat stores and the synthesis of new proteins from amino acids while conserving proteins already present in cells. The resulting increased level of new proteins and glucose feed back to reduce GHRF and, ultimately, the blood load of GH. GH sensitizes the beta cells of the pancreas, causing them to overreact to normal blood sugar loads with excessive insulin production. Despite this heightened insulin response, GH excess results in a condition similar to diabetes mellitus. An excess of GH in a young person produces gigantism; in an adult it produces acromegaly. In an adult the bones can no longer increase in length but continue to grow in width, producing enlarged, craggy facial features. Deficiency in a

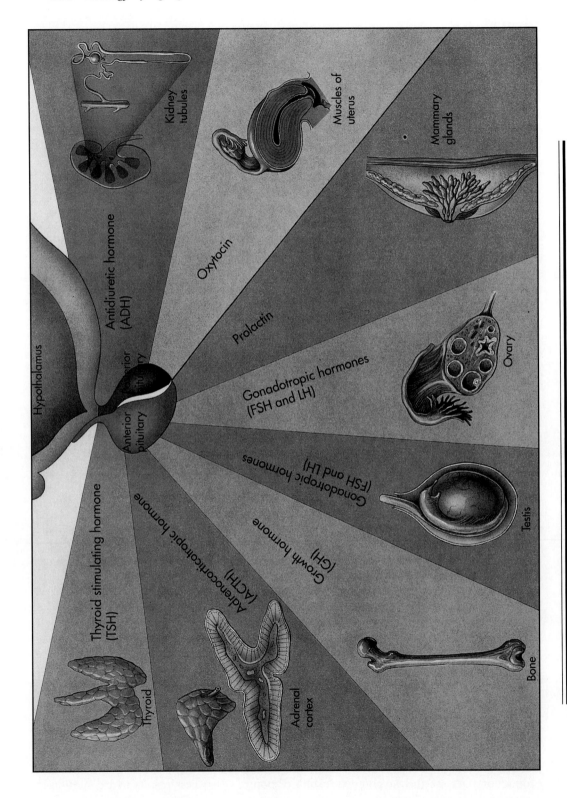

Figure 12-3 The principal anterior and posterior pituitary hormones and their target organs.

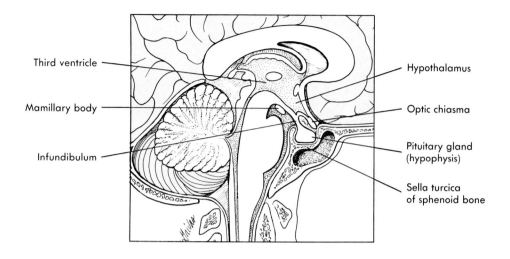

Figure 12-4   The brain stimulates the hypothalamus, which in turn stimulates the pituitary. The posterior pituitary is an extension of the neurosecretory cells of the hypothalamus, but the anterior pituitary is stimulated by hormones released by the hypothalamus.

child produces dwarfism that ultimately leads to a miniature adult. The results of excesses and deficiency of GH are shown in Figure 12-5.

Adrenocorticotropic Hormone. Also known as ACTH, this hormone is released when serum cortisol levels fall and under conditions of stress. The release occurs in response to corticotropin releasing factor (CRF) produced by the hypothalamus. ACTH acts on the cortex of the adrenal glands to cause an increase in cortisol production. The cortisol then feeds back on the hypothalamus, reducing CRF, and also acts directly on the adenohypophysis, reducing ACTH production. The adenohypophysis increases ACTH production in the early morning, normally reaching a peak at about 8 AM. Because the adrenal response is almost immediate, cortisol peaks at about the same time. These hormones decline steadily, reaching a low at around 8 PM. This diurnal (or circadian) cycle is directly related to the wake/sleep cycle; it changes when the sleep cycle changes. Many other hormones have also been shown to have circadian fluctuations. The symptoms of excess ACTH appear primarily as the symptoms of hypercorticoidism, with secondary features of hyperaldosteronism. Aldosterone is a steroid released in small amounts with cortisol.

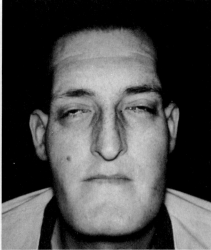

Figure 12-5   Growth hormone abnormalities. **(A)** Hypersecretion (far left) and hyposecretion (far right) of GH produce gigantism and dwarfism, respectively. **(B)** Hypersecretion of GH in adulthood produces acromegaly.

Melanocyte Stimulating Hormone (MSH). This substance causes increased deposition of melanin in the skin. A releasing or inhibiting factor for this hormone has not yet been found. MSH is often released from the adenohypophysis along with adrenocorticotropic hormone and often provides a telltale increase in pigment when the pituitary is releasing too much ACTH.

Thyroid Stimulating Hormone (TSH). This hormone, also known as thyrotropin, is released by the adenohypophysis in response to TSH releasing factor, also called thyrotropic releasing factor (TRF), secreted by the hypothalamus. TRF is released in response to low levels of circulating thyroxin. Exposure to cold and some emotional states apparently also stimulate TSH and subsequent thyroxin production. Thyroxin increases the body's rate of metabolism. High levels of circulating thyroxin, increased body temperature, and prolonged excitement ultimately result in reduced thyroxin secretion.

Follicle Stimulating Hormone **(FSH).** One of two gonadotropic hormones produced by the anterior pituitary gland, FSH and the other gonadotropin, luteinizing hormone, are produced by both sexes. In men, FSH in the presence of testosterone stimulates Sertoli cell activity that nourishes the sperm and contributes to sperm maturation. In women, FSH is responsible for growth of the ovarian follicle and the egg it contains, and for the production of estrogens. Like most other adenohypophyseal hormones, FSH production is stimulated by a hypothalamic releasing factor.

Luteinizing hormone. LH causes the interstitial cells of Leydig in the testes of mature men to produce testosterone. As with FSH, LH is released by a hypothalamic releasing factor (LHRF). The regulation of LH in women is similar, except the primary feedback hormone is progesterone. These hormones and their effects on the male and female reproductive systems are discussed in Chapter 11.

Prolactin. Stimulates lactation, milk production. Unlike other adenohypophyseal hormones, prolactin production is normally inhibited by a hypothalamic hormone, prolactin inhibiting factor (PIF). PIF is reduced at the end of pregnancy and its production continues to be inhibited for a period during lactation. If the hypothalamic-pituitary relationship is severed, prolactin increases while the other pituitary hormones decline.

Posterior Pituitary Hormones

Antidiuretic Hormone **(ADH).** Also known as vasopressin, this hormone, like the other posterior pituitary hormone, oxytocin, is actually produced by the hypothalamus and is released through the posterior pituitary. The bulbous ends of secretory nerves originating in the hypothalamus terminate in capillary beds of the pituitary, where they release their secretions.

The effect of ADH on kidney function was discussed in Chapter 8. Extremely small amounts of this hormone can cause antidiuresis, or greatly reduced urine volume by greatly increasing the permeability of the collecting ducts to water. This process allows increased reabsorption of water from the urine into the blood, lowering blood salinity. In response to high blood salinity, the hypothalamus is stimulated to release ADH. When enough water has returned to the blood, thus diluting the blood salts and changing the osmotic gradient at the hypothalamus once again, ADH secretion is terminated. Complete deficiency of ADH produces diabetes insipidus. Patients who have this disease will produce 4 to 15 liters of urine per day, depending on fluid intake. This condition can be controlled by nasal inhalation of small amounts of synthetic vasopressin.

Oxytocin. Like vasopressin (ADH), OXT is actually produced in the hypothalamus and released by neurosecretory cells into a capillary bed in the neurohypophysis. This again illustrates that the posterior pituitary is not a gland at all but rather a contact point for neurosecretory cells and capillaries. The actual gland consists of the neurosecretory cells of the hypothalamus.

Release of oxytocin initiates and promotes parturition, or the birth process by causing strong uterine contractions. It also promotes lactation. In addition, it has been suggested that the release of oxytocin during intercourse may promote fertilization by causing uterine contractions that carry semen toward the fallopian tubes.

ADRENAL GLANDS

The adrenal glands, like the kidneys from which they are embryologically derived, are composed of an outer cortex and an inner medulla (Figure 12-6 shows a diagram of adrenal structure; see also Color Plate 15; *B*). The adrenal cortex secretes steroid hormones called corticosteroids. The adrenal medulla secretes epinephrine and norepinephrine, functioning in parallel with the sympathetic nervous system.

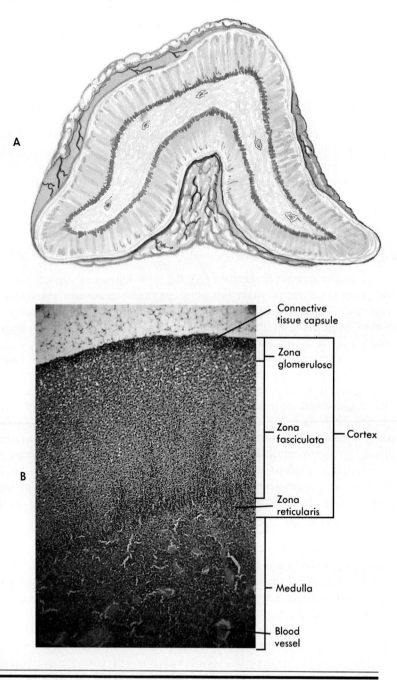

Figure 12-6   **(A)** Anatomy and **(B)** histology of the adrenal gland.

Adrenal Cortex

Of at least thirty adrenal cortical steroids, only three of them have clinical significance. These are aldosterone, cortisol, and androgenic steroids.

Aldosterone. Discussed previously with the excretory system in Chapter 8, this adrenal hormone increases reabsorption of salt from the urine. As long as the individual has access to water, an accumulation of excess salt in the blood due to aldosterone, hypernatremia, has little effect other than to slightly increase the blood volume and pressure. Water will be reabsorbed with the sodium, and activation of the thirst reflex will further dilute the level of blood salinity.

Aldosterone also causes the secretion of potassium into the distal convoluted tubules and collecting ducts as a part of the sodium reabsorption mechanism. Excess aldosterone secretion can therefore cause hypokalemia, decreased blood potassium, and result in hyperpolarization of nerve and muscle fiber membranes, leading to paralysis. The same mechanism may cause heart conduction system problems. The latter can produce weakness and arrhythmia. Finally, as sodium is reabsorbed, hydrogen ions are secreted. This absorption leaves alkaline bicarbonate ions rather than carbonic acid predominating in the blood. Excessive aldosterone can thus produce alkalosis, and inadequate aldosterone can produce acidosis. Aldosterone secretion can be stimulated by high potassium concentrations in the intercellular fluid, low sodium concentrations, the renin-angiotensin mechanism, and to a lesser degree by ACTH.

Cortisol. Also known as hydrocortisone, cortisol is an important regulator of carbohydrate, protein, and fat metabolism. It mobilizes amino acids from the tissues, increases conversion of amino acids into glucose, aids GH or ACTH mobilization of fatty acids from adipose tissue, decreases inflammation, and in very large (therapeutic) dosages, inhibits and causes atrophy of the immune system. As mentioned earlier, this hormone is secreted in response to stress, both physical and emotional, through the ACTH association with the hypothalamus and brain. High levels of cortisol feed back to reduce corticotropin releasing factor (CRF) from the hypothalamus, thereby reducing ACTH secretion and subsequently reducing cortisol secretion.

Failure of the adrenal cortex to produce enough cortisol and aldosterone usually results from primary atrophy of the adrenals but sometimes is caused by destruction of the glands by cancer or tuberculosis. This produces Addison's disease, resulting in electrolyte problems, low blood pressure, weakness, and inability to survive a mild stress such as an otherwise nonfatal disease. Hypersecretion of the adrenal cortex produces Cushing's disease (Figure 12-7). This is usually caused by excessive secretion of ACTH or by adrenal tumor, and results in a complex of problems associated with excessive cortical secretions.

Adrenal Medulla

When stimulated by the sympathetic nervous system, the adrenal medulla releases epinephrine, also known as adrenaline, and norepinephrine. Both of these hormones cause glucose formation, vasoconstriction, and acceleration of the heartbeat. Epinephrine has a much greater effect on the heart; norepinephrine has more effect on the blood vessels. Epinephrine also accelerates the metabolism of the body to a greater degree than does norepinephrine. The stimulating effect of these hormones lasts about ten times longer than does direct stimulation by the sympathetic nervous system. The sympathetic nerves stimulate release of these hormones in response to panic situations. This

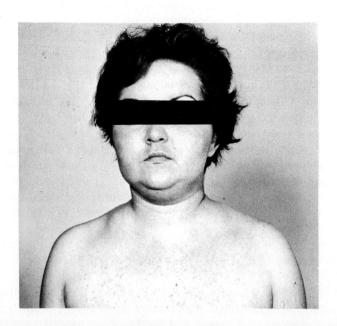

Figure 12-7   Cushing's disease showing facial edema.

release results in increased energy availability (glucose), rapid pulse, increased blood pressure, and generalized smooth muscle constriction, all of which combine to create a "flight or fight" mode.

THYROID GLAND

The thyroid gland secretes two major hormones, thyroxin and calcitonin. The secretory cells of the thyroid are arranged to form follicles, each filled with thyroglobin, a protein (Figure 12-8).

Thyroxin. Thyroxin, tetraiodothyrnine (T4), is manufactured in the follicle-lining cells and stored bound to thyroglobin. When needed, it splits off from the thyroglobin molecule and acts to increase the metabolic rate of most of the body's tissues. At the cellular level, T_4, a prohormone, is converted to triiodothyronine (T_3), the active hormone. Although T_3 has a much shorter life than T_4, it is about four times more potent. The effect of a single dose of thyroxin begins to be felt in 2 or 3 days and reaches a peak in 10 to 14 days. All of the hormone is eventually expended in about 40 days.

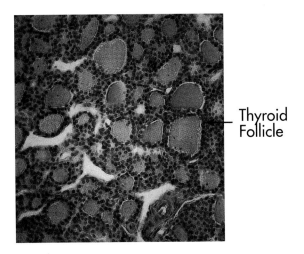

Thyroid Follicle

Figure 12-8 Thyroxin is produced by the epithelium and stored in the follicles of the thyroid gland.

Iodine is a part of the thyroxin molecule. A deficiency of this mineral produces a deficiency of thyroxin. This results in increased TSH releasing factor, increased TSH production, and hyperplasia of the thyroid gland. As the gland enlarges with still insufficient iodine and without thyroxin production, it becomes visible as a swelling in the front of the neck, a condition known as goiter. Thyroxin deficiency produces cold intolerance, slow thought processes, depression, obesity, decreased metabolism, and lethargy. Thyroxin deficiency also contributes to loss of libido and may produce complete failure of sexual function.

Hyperthyroidism is caused by excess production of thyroxin. This can occur as a result of tumor or thyroid gland hyperplasia or by production of a thyroid stimulator called long-acting thyroid stimulator (LATS). LATS, an immunoglobulin that functions as TSH, is often found in hyperthyroid patients. The levels of TSH in these patients is usually low because of the feedback effect of high levels of circulating thyroid hormones and high metabolic rates. Hyperthyroidism often produces bulging of the eyes, exophthalmos, as well as increased metabolism and nervousness (the Don Knotts syndrome). See Figure 12-9.

Calcitonin. Also called thyrocalcitonin, this hormone is secreted by the parafollicular cells of the thyroid gland (See Figure 12-8.) It decreases blood calcium by initially slowing the activity and production of osteoclasts, thus slowing the dissolving of calcium from bone. About 1 hour after a calcitonin injection, osteoblastic activity increases, with an increase in the removal of calcium from the blood and deposition of calcium in bone. (See the discussion of osteoporosis in Chapter 4.)

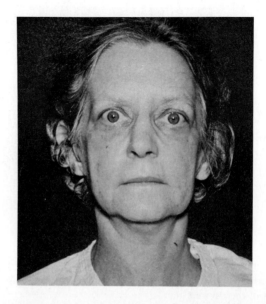

Figure 12-9   Protruding eyes are characteristic of Graves' disease.

PARATHYROID GLANDS

Parathyroid Hormone. Parathyroid glands are a set of four tiny organs usually adjacent to the thyroid. They secrete parathyroid hormone in response to low blood calcium. This hormone increases gastrointestinal absorption of calcium, increases tubular reabsorption, decreases renal excretion of calcium, and activates osteoclasts, causing them to dissolve calcium from the bone. Parathormone also causes rapid formation of new osteoclasts. This formation increases the levels of circulating calcium, which decreases parathormone secretion.

THE THYMUS

The glandular function of the thymus remained unknown long after the functions of other endocrine glands had been discovered. As discussed in Chapter 10, the most important function of this basically lymphoid organ is serving as a center for the maturation of T lymphocytes. The primary endocrine secretion of the thymus is a collection of hormones called thymosins. At least some thymosins are responsible for the immune competence of T cells. The decline or involution of the thymus is clearly an aspect of senescence and probably contributes to the overall decline in immune function seen with age.

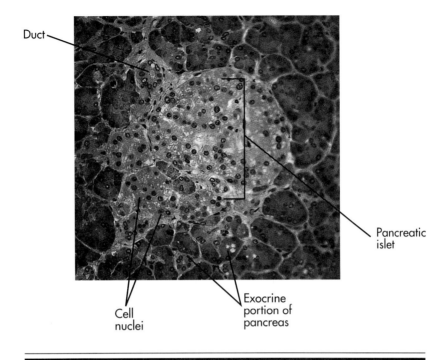

Duct

Pancreatic islet

Cell nuclei

Exocrine portion of pancreas

Figure 12-10   The islets of Langerhans are endocrine tissues that secrete insulin and glucagon. They are embedded in the pancreas.

PANCREATIC ISLETS OF LANGERHANS

The islets of Langerhans are endocrine cells located in an otherwise exocrine gland, the pancreas (discussed in Chapter 7). They secrete insulin and glucagon, two hormones that are important factors in metabolism. The endocrine components of the pancreas are shown in Figure 12-10 and Color Plate 15, *B*.

Insulin. The beta cells of the islets of Langerhans secrete Insulin in response to high blood glucose concentration. As this hormone circulates through the bloodstream and diffuses into the interstitial fluid, it binds with the membranes of the body's cells where it facilitates the transport of glucose across the cell membrane. This enables the glucose to move into the cells where it can be metabolized. There it will either be combined with oxygen in the process of respiration, releasing energy for cellular activity, or it will be converted to glycerol and fatty acids and stored as fat.

Insulin also increases glycogen formation and storage in the liver, and decreases glycogen utilization. Lack of insulin effect after the intake of sugars, or other food that is converted to sugars, results in the disease diabetes mellitus (to be discussed later this chapter). Inadequate insulin can also contribute to atherosclerosis. When insulin is lacking, fat cells break down, releasing glycerol and fatty acids into the blood. Increasing concentrations of free fatty acids circulate in combination with albumin. The increased level of fatty acids further depresses glucose utilization. Excess fatty acids are synthesized by the liver into triglycerides, cholesterol, and phospholipids, all of which also are released into the blood. Their levels may increase to as much as five times normal. They may then accumulate in the tunica intima of arteries where they are involved in the formation of atheromas, as shown in Figure 9-6.

Glucagon. Produced by the alpha cells of the islets of Langerhans, glucagon is released in response to low blood glucose and causes the liver to convert glycogen to glucose, which is released into the bloodstream. Glucagon also increases the formation of glucose from amino acids in the liver. The effect of glucagon (raising blood glucose) is opposite to that of insulin. Glucagon is secreted in response to low blood sugar and has the effect of raising it. Insulin is secreted in response to high blood sugar and has the effect of lowering it. Glucagon is also secreted in response to exercise and high levels of circulating amino acids. A person deficient in blood glucose, is said to be hypoglycemic. A person with an excess of blood glucose has hyperglycemia.

THE TESTES, OVARIES AND PLACENTA

The endocrine function of these organs was discussed earlier in this chapter. *Testosterone,* produced by the Leydig cells, is responsible for male secondary sexual characteristics. The follicular epithelium of the ovaries produces *estrogens* responsible for female secondary sexual characteristics. The corpus luteum produces *progesterone.* The placenta, a temporary endocrine tissue, exists only during pregnancy to nourish the developing baby. The feedback mechanisms and detailed effects of effects of androgens and estrogens were discussed in Chapter 11.

THE GASTRIC AND INTESTINAL MUCOSA

The hormones of the stomach and intestine were discussed in Chapter 7. Most of them, such as gastrin, secretin, and cholecystokinin, have a very local sphere of influence. Other examples are *vasoactive peptide,* which increases intestinal motility, and *somatostatin* which suppresses intestinal activity.

Gastrin is a hormone secreted by the gastric and duodenal mucosa in response to the ingestion of food, especially protein. This hormone circulates through the bloodstream and causes the gastric glands to secrete digestive enzymes and acid. *Secretin* is secreted by the duodenal mucosa in response to acid entering from the stomach. Secretin causes its target cells in the pancreas to release trypsinogen and other enzymes into the duodenum. Secretin may also facilitate the release of enzymes from the intestinal glands. *Cholecystokinin* is secreted by the duodenal mucosa in response to the ingestion of fat. It stimulates the pancreas and gall bladder to contract, releasing their contents into the duodenum.

Table 12-1   Age-Related Changes in the Endocrine System

STRUCTURE GLAND AFFECTED	CHANGE
Most glands	Some degree of glandular atrophy and fibrosis
	Decreased rate of secretion
	Decreased rate of metabolic destruction of hormones produced. Circulating hormone levels remain fairly constant because of this decrease, or decreased excretion through the kidneys
Target tissues of most glands	Changes in sensitivity
Hypothalamus, pituitary	Progressive loss of sensitivity to feedback control

Adapted from Eckel and Hofeldt.

❖ *The Aging Endocrine System*

GENERAL AGE-RELATED CHANGES

Table 12-1 summarizes age-related changes in the endocrine system.

The pituitary gland begins to atrophy after middle age but shows no decrease in growth hormone secreting cells or prolactin secreting cells. Growth hormone fails to be suppressed by nutrients as it was in earlier years. In women, follicle stimulating hormone increases by ten to fourteen times after estrogen levels begin to decline as a result of decreased ovarian responsiveness to FSH and subsequent decreased estrogen production. Experimental studies have shown that even if estrogen production were increased, the feedback response of the pituitary gland would be greatly reduced. Luteinizing hormone follows a similar pattern to that of FSH.

Growth hormone declines to about 50% of the levels of early adulthood by age 65. Recent preliminary studies have indicated that replacement of this growth hormone may have favorable effects. Growth hormone treatments given to elderly males appeared to increase lean body mass, skin thickness, and bone density. The experiment suggested that growth hormone decline could be a significant feature in the aging process. Growth hormone may accelerate the progress of already existing cancers. In addition, excesses of growth hormone have been associated with lipid and pancreas problems. Much more work needs to be done before growth hormone can be considered as therapy for senescence.

A person's ability to concentrate urine decreases with age. This may be due in part to increasing renal tubular resistance to antidiuretic hormone. However, when there is excessive ADH production, decreased tubular sensitivity may result in fairly normal urine concentration.

The aging thyroid shows an infiltration of lymphocytes and a decrease in glandular cells. Some of this may be associated in part with autoimmune destruction of the gland. There is often an increased prevalence of antithyroglobin antibodies with age, especially in women. Nodularity of the thyroid also increases with age. Postmortem studies found nodules in 27% of the elderly patients examined.

Hypothyroidism occurs in 3% to 4% of elderly hospital admissions, but hyperthyroidism occurs in less than 1%. Thyroid disorders are more common in women than men. Diagnosis of thyroid dysfunction in the elderly is difficult. Although symptoms such as intolerance of cold, obesity, slow thought processes, and depression can be caused by hypothyroidism, they are often caused by other factors. The symptoms of thyroid problems are often atypical in the elderly. Some of these are discussed later in this chapter (see thyrotoxicosis).

Iodine uptake by the thyroid shows little age-related change. When hypothyroidism occurs in elderly patients, it is usually in association with increased thyroid stimulating hormone, indicating that the problem lies in lack of thyroid responsiveness rather than with pituitary or hypothalamus insufficiency. Drug interactions with the thyroid are common, and many drugs distort thyroid function tests.

A recent study demonstrated that production of cortisol by the adrenal cortex declines by 25% in elderly men, but plasma cortisol levels are unchanged. As with most other hormones, renal clearance of cortisol is diminished. Responsiveness of the adrenal cortex to ACTH apparently does not decline; nor does responsiveness of the pituitary to cortisol feedback. Several other cortical steroids decrease with age, including progesterone and aldosterone. Cortical androgens decrease in both sexes with age. These reductions can affect attitudes, behavior, and related physical factors. The response to the renin-aldosterone mechanism also declines with age.

The adrenal medulla may increase its catecholamine and norepinephrine production in elderly subjects, but the cardiovascular response to norepinephrine may decline. Nerve ending production of norepinephrine may decline in some patients, producing a delayed blood pressure response to moving to an upright posture (orthostatic hypotension).

Changes in the ovary and testes were discussed in Chapter 11.

The islets of Langerhans in the pancreas show little age-related change. In spite of this, there is usually a substantial decline in glucose tolerance. This could be caused by decreased islet response to high blood glucose, resulting in inadequate insulin production. An alternative explanation involves decreased cell membrane responsiveness to insulin. The latter is supported by recent evidence of increased insulin levels in response to oral glucose in some affected elderly. Yet another explanation is production of an ineffective form of insulin.

We can find little to show that any change occurs in gastrin and secretin levels in the elderly. Gastrin levels increase in cases of atrophic gastritis as a result of lack of feedback.

Diabetes mellitis and thyroid dysfunction are the two most important general categories of endocrine/metabolic disorders in the elderly. They are followed by the consequences of menopause in women, hypocalcemia and hypercalcemia (either dietary-absorptive or parathyroid in origin), electrolyte problems related to adrenal or renal changes, malignancy-generated imbalances, and drug-related endocrine problems. One researcher has observed that there is likely to be, on average, at least one endocrine-related problem in each new elderly patient.

ENDOCRINE DISORDERS ASSOCIATED WITH ADVANCED AGE
Diabetes Mellitus

This ailment is by far the most common endocrine disease in the elderly. It is diagnosed when hyperglycemia occurs and may be associated with ketosis, the production of ketones that give the breath a distinctive odor. The diagnosis is made when the fasting blood glucose level is above 140 μgm/dl or when there is an abnormal response on a glucose tolerance test. This disease is associated with inadequate insulin effects and with degenerative changes in the nerves and blood vessels. Oral glucose tolerance tests show that many elderly people have inappropriately high blood sugar levels without having diabetes mellitis. This condition is known as carbohydrate intolerance. These individuals have inadequate insulin response due to ineffective forms of insulin or decreased numbers of insulin receptors in the body's cells.

Diabetes may be juvenile onset (type I, or insulin dependent) or adult onset (type II). In this discussion we are concerned only with type II. This form of diabetes is often found in overweight patients. Symptoms of type II diabetes may be controlled by diet. Often a program for obese people designed to reduce weight with restriction of sugar intake is adequate and eliminates all the symptoms of the disease. With others a combination of diet and perhaps oral agents to stimulate insulin production and effectiveness are adequate, although the use of the latter has recently been questioned. The primary treatment for severe diabetes, even type II, remains insulin.

Because diabetics do not handle infections normally and they often have poor circulation in their feet, foot care becomes extremely important in the management of this disease. Many develop foot infection as the disease progresses. The infections heal slowly and are usually associated with ischemic vascular problems. Nerve problems, atherosclerosis, and small vessel disease are likely occurrences with advanced, long-standing diabetes. Eye problems develop due to the vascular changes in the retina and are called diabetic retinopathy. Diabetic retinopathy usually develops from small vessel hemorrhages obscuring the macula lutea of the retina. Cataracts can also develop as a result of sustained hyperglycemia. While eye and kidney problems (diabetic nephropathy) are more common in type I disease, they do occur in the type II patient. Such problems in both types of cases can result from either abnormal protein deposition or vascular changes.

Thyrotoxicosis

The classic symptoms of Graves' disease, or hyperthyroidism, (exophthalmos, large goiter, and systemic symptoms) are rarely seen in the elderly. The causes of thyrotoxicosis include:

1. Autoimmune disease—The typical case of Graves' disease is caused by an immunoglobulin that stimulates the thyroid gland and leads to proliferation of soft tissue behind the eyes, causing them to bulge. See Figure 12-9.
2. Thyroid tumors—Thyroid hormone production can arise from thyroid tumors that become autonomous and do not require TSH stimulation.
3. Excessive consumption of thyroid medication—This can result from accidental ingestion or overprescription. The latter is fairly common due to failure to recognize decreased renal clearance and lowered metabolic demand.
4. Physical and psychologic stress can precipitate hyperthyroidism.
5. Radioactive [25]iodine-labeled ([25]I) fibrinogen injected to study venous thrombosis may precipitate thyrotoxicosis.

Thyrotoxicosis is often manifested by heat intolerance, weakness, and muscle wasting, sometimes with tremor, or by weight loss, anoxia, and either diarrhea or constipation. It may be associated with either depression or agitation, both of which symptoms can be confused with senile dementia. The prognosis for these patients is usually excellent with treatment consisting of radioactive iodine or sometimes thyroid blockade with medication. Surgery is usually required only for treating neoplasms.

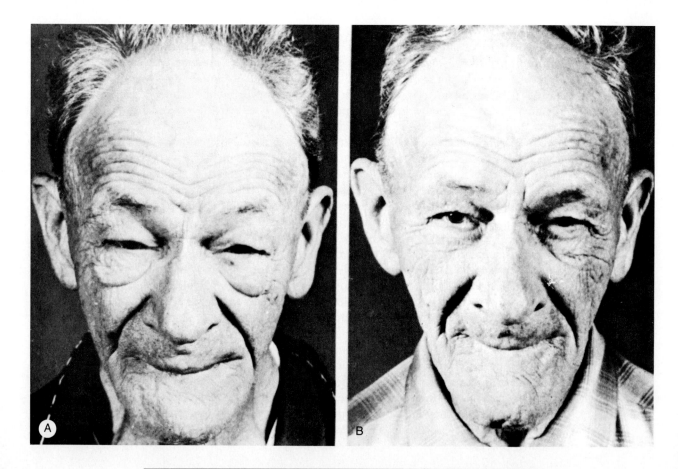

Figure 12-11   A patient with myxedema caused by hypothyroidism, **(A)** before, and **(B)** after treatment with thyroxine.

Hypothyroidism

Hypothyroidism (myxedema) is more common than thyrotoxicosis in the elderly. Like hyperthyroidism, it is probably most often the result of autoimmune disease. It is also a common aftermath of treatments of thyrotoxicosis and may result from pituitary or hypothalamus failure with inadequate secretion of thyroid stimulating hormone. Elevated TSH with low thyroxin levels indicates primary hypothyroidism, usually resulting from autoimmune disease.

Hypothyroidism in the elderly is associated with many failings that typically accompany aging. It produces cold intolerance, lethargy, immobility, increased likelihood of falling, slow thought processes, and constipation. The condition may progress if not treated and the progression is sometimes explained away as "rapid aging." Many other factors, such as a cold house or prolonged inactivity, may contribute to hypothermia, but a hypothyroid condition can blunt the response to hypothermia and make it more severe and even life-threatening. Hypothyroidism is usually treated with (T_4) thyroxine. Figure 12-11 shows the response to replacement therapy with thyroxine.

Cushing's Disease

Cushing's disease is a syndrome resulting from hypersecretion of the adrenal cortex. Cortisol, aldosterone, and androgens are commonly all overproduced. The cause may be an adrenal tumor or hyperplasia of both adrenal glands, either as a result of excessive ACTH from the anterior pituitary or from an "ectopic" source of the hormone such as certain carcinomas.

The effects of excesses of these hormones include mobilization of fat from the lower parts of the body with deposition of fat in the upper abdomen and thorax. The antiinsulin effects of cortisol may induce the symptoms of diabetes mellitus. The excess of aldosterone contributes to an elevated blood pressure and facial edema. The androgens may cause acne and growth of facial hair in women. The condition is treated by removing parts of the adrenal glands or destroying parts of the pituitary, depending on the cause.

Addison's Disease

Addison's disease involves deficiency in adrenocortical hormones. This is usually caused by atrophy of the adrenal cortices, usually as a result of destruction of the glands by autoimmune disease. The glands are sometimes damaged by tuberculosis infection or by cancer.

With the decline in cortisol, anterior pituitary secretions increase. These include ACTH and melanocyte stimulating hormone. The latter causes increased melanin deposition, especially around the lips and nipples. The primary effect of the disease is the loss of sodium and chloride ions and water because of insufficient aldosterone. Hyperkalemia (potassium excess) develops and acidosis results from failure of the renal tubules to exchange hydrogen ions for sodium ions in the urine. Death from shock may occur within 4 days to 2 weeks after total cessation of aldosterone secretion. The disease is treated by daily replacement of the missing hormones. However, the body is still unable to produce the large amounts of cortisol needed in times of stress.

CASE HISTORIES

❖ Diabetes Mellitus

D.H. is a 32-year-old single woman who at the age of 14 had a "flu-like" illness that degenerated into coma. In the hospital a random blood sugar was found to be 1080 µg/dl (normal less than 110), and she had profound acidosis with both acetones and ketones present. This resulted in a diagnosis of diabetic coma. When the sugar in the blood cannot be utilized, alternate metabolic pathways occur, leading to formation of these acidic waste products. These inhibit cellular functions, contributing to diabetic coma. During her hospitalization she was stabilized using subcutaneous insulin and advised about proper diabetic care, including insulin injections and the need for careful monitoring of her diet. She did not accept her illness and was noncompliant with her diet and insulin administration. This resulted in repeated hospitalizations for management of uncontrolled diabetes.

Despite her medical problems, she was a good student, finished high school, and obtained a degree in medical technology. While employed as a medical technologist she developed a limp and was found by her local doctor to have a severe peripheral neuropathy with an open, infected ulcer involving her lateral malleolus (outer ankle), extending into the bone. She was hospitalized for 3 weeks for good diabetic control, IV antibiotics, and debridement of the wound. It partially healed and she returned to work doing limited walking. At this time her vision rapidly deteriorated, with almost complete blindness in one eye and reduced vision in the other. Examination revealed severe diabetic retinopathy with superimposed large hemorrhages. Laser ablation (distruction) of the abnormal blood vessels resulted in stabilization of her vision at a barely useable level.

During this period, great emphasis was placed on her need for good diabetic control. However, follow-ups demonstrated blood sugars in the 200 to 300 range and her high glycohemoglobin levels indicated poor blood sugar control. When she attempted to control her diabetes, she experienced four episodes of insulin-induced hypoglycemia and coma (insulin shock). Two of the episodes resulted in car accidents and her driver's license was suspended. During the insulin reactions she also became confused, her mentation slowed, and she developed an unexplained weight gain. She was returned to the hospital.

Laboratory tests revealed a blood urea nitrogen (BUN) reading of 98 and a creatinine measurement of 6.9 μg/dl, suggesting diabetic nephropathy. She was started on peritoneal dialysis and experienced improved mentation, fluid balance, and sense of well-being. She continued to have insulin reactions, however, and experienced four episodes of peritonitis. She was then switched to hemodialysis with a marked improvement in her glucose control, mentation, and outlook. She has been able to stabilize her blood sugars and is waiting for a kidney transplant.

Comment. While this case deals with a young person, it is presented here because it demonstrates nearly all the complications of diabetes mellitus. She had diffuse small vessel disease affecting her skin, eyes, kidneys, and many other organs. Her ischemic ulcers and retinopathy are typical of uncontrolled diabetes. Medical observers currently believe diabetic complications are caused by exposure of the tissues to high glucose levels. Good control of blood sugars is believed to substantially delay these complications, or even avoid them completely.

It is unknown to what extent this patient failed to control her blood sugar because she refused to recognize the dangers involved, or whether she had "brittle" diabetes. "Brittle" diabetics have unexplained wild swings in blood sugar and exaggerated insulin effects. This causes the tissues to be bathed in high sugar solution, subsequently damaging their proteins. Some believe, however, that the complications of diabetes are independent of blood sugar levels.

The sensory peripheral neuropathy (loss of feeling) in diabetes often has a "glove and stocking" distribution. This can allow significant physical damage or infection because the warning pain is absent. Similarly, the renal and small vessel damage can become severe before the patient is aware of it. Unfortunately, this allows many patients to believe that they can ignore diabetes without consequences.

❖ Hypothyroidism

E.F. was a 76-year-old retired plumber who developed shaking chills, dizziness, weakness, and headache. The symptoms began abruptly. He was hospitalized the same day with a reduced BP (60/30), a rapid pulse (118 beats per minute), a temperature of 37.2° C, and an RR of 20. His skin was warm and dry. He knew who he was, where he was, and what day it was, but was slow to respond and slow in speech with a deep voice. His physical examination was unremarkable with the exception of a 6 cm pulsatile mass in the abdomen. A complete blood count showed no anemia, elevated WBCs (17,000) and increased bands (20%), suggesting inflammation. Blood Chemistry assays were abnormal for glucose at 122 (normal when fasting is 110), and a CK of 505 (<232 IU/L normal). CK is a muscle enzyme and fractionation did not reveal an increase in the MB band, the heart fraction. The ECG was abnormal with T wave changes and a prolonged Q-T interval, nonspecific changes. Because of the abnormal ECG and the elevated CK, the patient was admitted to the coronary care unit (CCU) to rule out myocardial infarction (MI).

In the CCU he quickly became lethargic. Urine cultures grew *E. coli* sensitive to ampicillin. He was treated for a urinary tract infection (UTI) with possible sepsis. Follow-up ECGs and CK and LDH enzyme readings were stable, indicating that E.F. did not have an evolving MI. Consultation was obtained.

The consultant found the patient to be lethargic with a slow response to questions, and drifting off to sleep without finishing sentences. His voice was raspy and hoarse and his skin was dry and scaly, especially over the tibia. His hair was coarse and dry with thinning of the eyebrows laterally. There was mild periorbital edema, and the deep tendon reflexes (DTRs) were brisk but with a slow return phase. A stat T_4 measurement was found to be low, less than 1.0 mg/ml with an elevated thyroid stimulating hormone (TSH) of 50. E.F. was diagnosed as a typical case of hypothyroidism and treated with thyroxin. He made an uneventful recovery.

With a return to normal mentation, the patient revealed that he had quit taking all of his medications 3 to 4 months prior to the episode. Forty years earlier he had been hyperthyroid and was treated with radioactive iodine, destroying his thyroid gland. One of his medications was replacement thyroxin.

Comment. Profound hypothyroidism (myxedema) often leads to coma and can be confused with many disease processes. In this case, the CK elevation that often accompanies hypothyroidism led the admitting physicians into thinking the patient was having a heart attack. The alteration of consciousness brought on by the lack of thyroxin precluded obtaining an accurate past history, and some myxedematous patients do not have a history of prior thyroid problems. Luckily, the classic findings (dry skin, coarse hair, deep voice, slow mentation and action, and elevated CK) were recognized by the consultant and appropriate therapy was begun.

❖ Hyperthyroidism

C.E. is an 81-year-old widow who developed fainting spells. Because of her mental incompetence she had been a patient in a nursing home for the last 2 years. During the last week the staff had observed her falling several times, after which she maintained consciousness. Medical attention was sought when she could not be aroused after the last episode. The patient was an unreliable historian, so most of the history was obtained from the nursing home staff. They reported that she had experienced mild weight loss, diarrhea, and tended to wear her sweater less often than she had earlier.

Her physical examination showed her to have mild systolic hypertension (BP 190/68) and an increased pulse (98/min). Her skin was moist with a fine velvety feeling. Her eyes appeared slightly protuberant and had a mild horizontal nystagmus (fine, quick back and forth movements). Her tongue was tremulous, as were her fingers. The tremors were fine and occurred regularly. A diffusely enlarged, symmetric, nontender thyroid was palpable. The chest, heart, and abdomen were unremarkable. Her DTRs were brisk and exaggerated bilaterally. Pertinent laboratory work included an elevated total thyroxine and T_3 uptake with a high free thyroxin index. The chemistry profile showed an elevated amount of alkaline phosphatase and a low cholesterol level; both are nonspecific findings that can be seen in hyperthyroidism. A screening ECG was normal but her heart rate became irregular and rapid during a fainting attack. An ECG conducted during an attack showed atrial fibrillation which spontaneously reverted to normal rhythms.

A diagnosis of hyperthyroidism with paroxysmal atrial fibrillation was made. She was treated with radioactive iodine (I^{131}) followed by thyroid replacement.

Comment. Thyroid conditions are the most common endocrine abnormalities after diabetes mellitus. In the elderly, both hypothyroidism and hyperthyroidism can be difficult to diagnose because many of the symptoms can be attributed to aging. This is especially true of hypothyroidism. Because thyroid disease is easy to treat, it is important to make a correct diagnosis. This case, with weight loss, diarrhea, heat intolerance, goiter, suggestion of exopthalmos, tremor, fine and moist skin, hyperreflexia, and elevated serum thyroxin, included most of the classic symptoms of hyperthyroidism.

❖ Cushing's Disease

R.C. was a 58-year-old homemaker who noticed a gradual onset of weakness. She usually walked for a half hour in a local park in the morning but had recently stopped doing this. Recently she has not had enough energy to go shopping. Over the same time period she had noticed a 10-pound weight gain. On examination she was an obese, middle-age female in no acute distress with a round, pink facial appearance. Her obesity was most pronounced in her abdomen, back, and thorax with relatively thin arms and legs. Her skin was fine and moist with many "stretch marks." The remaining portions of the exam were normal with the exception of hypertension and weakness. Preliminary laboratory evaluation showed no evidence of anemia, but 2+ glucose was found in her urine with no ketones. Glucose is normally not found in the urine unless blood sugar levels are abnormally high.

A working diagnosis of adult onset diabetes mellitus was made. She was advised to follow an 1800-calorie ADA (American Diabetic Association) diet. She also was scheduled for a 3-hour glucose tolerance test that subsequently showed glucosuria on all but the fasting specimens and revealed blood sugar levels diagnostic of diabetes mellitus. The physician added an oral hypoglycemia agent to the management and scheduled her for a consultation with a diabetologist.

On her referral visit, the consultant was impressed with the obesity and stretch marks. Further laboratory work demonstrated high AM and PM serum cortisols that did not suppress with dexamethasone and were associated with high urinary cortisol levels. Cortisol metabolism shows circadian variation with high levels in the morning and operates through a feedback mechanism with ACTH. R.C. lacked diurnal variation and high cortisol levels did not suppress future cortisol production. She therefore had uncontrolled excess production. CT scans of the adrenal glands showed an irregular tumor mass in the right adrenal that was surgically removed and found to be a benign adenoma. Her glucoses rapidly returned to normal and she experienced mild weight loss associated with a significant increase in exercise tolerance.

Comment. Cushing's syndrome is caused by an increased secretion of cortisol. This hormone has an antiinsulin effect that is associated with a diabetes-like carbohydrate intolerance, unusual weight distribution, catabolic activity leading to loss of muscle mass, and weakness. In addition, the mineralocorticoid effects (aldosterone) lead to expanded blood volume and hypertension. An associated problem that can lead to a nursing home admission is osteoporosis caused by the catabolism of bone.

❖ Addison's Disease

D.H. is an 82-year-old homemaker whose husband was suffering from the late stages of Alzheimer's disease. While caring for him she noticed that she tired easily. Her prior medical history included hypothyroidism due to Hashimoto's thyroiditis, for which she had been taking replacement thyroid medication. Because fatigue had been a symptom of her thyroid deficiency, she increased the amount of thyroid she was taking but this produced no improvement. In fact, the weakness increased over a 1-month period to the point that she could no longer care for her husband.

When seen by her physician she was noted to have recently lost weight and to appear well tanned. This was surprising because she had been almost homebound in caring for her husband. In addition, she had lost muscle strength, especially noticeable in her hands and legs. Her blood pressure, which was usually near normal at 130/75, was found to be 96/70. Her thyroid was still enlarged but unchanged from her prior examination. Routine testing showed normal ECG readings, chemistry profile, blood count, and electrolyte levels. Her T_4 (thyroxin) level was mildly elevated at 15.1 µg/dl (normal to 12.5) and consistent with her increased intake.

Serum cortisol was ordered because of the hypotension and her pigmentation. Her cortisol was low (6 µg/dl) and lacked the normal diurnal variation (AM 7 to 25; PM 2 to 14). Her ACTH level was found to be markedly elevated at over 200 picograms/ml (normal less than 60). After 3 days of ACTH infusion, her serum cortisol remained at 6. A CT scan of the abdomen revealed small adrenal glands. A diagnosis of idiopathic adrenal atrophy was made. She was treated with corticosteroids with resultant complete recovery of her strength.

Comment. Addison's disease can be difficult to diagnose in the elderly because of its nonspecific symptoms. Fortunately, many of these patients have increased production of melanocyte stimulating hormone, resulting in unexpected pigmentation that alerts the physician. The melanocyte stimulating hormone increase is because of excessive pituitary activity resulting from lack of corticosteroid feedback.

The most common cause of adrenal atrophy is prolonged corticosteroid use. This case is an example of the second most common cause—"idiopathic" atrophy. It is often associated with problems such as diabetes or Hashimoto's thyroiditis and may be secondary to an immune destruction of the gland. The disease is important to recognize because these patients can not respond well to stress and may be unable to combat another illness such as pneumonia.

BIBLIOGRAPHY

Davis PJ: Aging and endocrine function, *Baillieres clin Endocrinol Metab* 8:603-619, 1979.

Eckel RH, Hofeldt FD: *Endocrinology and metabolism in the elderly.* In Schrier, ed: *Clinical internal medicine in the aged,* Philadelphia, 1982, WB Saunders.

Green MF: *The endocrine system.* In Pathy MSJ, ed: *Principles and practice of geriatric medicine,* Philadelphia, 1985 WB Saunders.

Greenspan FS, Forsham PH: *Basic clinical endocrinology,* ed 2, New York, 1986, Lang.

Kjeldsen SE, and others: Renal contribution to plasma catecholamines—effect of age, *Scand J Clin Lab Invest* 42:461-466, 1982.

Rudman D, and others: Effects of growth hormone in men over 60 years old, *N Engl J Med* 323(1):1-6, 1990.

Vance ML: Growth hormone for the elderly?, *N Engl J Med* 323(1):52-54, 1990.

13 *The Nervous System*

RELEVANT ANATOMY AND PHYSIOLOGY
General Functions

The nervous system has three functions: sensory, integrative, and motor. The sensory function involves receiving stimuli from both inside and outside the body and transmitting impulses caused by those stimuli from the sensory receptors to the integrative areas. The integrative areas process the sensory stimuli, combining them with data already in the memory. This combination of information creates a new series of stimuli that we perceive as a new concept or thought. The profusely branching integrative nerves involved in this process are called interneurons. If this integration results in a decision to take action, the nerves involved in motor function will be stimulated. Motor nerve tracts extend from the integrative areas to muscles, and when stimulated, cause them to contract. Rarely, with some reflex actions, sensory nerves stimulate motor nerves directly, without involving interneurons.

All of these functions are affected by the aging process. Aging impairs the reception of such stimuli as taste, smell, touch, hearing, and sight. The processing and integration of information is disrupted, resulting in errors in judgments and loss of memory. The execution of ideas by the motor nerves may be compromised by tremors, delays, and varying degrees of paralysis.

The nervous system is separated anatomically into central and peripheral divisions (Figure 13-1). The central nervous system (CNS) consists of the brain and spinal cord. The peripheral nervous system (PNS) is made up of the nerves and ganglia residing outside of the CNS. These nerves are either sensory, (receiving information about the body or its environment), or motor (stimulating muscle action). The motor nerves are further subdivided into somatic nerves that operate under conscious control, and autonomic nerves that generally operate without conscious thought. Both kinds of motor nerves function under the direction of the CNS as it integrates a tremendous volume of sensory impulses with an even more overwhelming load of stored information.

The Structure and Operation of Nerves

Nerve cells, called neurons, are of many shapes and sizes (Figure 13-2). They receive impulses through the many branches of their dendrites, and conduct them through the cell body to the axon. The axon delivers the impulse to a muscle, another nerve, or a gland. The longest neurons extend from the tip of the great toe to the base of

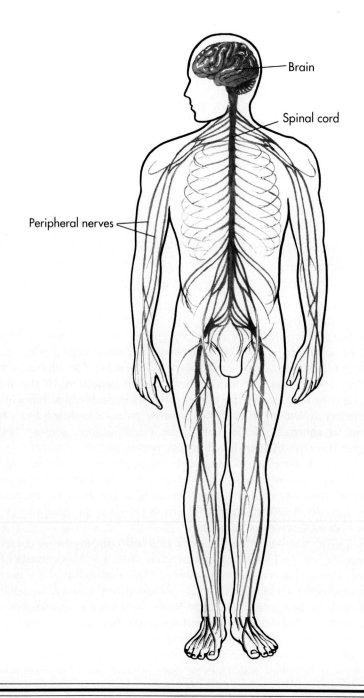

Brain

Spinal cord

Peripheral nerves

Figure 13-1   The central nervous system includes the brain and the spinal cord. The peripheral nervous system includes the nerves extending from the CNS into the body.

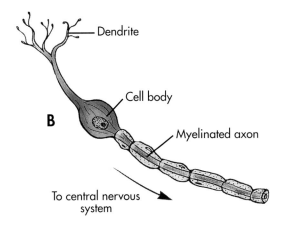

Dendrite

Cell body

B

Myelinated axon

To central nervous system

Figure 13-2   A neuron is a nerve cell. This one is bipolar, having one dendrite and one axon. Many neurons are much more complex.

the spine, a distance of 3 feet or more. Even so, they are microscopic cells. Neurons outside of the CNS are grouped into bundles called nerves.

Nerve cells usually do not reproduce after the first 6 months of life. When reproduction does occur, the process is slow and often ineffective. For this reason any damage to or death of nerve cells may produce permanent impairment. Although neurons and the nerves they make up have excellent support and maintenance from certain surrounding tissues, it is not surprising that decreased nervous function is one of the markers of advancing age.

Different neurons respond to a variety of different stimuli, including sudden temperature change, pressure, light, various chemical agents, and neurotransmitters. These stimuli change the permeability of the neuron membrane, causing positively charged sodium ions to temporarily pass into the cell and leaving a negative charge on the outside of the membrane (Figure 13-2). As soon as this depolarization occurs, cellular energy returns the sodium to the outside of the membrane (repolarization). This movement of sodium displaces potassium ions. The next sodium ion in line then moves into the cell and the process is repeated. Consequently, a wave of depolarization moves from the point of stimulus along the axon or dendrite. Nerve impulses move in only one direction along an axon or dendrite because the cell membrane requires time to recover from depolarization. Impulses normally move from dendrites toward axons because only dendrites have neurotransmitter receptors and only axons can release neurotransmitters. The role of neurotransmitters in nerve stimulation is discussed in the following passage

Synapses and Neuromuscular Junctions

The associated ends of two connecting nerves are collectively called a synapse (Figure 13-3). When one nerve stimulates another, it does so by releasing a neurotransmitter from synaptic vesicles at the end of an axon. The neurotransmitter released from the synaptic vesicles diffuses across the synaptic cleft to neurotransmitter receptors located in the dendrite of the adjacent neuron. When the concentration of neurotransmitter in the receptors is great enough to stimulate the dendrite of the receiving neuron, the neuron is said to have reached threshold. Threshold concentrations of neurotrans-

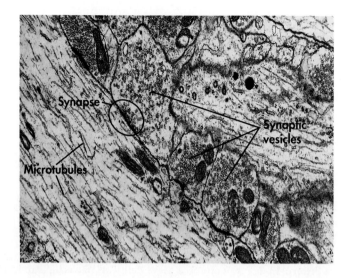

Figure 13-3 Impulses can move only one direction across a synapse, the point of contact between two nerves. Only dendrites have neurotransmitter receptors and only axons have neurotransmitter releasers.

mitter on the receiving dendrite cause membrane changes, resulting in depolarization of the dendrite.

If a neurotransmitter such as acetylcholine (present in most neurons) were allowed to remain in the receptors of the dendrite in threshold concentrations, the dendrite would be continuously depolarized. However, an enzyme produced by the dendrite splits acetylcholine into its components, which cannot separately stimulate a neuron. These components are then returned to the axon, where they can be reused. In sympathetic neurons, norepinephrine is returned intact to the axon terminal for reuse.

A motor neuron stimulates a muscle at the point of contact, which is called a neuromuscular junction. See Chapter 5 for information on neuromuscular junctions.

Many factors can affect synapses. For example, when the blood becomes too acidic (pH below 7.35), as occurs with a variety of medical conditions, conduction is inhibited and nerve activity is suppressed. On the other hand, when blood becomes too alkaline (pH above 7.45) the nerves can become too excitable. A variety of age-related conditions can threaten the maintenance of a pH level in the 7.35 to 7.45 range. For example, emphysema commonly increases blood acidity (causing acidosis), and liver failure can increase blood alkalinity (causing alkalosis).

Supportive Tissues

Axons of peripheral nerves are often covered with a myelin sheath (Figure 13-4), which is an extension of a supportive cell called a Schwann cell. Similar myelin sheaths are produced around axons in the central nervous system, but it is done by a different cell called an oligodendrocyte. Each Schwann cell or oligodendrocyte wraps an axon in layers of membrane (myelin). A small area between each of the supportive cells, the node of Ranvier, remains unmyelinated. The myelin sheath greatly increases the speed of nerve impulse conduction by permitting the depolarization to jump from one node of

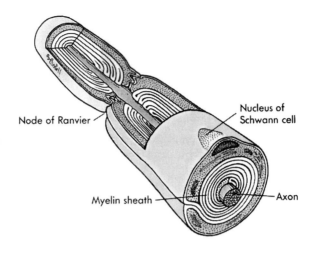

Node of Ranvier

Nucleus of Schwann cell

Myelin sheath

Axon

Figure 13-4   Myelinated nerves conduct impulses much more rapidly than unmyelinated nerves. The myelin sheath is laid down by Schwann cells in peripheral nerves and by oligodendrocytes in the central nervous system.

Ranvier to the next. It also contributes to the insulation and maintenance of the nerve. The white matter of the brain and spinal cord is composed of myelinated nerves; the gray matter consists mostly of cell bodies with profusely branching unmyelinated fibers, as well as the neurons for some fibers that become myelinated when they leave the gray matter.

Supporting cells in the central nervous system are called neuroglia, which means "nerve glue." Schwann cells have sometimes been called peripheral neuroglia. Oligodendrocytes are a form of glial cell that provides myelin for certain nerves in the CNS. Astrocytes attach nerve cells to each other and to blood vessels. Microglia are small phagocytic cells that consume bacteria and parts of damaged cells. Ependymal cells line the ventricles of the brain and circulate in the cerebrospinal fluid.

Regeneration and Repair of Nervous Tissue

One function of the Schwann cells (also called neurilemma cells) involves aiding in the repair of damage. For example, when peripheral axons are severed, the part of the axon distal to the injury degenerates and dies. The remains of the dead axons are consumed by phagocytes. The Schwann cells then multiply and form a tube where the axon was. However, this formation will occur only if the injured area is not too large and the deposition of scar tissue is not too great. If the tube is satisfactory, the living axon will grow a new process through the tube and perhaps reinnervate the structure. Unfortunately, the oligodendrocytes of the central nervous system cannot perform similar repairs. Because of this, severing of the spinal cord results in permanent loss of both sensory and motor function of the limbs and tissues dependent on the severed nerves. However, spinal reflexes usually return after a period of several days or weeks.

Injury to the spinal cord, the spinal nerves, or their branches can cause loss of function or feeling. Severing all or part of the spinal cord will cause paralysis. The types of paralysis produced are classified as follows:

Monoplegia (one limb only)
Diplegia (two limbs)
Paraplegia (both upper or both lower limbs)
Hemiplegia (both limbs on one side)
Quadriplegia (all four limbs)

The Spinal Cord and Spinal Nerves

The spinal cord is that portion of the CNS surrounded by the vertebrae. It is composed of gray matter and white matter (Figure 13-5). Two spinal nerves extend outward from the spaces between each of the vertebrae, with one nerve on each side of the spinal column. Each spinal nerve has a dorsal sensory component and a ventral motor component. The dorsal root receives impulses from the organs and tissues of the body and the ventral root stimulates the muscles associated with those organs and tissues. The cell bodies of the sensory nerves are located in a ganglion, an expanded area of each dorsal root. The cell bodies of the motor nerves are located in the ventral horn of the gray matter within the spinal cord.

When certain sensory nerves are stimulated, the impulse enters the spinal cord and usually passes through one or more interneurons to a motor neuron. This results in almost instant contraction of flexor muscles of the limb that was stimulated. This rapid response by the spinal cord is known as a reflex arc. Spinal reflexes may prevent damage when immediate action is required and there is not enough time to produce a more

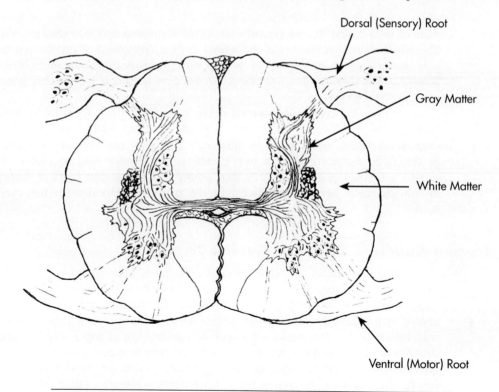

Dorsal (Sensory) Root

Gray Matter

White Matter

Ventral (Motor) Root

Figure 13-5 A cross section of spinal cord. The spinal cord has gray matter in the center and white matter on the outside. The dorsal root of a spinal nerve is always sensory and the ventral root is always motor.

complex decision from the brain. For example, when a person accidentally places a hand on a hot stove burner, the hand will usually be withdrawn before any pain is felt or even before the brain is aware that a problem exists. Tests of spinal reflexes such as the knee-jerk response can be a useful diagnostic tool when certain nerve problems are suspected.

Spinal nerves are designated according to the name and number of the vertebra with which they are most associated. For example, the second thoracic vertebra gives rise to the T2 spinal nerves. Each spinal nerve supplies a segment of the body; these segments can be mapped on the surface of the skin. Each segment is called a dermatome. (See Figure 3-7, *B*).

The Peripheral Nervous System

Most of the nerves we think about are large nerves that innervate skeletal muscles or receive stimuli from the sense organs. These are channeled to or from the conscious mind and are referred to as the somatic peripheral nervous system. They originate from the spinal cord and from nerves arising from the brain. The 12 pairs of cranial nerves and 31 pairs of spinal nerves are responsible for all purposeful movements. Some of these nerves are sensory nerves extending from receptors to the spinal cord; others are motor nerves extending from the spinal cord to the muscles they innervate. Still others are mixed, having both sensory and motor components. The cranial nerves and their functions are explained in Table 13-1.

The first branches of the spinal nerves after they exit the vertebrae are called rami. These sometimes join with other nerves to form networks called plexuses, which

TABLE 13-1   Cranial Nerves

NERVE	IMPULSE CONDUCTS	FUNCTIONS
I Olfactory	Nose → brain	Smell
II Optic	Eye → brain	Vision
III Oculomotor	Brain → eye muscles	Eye movements
IV Trochlear	Brain → external eye muscles	Eye movements
V Trigeminal	Skin and mucous membrane of head and teeth → brain; also brain → chewing muscles	Sensations of face, scalp, and teeth chewing
VI Abducens	Brain → external eye muscles	Outward eye movements
VII Facial	Taste buds → brain; brain → face muscles	Taste; contraction of muscles of facial expression
VIII Acoustic	Ear → brain	Hearing, balance
IX Glossopharyngeal	Throat and taste buds → brain; Brain → throat muscles and salivary glands	Sensations of throat, taste, swallowing, secretion of saliva
X Vagus	Throat, larynx, and organs in thoracic and abdominal cavities → brain; Brain → muscles of throat and organs in thoracic and abdominal cavities	Sensations of throat and larynx and of thoracic and abdominal organs; swallowing, voice production, slowing of heartbeat, acceleration of peristalsis (gut movements)
XI Spinal accessory	Brain → some shoulder and neck muscles	Shoulder movements; turning of head
XII Hypoglossal	Brain → tongue muscles	Tongue movements

TABLE 13-2   Spinal Nerve Plexuses

PLEXUS	ORIGIN	MAJOR NERVES	MUSCLES INNERVATED	SKIN INNERVATED
Cervical	C1-C4		Several neck muscles	Neck and posterior head
	(Neck)	Phrenic	Diaphragm	
Brachial	C5-T1	Axillary	Two shoulder muscles	Part of shoulder
	(Shoulder)	Radial	Posterior arm and forearm muscles	Posterior arm, forearm, and hand
		Musculocutaneous	Anterior arm muscles	Radial surface of forearm
		Ulnar	Two anterior forearm muscles, most of the intrinsic hand muscles	Ulnar side of hand
		Median	Most anterior forearm muscles, some intrinsic hand muscles	Radial side of hand
Lumbosacral	L1-S4	Obturator	Medial thigh muscles	Medial thigh
		Femoral	Anterior thigh muscles (extensors)	Anterior thigh, medial leg and foot
		Sciatic		
		Tibial	Posterior thigh muscles (flexors), anterior and posterior leg muscles, most foot muscles	Sole of foot
		Common peroneal	Lateral thigh and leg, some foot muscles	Anterior and lateral leg, and dorsal foot

are named according to their locations within the body. Table 13-2 lists the spinal nerves and their functions.

The Autonomic Nervous Systems

Nerve tracts that originate from certain spinal and cranial nerves do not involve the conscious mind, although they are regulated by the brain. These tracts comprise the autonomic nervous system (ANS), which is made up entirely of motor nerves. The ANS responds to sensory impulses received by the PNS and processed through the brain. The ANS is regulated by the hypothalamus. It stimulates both smooth and cardiac muscle, as well as glandular epithelium.

The ANS is divided into sympathetic and parasympathetic components, each of which has generally opposite effects. Figure 13-6 shows the target organs and effects produced by the sympathetic and parasympathic divisions of the ANS.

Several health problems appear to arise in part because of unwanted action by the autonomic nervous system. Asthma attacks may be initiated by an allergen or other unknown cause, but parasympathetic stimulation of the bronchioles is known to be a major contributor. The parasympathetic response may greatly magnify vascular shock, critically reducing blood pressure. Heart attack may be initiated by parasympathetic

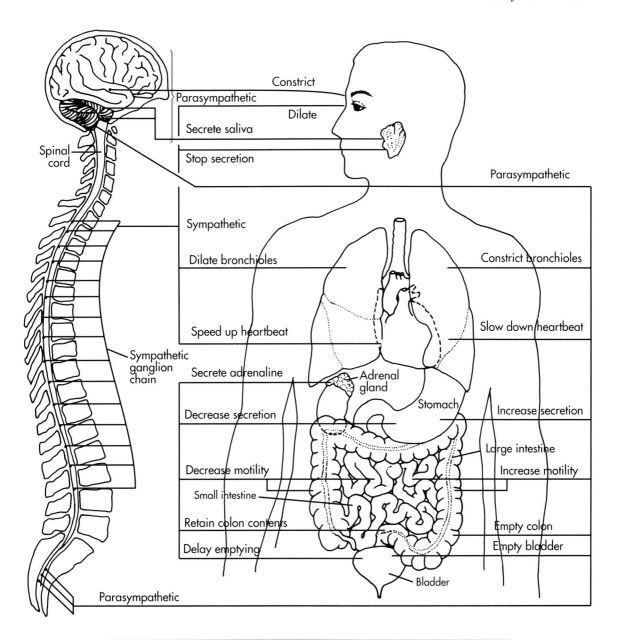

Figure 13-6 Innervation of the major target organs by the ANS. Sympathetic fibers arise through the cranium and sacrum; parasympathetic fibers are associated with thoracic and lumbar vertebrae.

constriction of the coronary arteries, depriving the heart of adequate oxygen. There is good reason that a person who has just lost a spouse or even a business has an increased chance of heart attack. Sympathetic constriction of arteries can contribute to hypertension. Evidence is mounting that the conscious mind may be able to reverse these responses to some extent through "biofeedback." For example, one may slow excessive heart rate by thinking of something pleasant and relaxing.

The sympathetic nerves leave the CNS through the spinal nerves of the thorax and lumbar regions of the spinal cord. The parasympathetic system exits through certain cranial nerves and spinal nerves exiting through the sacral vertebrae. Both systems synapse with other nerves by utilizing ganglia outside the spinal column. The sympathetic system utilizes a chain of connected ganglia lying on each side of the spinal column. These sympathetic trunks act as relay centers for sympathetic nerve transmission. The parasympathetic system utilizes ganglia near the visceral muscles they innervate. If a sympathetic trunk is damaged, that side of the body may exhibit the symptoms of parasympathetic stimulation.

The Brain

The brain could be described as a greatly expanded portion of the spinal cord, arranged inside out. Unlike the spinal cord, the gray matter is on the outside and the white matter is inside. In addition, with a few exceptions, each side of the brain controls the opposite side of the body. Consequently, severe damage to the right cerebral hemisphere will cause loss of function on the left side of the body. A longitudinal section of a human brain is shown in Figure 13-7, *A,* and a magnetic resonance image of a normal brain is shown in Figure 13-7, *B.* Some of the functions of the labeled structures in Figure 13-7, *A* are listed below.

Cerebrum. The cerebral hemispheres are the largest part of the mammalian brain. The gray matter of the cerebrum, the cerebral cortex, is a layer of interneurons deposited on a thick mass of white matter. The cerebral cortex of mammals is usually folded, forming ridges called gyri and shallow grooves called sulci. The two cerebral hemispheres are connected internally by the white matter of the corpus callosum.

Each cerebral hemisphere is considered to have four lobes, with each performing a variety of functions (Figure 13-7, *A*). The primary function of the frontal lobe is the elaboration of thought and emotion. Part of the frontal lobe, Broca's area, which develops in only one hemisphere, translates thoughts into words. The most posterior part of the frontal lobe, the area adjacent to the central sulcus, is the main motor area. It translates thoughts into actions and controls all voluntary motion.

The parietal lobe interprets sensations, (except for light and sound) received from the body. The temporal lobe receives and interprets sound. Large parts of both the parietal and temporal lobes are concerned with speech. The temporal lobe also interprets smell. The occipital lobe receives visual stimuli from the eyes. The more anterior parts of the occipital lobe interpret these stimuli.

Cerebellum. The cerebellum lies immediately below the occipital lobes of the cerebrum. Like the cerebrum, it is separated into two hemispheres that are subdivided into lobes. Also like the cerebrum, the cerebellum has a cortex of gray matter and a core of white matter. The white matter is a tree-like structure called the arbor vitae. It surrounds and radiates away from a cavity called the fourth ventricle. The overall function of the cerebellum involves coordinating the skeletal muscles and maintaining balance.

The cerebellum ensures the relaxation of opposing muscles while a muscle is being contracted. It aids in determining how many muscle fibers in a muscle will be stimulated to carry out a smooth movement as a result of impulses received from the motor areas of the cerebrum. Damage to the cerebellum may cause ataxia, a loss of coordination associated with a lack of awareness of limb position. Because some nerves from the

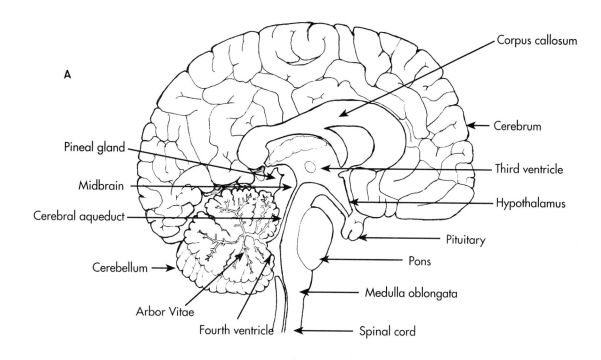

A

- Corpus callosum
- Cerebrum
- Third ventricle
- Hypothalamus
- Pituitary
- Pons
- Medulla oblongata
- Spinal cord
- Pineal gland
- Midbrain
- Cerebral aqueduct
- Cerebellum
- Arbor Vitae
- Fourth ventricle

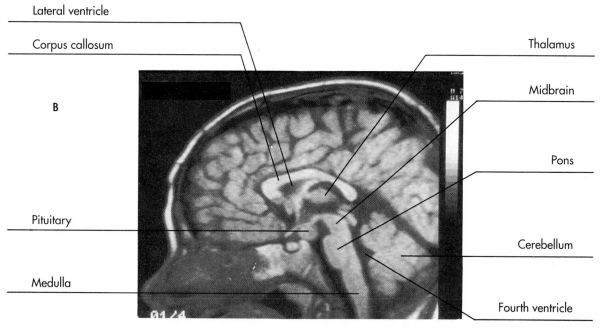

B

- Lateral ventricle
- Corpus callosum
- Pituitary
- Medulla
- Thalamus
- Midbrain
- Pons
- Cerebellum
- Fourth ventricle

Figure 13-7   A, This longitudinal section of brain shows many of the major parts. The central cavity, the third ventricle, is connected to the fourth ventricle in the cerebellum by a canal called the cerebral aqueduct. Also connected to the third ventricle are two lateral ventricles extending into the sides of the brain, the temporal lobes. Dotted lines indicate the four lobes of the cerebral hemisphere. **B:** MR image of a normal brain.

cerebellum cross to the opposite side of the body, then back again, coordination failure may be on the same side of the body as the brain damage.

The Brain Stem

The brain stem is divided into the medulla, pons, midbrain, and diencephalon (Figure 13-7, *A*). Although it contains islands of gray matter (nuclei) that regulate important functions, the brain stem is basically a two-way communication network. Sensory and motor impulses flow through the brain stem.

Medulla Oblongata. The medulla transmits motor and sensory impulses between the spinal cord and the rest of the brain. Most of the motor fibers from the cerebral cortex pass through this structure to the opposite sides of the spinal cord, therefore stimulating muscles on the side of the body opposite to the side of the brain where they originated. Sensory impulses coming in from each side of the body pass to opposite sides of the brain through the medulla, then on to the thalamus and cerebral cortex.

The medulla contains the vital centers that control heart rate, breathing, and blood vessel diameter. Together with the pons and midbrain of the brain stem, as well as the spinal cord and thalamus, it contains reticular fibers. The medulla contains fibers that are part of the reticular activating system, process that maintains consciousness.

Pons. The pons is a continuation of the medulla that ends in the midbrain. Nerve fibers pass through the pons into the cerebellum. The sensory and motor fibers in the medulla also pass through the pons into the midbrain before connecting with the cerebrum. Fibers from the spinal cord, cerebral cortex, and various nuclei pass through the pons into the cerebellum, become coordinated there, and return to exit through the pons.

Midbrain. The midbrain is formed from the motor and sensory fibers flowing between the pons and cerebrum. In addition, it contains nuclei responsible for eye reflexes. Two pigmented nuclei, the substantia nigra, and red nucleus, both important in control of tremor and undesired muscle contraction, are also located in the midbrain.

The substantia nigra filters out unwanted impulses from motor commands. Damage to this nucleus is associated with Parkinson's disease. The red nucleus is the center of the reticular formation of the midbrain. It receives fibers from the cerebral cortex and cerebellum.

Diencephalon. This structure makes up the walls of the third ventricle, the fluid-filled central cavity of the brain. It contains the thamalmus and hypothalamus. The thalamus is a distribution center for sensory impulses entering from the cerebrum, cerebellum, spinal cord, ears, eyes, and tongue. It performs a similar function for voluntary motor nerve tracts. It is also important in the arousal process and in emotional expression.

The hypothalamus is located under the thalamus on each side of the third ventricle and at its base. This part of the brain is the link between the pituitary gland and the cerebral cortex. The hypothalamus secretes neurochemicals that act as hormones operating directly on the body. Others hypothalamic products stimulate the anterior pituitary to release its hormones. The hypothalamus also regulates the autonomic nervous system. Because of the effect of the hypothalamus on the endocrine system through the pituitary and on behavior and ability to perform through the autonomic nervous system, this organ may play an integral part in the establishment of psychosomatic illness.

The hypothalamus, in conjunction with the thalamus, regulates sleep cycles, appetite, water balance, and body temperature.

The Basal Ganglia. The basal ganglia are collections of gray matter (nuclei) located around the lateral and third ventricles in close association with the thalamus and hypothalamus. They are summarized in Figure 13-8 and include nuclei important in coordination or refinement of motion. In conjunction with the substantia nigra, these nuclei help to control and refine movement and allow for the fine facial movements that give us our expressions. Along with parts of the temporal lobe of the cerebrum, one of these nuclei, the amygdaloid nucleus, is important in emotion.

The Limbic System. The limbic system uses some parts of the brain that have already been described. Many of its tissues are associated with the lateral and third ventricles (the large cavities of the brain). The limbic system consists of a cerebral component known as the limbic lobe, a portion of the floor of the lateral ventricles making up the hippocampus, the amygdaloid nucleus near the anterior tip of the lateral ventricles, and the hypothalamus and thalamus. This system influences the perception and expression of intense emotion. It is the center for anger, terror, fear, happiness, pleasure, and sexual arousal. Nervous disorders affecting the limbic system can produce excesses or deficiencies in any of these emotions.

The Meninges, Ventricles, and Cerebrospinal Fluid

The meninges. The brain and spinal cord are fragile organs and are encased in bone. This provides protection from most trauma. In addition, between the nervous tissue and the bone are three encasing membranes called meninges (Figure 13-8). Immediately beneath the bone lies a tough connective tissue layer, the dura mater. This usu-

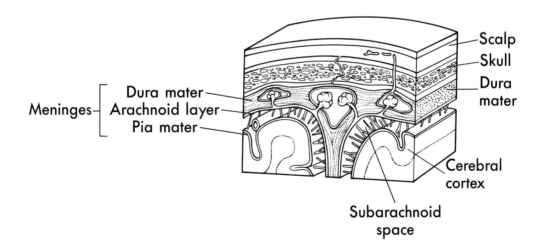

Figure 13-8 The meninges. Between the cranium and the brain are three protective tissues: the tough dura mater, the vascular arachnoid layer, and the fragile pia mater.

ally protects the brain from bone fragments in case of skull fracture. Under the dura mater lies a membrane closely associated with a web of blood vessels. This membrane and its associated blood vessels are called the arachnoid layer. The subarachnoid space is filled with cerebrospinal fluid (CSF) and separates the arachnoid layer from the fragile, inner membrane, the pia mater. The pia mater adheres to the brain along its contours.

The Ventricles and Canals of the Brain. The spinal cord and brain are hollow and contain cerebrospinal fluid (CSF). Extending the length of the spinal cord is a narrow tube called the central canal. At the brain, this canal enlarges, forming the fourth ventricle that extends into the cerebellum. Cerebrospinal fluid can enter or leave the fourth ventricle through numerous openings and pass into the subarachnoid space outside of the brain.

One opening of the fourth ventricle leads to the third ventricle, the central cavity of the brain. Two openings in the third ventricle permit passage of CSF from the first and second ventricles (usually called lateral ventricles) located in the cerebrum.

Cerebrospinal Fluid. The bone and dura mater usually protect the brain from penetration but do little to protect against damage from sudden acceleration or deceleration. The CSF provides this kind of protection. The CSF is a filtrate of blood plasma produced by the choroid plexus, located in the ventricles of the brain.

CSF is normally absorbed by vascular tissue in a CSF-filled space in the duramater called the superior sagittal sinus. It is normally absorbed by this tissue as rapidly as it is produced. Hemorrhage into the fluid however, can increase CSF pressure to the point that brain cells are killed. Likewise, inflammation of the meninges (meningitis) can decrease the rate of absorption of CSF to the point that CSF pressure is increased. Measurements of CSF pressure, types of cells present, and the presence of bacteria are often important tests when some classes of brain disease are suspected. These data are obtained by conducting a "spinal tap," a procedure that requires inserting a needle between the lumbar vertebrae and withdrawing a small amount of cerebrospinal fluid.

The Blood-Brain Barrier

Brain tissue is protected by specialized capillaries possessing walls that are highly selective to the passage of materials through their membranes. This selectivity creates a barrier against many toxic substances. While freely permitting the flow of oxygen, carbon dioxide, and sugar to and from the brain, the blood-brain barrier prevents the passage of proteins and antibiotics. This creates a problem in the treatment of infection in the brain and in the use of some chemotherapeutic agents to treat brain cancer.

THE AGING NERVOUS SYSTEM
General Observations

The nervous system, like many of the organ systems, exhibits pathologic conditions that emerge gradually from the progressive senescence of the body. Therefore it is extremely difficult to distinguish changes that are caused by aging from those resulting from many other, often poorly understood disease processes. In addition, nervous system performance can be affected by disorders arising elsewhere in the body. Some of these are summarized in Table 13-3.

The number of seemingly internal factors affecting the nervous system is much smaller. They include genetic disease such as amyotrophic lateral sclerosis, tumor arising within the brain, and conditions such as Alzheimer's disease and Parkinson's disease.

TABLE 13-3   Factors Outside the Nervous System that Can Affect Its Function

FACTOR	EFFECT
Vascular disease	Interrupts supply of oxygen (atherosclerosis)
Autoimmune disease	Immune reactions to antigen-antibody complexes deposited on nervous tissue (multiple sclerosis)
Skeletal disease	Bone encroaches on neural pathways (Paget's disease)
	Degeneration of neck vertebrae (cervical spondylosis)
Tumors	Most are metastases from outside the CNS
Trauma	Hemorrhage pressure can damage nervous tissue
	Falls can damage the brain
Viral, fungal, bacterial infections	Damage to CNS (encephalitis, meningitis, dementia of syphilis)
Viral, bacterial agents	Damage to PNS (herpes zoster, leprosy, syphilis)

The Aging Peripheral Nervous System

Aging brings a decline in many of the peripheral senses (touch and taste for example, which will be discussed in Chapter 14). Beyond the changes in receptors, however, degenerative changes occur in the peripheral nerves. Some myelinated nerves lose myelin, and Schwann cells may cease to function. This may slow peripheral nerve conduction.

Pain can occur in the peripheral nerves for a number of reasons. Nerve inflammation, or neuritis, can be caused by injury, repeated stress or pressure, inflammation, or infection. Changes in body proportion combined with losses of tissue elasticity may produce pressure or friction against a nerve. Sciatica an example of nerve pain that often originates in this way. This condition involves severe pain along the sciatic nerve extending from the lower back into and along the leg. It most commonly results from a slipped vertebral disk but may result from entrapment or injury anyplace along the nerve.

The Aging Spinal Cord

Nerve cells and processes in the spinal cord are lost as a person ages. They are replaced with glial cells during a process called gliosis. Although lipofuscin granules are seen in some spinal nerve cells, many age-related problems seen in the brain, such as neurofibrillary tangles, senile plaques, and amyloid deposits (all discussed later in this chapter) have not been found. Deep tendon (jerk) reflexes decline with age; most such reflexes are lost by age 90. It is suspected that a loss of neurons is responsible for this.

Damage to the spinal cord or spinal nerves most often results from displaced vertebral disks or vertebral fractures associated with osteoporosis. Paget's disease sometimes deposits bone in the neural canal or where nerves exit from the spinal cord, and pinches a nerve.

The Aging Brain

General Changes in the Brain and Meninges. The meninges thicken with age. This is especially true with the dura mater; perhaps less so with the pia mater and the arachnoid layer. The dura mater adheres tightly to the skull in the very old. The blood vessels of the arachnoid layer reflect the vascular condition of the rest of the body. Hypertensive hemorrhage of vessels in the white matter of the cerebrum is a common cause of stroke.

Brain weight declines from about 1400 g in a typical young male adult to 1200 g to 1300 g in the very old. However, these individuals do not show brain dysfunction. While a small loss is apparent by age 50, the loss accelerates after age 60. An increase in size of the ventricles is usually associated with this. Much of this atrophy involves the frontal cerebral cortex, producing an increase in the size of the sulci and a decrease in the size of the gyri. The number of neurons declines from the approximately 20 billion present at birth, but the loss varies greatly in different parts of the brain and with different tissues. For example, some specialized neurons called Purkinje's cells decline by about 2.5% per decade between birth and 100 years of age. Astrocyte increases accompany an increase in glial fiber as a person ages, but no change is seen in oligodendrocytes or microglia. Net loss of cells in the cerebral cortex may be as much as 40% in a normal brain. Reflecting the cellular loss, total brain protein declines about 15% between ages 35 and 70. Extracellular water increases, probably as a result of the increased extracellular space that develops as cells are lost.

Extracellular Senile Plaque and Intracellular Pigment. Extracellular deposits of amyloid polypeptides increase with age. Amyloid deposition is involved with the formation of senile plaque, a mixture of amyloid, various cell organelles and degenerating nerve process. It is found in the cerebral gray matter. The protein forms in association with arterioles and capillaries, and is increased in some dementias such as Alzheimer's disease. Recent studies have shown subtle differences between the amyloid present in normal aging and that found in Alzheimer's disease.

Lipofuscin pigments within the cytoplasm in old neurons increase with age, as they do in many other cells, especially muscle cells. Although the pigments are deposited heavily in some parts of the brain, they may be completely absent in other sections. They have not been associated with any pathologic condition.

Cellular Changes. The axon terminals swell in certain areas of old brains. This swelling often precedes a degenerative loss of function. This phenomenon occurs in other parts of the brain and even in the peripheral nerves after trauma, or with certain pathologic conditions. Cortical dendrites lose cytoplasmic spiny extensions throughout life. Eventually branches and processes are lost; this is accompanied by swelling of the tissue. Finally the entire nerve is lost and replaced by neuroglia.

Neurofibrillary tangles were first demonstrated by Alzheimer, who discovered them in the gray matter of the cerebral cortex (Figure 13-9). These fibrils are especially prominent in the hippocampus and appear to be composed of abnormal neurofilaments, small twisted strands of a nerve protein. They become more abundant and thicken with age, forming a variety of shapes. Eventually, they may interfere with normal cellular metabolic and transport mechanisms.

Cerebrovascular Changes. Cerebral arteries and capillaries differ from those in the rest of the body in that both have a reduced tunica externa. The arteries also have a somewhat thinner tunica media. Some of the function of the tunica externa may be filled by the surrounding extensions of astrocytes. These arteries are subject to the ravages of atherosclerosis to at least as great an extent as arteries located elsewhere in the body. The brain is less tolerant of vascular failure than the rest of the body, so even minor hemorrhages or emboli are likely to have significant effects on function.

Neurochemical Changes. Acetylcholine transferase, an enzyme essential for the synthesis of acetylcholine, declines in the brains of patients who have Alzheimer's

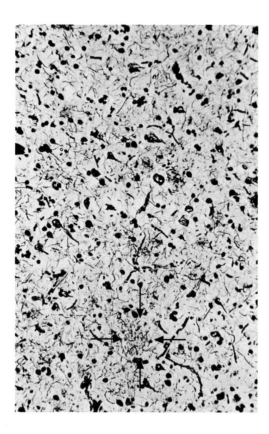

Figure 13-9   This section of cortex cell contains many senile plaques and neurofibrillary tangles. They are characteristic of an aged brain and of some brain diseases such as Alzheimer's disease. (Courtesy James S. Nelson. From Kissane JM: *Anderson's Pathology,* vol 2, ed 9, St. Louis, 1990, Mosby.)

disease. This suggests that acetylcholine may decline as well. A decline in norepinephrine with age has been found in parts of the brain, and lower levels of dopamine have been measured, especially in the basal ganglia of the elderly. Lower levels of norepinephrine have been associated with depression, and dopamine deficiency is a primary factor in Parkinson's disease. A number of studies have noted changes in the mineral content of aged brain tissue.

Diseases and Conditions Associated with Aging Nervous Systems

Peripheral Neuropathy. Peripheral neuropathy is often manifested in the elderly, although it is not the result of aging. Common causes include diabetes mellitus and nutritional problems such as vitamin B_{12} deficiency. Typically diabetes causes loss of sensation starting at the tips of the limbs and slowly progressing toward the trunk. Vitamin B_{12} nerve damage is associated with position (kinesthetic) sense and the ability to feel vibration. Other neuropathies in the elderly can be associated with toxins or tumors. They can also be idiopathic and involve both movement and sensation.

Amyotrophic Lateral Sclerosis. Also known as Lou Gehrig's disease, this condition results from degeneration of the ventral roots (motor neurons) of the spinal cord.

The deteriorating neurons are replaced by glial cells. The disease has its onset in middle age, usually affecting people between 40 and 70 years of age. It normally requires about 4 years to progress to death, although remissions may occur that extend the course of the disease over as many as 15 years. The first symptom is often weakness of a leg, resulting in an unexpected fall. As the disease progresses, muscle atrophy of disuse develops in the effected limbs. Eventually the muscles involved in swallowing and speech are affected, requiring insertion of a feeding tube.

Multiple Sclerosis (MS). Multiple sclerosis is a disease of the myelinated nerves of the spinal cord and brain. It often begins between ages 20 and 40 and may progress to death in 10 to 30 years. Early symptoms are often mild and transient, involving attacks of weakness or loss of feeling in a leg or hand, or irregularities in vision. As the disease progresses, the attacks become more severe, last for a longer period, and involve more of the body. Remissions may last a year or two early in the disease but become shorter and less complete as the disease progresses. Each attack characteristically leaves the patient more disabled. Death usually results from infection promoted by lack of motor activity. For example, inability to empty the bladder may lead to kidney infection; inability to breathe deeply may contribute to lung infection.

MS is believed to be caused by an autoimmune process, perhaps involving immunocomplexes deposited on oligodendrocytes. This deposition results in inflammation and destruction of the myelin sheaths the oligodendrocytes produce. Astrocytes proliferate and plaque is deposited. As the inflammation subsides, some nerves return to normal, restoring some of the lost function. With each attack, some function is permanently lost. Consequently, effects of the attacks progressively worsen. The extent and periods of the remissions decline as the white matter is replaced with sclerotic, plaque-encased fibers.

Traumatic Injury. Because of problems with balance or weakness, the elderly are more likely to fall than are younger people. These falls commonly cause injury; when this injury is to the cranium it is especially serious. Severe trauma to the head may produce subdural hematoma, creating a pool of blood between the dura mater and the cranium. The resulting pressure on the underlying brain can produce permanent damage or death.

Cerebrovascular Accident. Cerebrovascular accident (CVA), or stroke, refers to vascular events that produce acute brain dysfunction. CVA accounts for about 15% of all deaths; 60% of these are by cerebral infarction and 40% are caused by cerebral and subarachnoid hemorrhage. A major contributor to both types of stroke is atherosclerosis (see Chapter 9). Cerebral infarction usually involves a vascular obstruction resulting from an embolism arising elsewhere in the body, from a thrombus building on an atheroma located in one of the cerebral arteries, or from a bleeding vessel no longer capable of carrying blood to its normal destination. Cerebral hemorrhage is usually a product of hypertension affecting an artery weakened by atherosclerosis. The hemorrhage sometimes results from the rupture of an aneurysm.

Vascular obstruction or rupture with bleeding produces ischemia and death of the tissue dependent on the involved vessel (Figure 13-10 and Color Plate 16, *A*). The extent of the area damaged is related to the size of the vessel blocked, whether it is blocked partially or entirely, and the amount of collateral circulation available. Nearby collateral vessels will dilate and may resupply an ischemic area over time, sometimes restoring injured but still-living neurons to function. Dead nerves will not be replaced; their function, if regained, must be assumed by other parts of the brain.

Following ischemic infarction, the affected tissue dies. Additional small hemor-

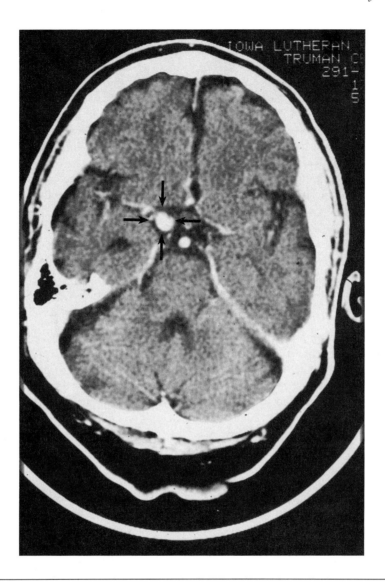

Figure 13-10 This MRI shows a threatening berry aneurysm (arrow). Rupture of this expanded blood vessel can be life-threatening.

rhages may follow necrosis. Inflammation occurs with tissue edema, swelling, and proliferation of astrocytes within the first 3 days. Microglia enlarge as they phagocytize debris and remove necrotic tissue. Ultimately, a scar is produced. Autopsies performed on these brains often reveal a cystic, softened area in the region of the infarction, usually brownish. It is not uncommon to find several such areas, indicating past CVAs.

Massive cerebral hemorrhage occurs most commonly due to rupture of a congenital aneurysm often located at the base of the brain. These berry aneurysms result in large hemorrhages that surround the brain and invade the ventricles, following the path of the cerebrospinal fluid. Most spontaneous hemorrhages do not involve aneurysms and occur in the region of the basal ganglia. Less commonly, hemorrhage occurs near the pons and cerebellum. Massive hemorrhage may flood the ventricles and extend into the subarachnoid space. Small hemorrhages produce small hematomas in or beneath the cortex.

Large vessel obstruction or large amounts of hemorrhage cause severe impairment of function of the part of the body controlled by the effected part of the brain. This is usually on the side of the body opposite the side of the brain where the CVA occurred. Large vessel problems are dramatic, sudden events, and are not associated with gradual dementia. Most dementias are not vascular in origin but when they are, they are most likely to be the result of numerous small vessel occlusions and are usually associated with hypertension. Primary dementias are those that arise from changes within the brain not known to be of external cause. Secondary dementias are caused by external factors such as vitamin B12 deficiency.

The variety of CVA-induced malfunctions is extensive. Examples include loss of ability to perform tasks, loss of memory, loss of reasoning ability, and slow mentation. An interesting example is aphasia, which involves the loss of comprehension of spoken or written words and sometimes the ability to comprehend a simple concept. A similar dysfunction involves loss of ability to ascribe the proper function to an object. This is sometimes called "object blindness." Even without demonstrable CVA, these conditions often appear with age-related dementias.

Transient Ischemic Attack. Many patients experience reversible small strokes prior to the large one resulting in permanent damage. When there is loss of a central nervous system function that persists for a short time but recovery occurs within 24 hours, the condition is known as transient ischemic attack (TIA). These are usually caused by small thrombocytic emboli arising from atherosclerotic large neck arteries. The small clot travels to the brain and causes function loss. When the clot is cleared, function is restored. Often surgical removal of the atherosclerotic material in the neck vessels is required. Anticoagulation therapy is used to prevent formation of more clots. Without appropriate therapy, many of these cases evolve into CVAs that cause permanent damage.

Senile Dementia and Senile Dementia of the Alzheimer's Type (SDAT). These terms refer to the gradual loss of brain function associated with advancing age. The condition involves slow mentation, decreased ability to concentrate, decreased reasoning ability, confusion, and sometimes varying amounts of belligerence. Aphasia and object blindness may occur in advanced cases. Senile dementias are primary degenerative diseases of the brain, but there are diseases that resemble senile dementias and are not due to aging. Such ailments often involve infarctions caused by small artery disease.

Typically, brains with primary senile dementia show a buildup of lipofuscin granules and neurofibrillary tangles. Often there are many senile plaque deposits. There is atrophy of the frontal and (sometimes) temporal lobes of the cerebral cortex and enlargement of the ventricles. Because these changes also are seen in cases of Alzheimer's disease, the latter may be a special form of senile dementia. Slowly progressing, late onset Alzheimer's disease is usually called SDAT. A surprising number of people who have demonstrated no significant loss of brain function in life show all of these brain changes at autopsy.

Alzheimer's Disease. This more rapidly progressing form of dementia classically begins earlier in life than the senile dementias just discussed and is often called a presenile dementia. The neurofibrillary tangles seen in parts of most aging brains are always present in Alzheimer's disease. This supports the hypothesis that Alzheimer's disease is an accelerated aging of the brain. However, some rapidly progressing dementias appearing as late as age 85 to 95 have all the anatomic and behavioral characteristics of Alzheimer's disease and have been described as such. The instance of Alzheimer's disease increases dramatically in older people and reaches a peak among those in their 80s. Alzheimer's disease accounts for 50% to 60% of dementias; multiinfarct (small vessel) dementia, the next most common type, accounts for 10% to 20%. The incidence seems higher in lower educational levels, which may suggest that higher education produces more "intellectual reserve."

There is a clear genetic connection, with Alzheimer's disease four times more common in families with a history of dementia than in other families. The brain changes in 40-year-old people with Down syndrome are nearly identical to those of Alzheimer's disease patients. Therefore, a connection possibly exists between early onset Alzheimer's disease and the 21st chromosome involved with Down syndrome. It is now suspected that the gene for late onset Alzheimer's disease may be on chromosome 19. Such findings may suggest that multiple genes produce the condition. Clearly, the increased incidence of Alzheimer's disease is due largely to the fact that people are living longer, allowing whatever factors that produce the late onset form of the disease to take effect.

A typical early onset Alzheimer's disease patient experiences forgetfulness or confusion as a first symptom. Getting lost on the way home from work is a fairly common example. As the disease progresses, the patient passes through a period of intense confusion, often with belligerence. This stage is followed by a period of greater confusion and hyperactivity, often with continual aimless walking. The final stages of the disease are decreased wakefulness, coma, and death.

At death, the brain is typically reduced to nearly two thirds of its maximum weight (1000 g vs 1400 g). The cortical atrophy is reflected in greatly reduced gyri with increased sulci and ventricular enlargement. Senile plaques and neurofibrillary tangles are abundant throughout the cortex and are especially common in the hippocampus and amygdaloid nucleus. The changes seen in the brain are similar to but more severe than those seen in a "normal" aging brain. Lipofuscin granules and amyloid deposits are more abundant than in young brains, but similar changes can be seen in healthy aged brain. However, the latter lacks the extensive neurofibrillary tangles and the amyloid deposits that appear in the hippocampus of Alzheimer's disease patients. As mentioned earlier, a decrease in acetylcholine transferase has been demonstrated in Alzheimer's disease patients. This decline suggests a lack of the neurotransmitter acetylcholine.

Parkinson's Disease. Sometimes called paralysis agitans, Parkinson's disease begins to produce symptoms after age 55. It usually begins with a resting tremor of one hand. With the passage of years, the tremor spreads to the other extremities. The tremor stops with deliberate movement but resumes when the movement is completed. As the disease progresses, movements become slower and the range of motion decreases, leading eventually to paralysis. The facial muscles are the first to show paralysis; as a result, Parkinson's disease patients develop a characteristic fixed facial expression. About 30% of Parkinson's disease patients develop dementia. The period from diagnosis to death may be as long as 20 years with the last 2 requiring nursing care.

The brain of Parkinson's disease patients appears grossly normal except for a loss of pigment in the substantia nigra. Microscopically, the substantia nigra shows a loss of pigment cells and replacement with glial cells. The neurons develop rounded cytoplasmic bodies in the substantia nigra and in other pigmented areas of the brain. The substantia nigra loses the ability to produce two locally important neurotransmitters, dopamine, and to a lesser extent, norepinephrine. Dopamine production has been enhanced by treatment with levodopa, and the side effects (reduced sleep, night terrors, finally psychosis) are reduced with carbidopa which inhibits dopamine formation outside of the brain. New treatment involving transplantation of fetal substantia nigra cells have shown greater success in reducing the symptoms of the disease; there are of course important ethical concerns about the use of fetal tissue. Another critical area involves implanting adrenal tissue in the brain. New drugs seem likely to make both these procedures unnecessary.

Tumors. Most brain tumors are metastatic from primary cancer elsewhere in the body, especially the lungs, breast, and prostate. Primary brain tumors are usually benign meningiomas but can be malignant neoplasms arising from astrocytes, or rarely, other glial cells. Surgery for glial tumors fails to provide a complete cure most of the time because of the difficulty of removing all of the malignant tumor. An astrocytoma is shown in Figure 13-11 and Color Plate 16, *B.* Meningiomas, being on the surface of the brain and benign, can be completely cured by surgery.

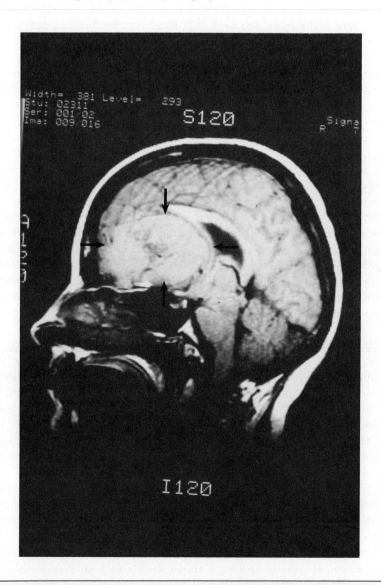

Figure 13-11   This MRI photograph shows an invading astrocytoma (arrows) that has replaced much of the brain and compressed much of the remainder. Compare with Figure 13-7 *B.*

Depression. Any elderly person who faces reality realizes that the future holds an accelerating degradation of body and mind. The effect of this recognition on an individual's ability to achieve life objectives is enough to produce a depressed state of mind. Fortunately, few people face this reality, or at least brood on it. Evidence suggests that body chemistry has as much and perhaps more to do with depression than do the challenges we face in life, including battling the aging process.

Depression may be either the cause or result of physical problems. An early study of 30, 50, and 70-year-olds found that 70-year-olds were only slightly more depressed than 50-year-olds. It follows that most elderly are not clinically depressed. A recent study found that major depressive disorders occurred in 12.6% of a large nursing home population. An additional 18.1% had depressive symptoms. Researchers also determined that major depressive disorders increased the death risk by 59%, but that depressive symptoms alone did not increase the likelihood of dying. Interestingly, the majority of the depressive disorders listed in this study were unrecognized by nursing home physicians and were untreated.

Unlike some dementias, depression is treatable. Counseling, psychoanalysis, and a variety of drugs can bring about improvement. Several antidepressive drugs are widely used, and electroconvulsive therapy (ECT) is used effectively in certain severe cases.

Insomnia. Elderly people (and many others) often complain about insomnia, or inability to obtain enough restful sleep at night. It has been reported that sleeping is a problem in 30% to 50% of elderly people. Elderly people tend to sleep lightly and for shorter periods than do the young. They tend to make up for this by napping during the day. The greatest problems are decline in the deepest levels of sleep and a tendency to awaken often during the night. Causes of insomnia include medical and psychologic illness, alcohol, medication, and poor sleep habits.

Complete rest seems to require adequate amounts of various levels of sleep. These include deep sleep and rapid eye movement (REM) sleep. Electroencephalograph (EEG) studies have shown regular cycles between these two sleep states. Damage to an area of the brain stem can decrease or destroy REM sleep, seemingly by causing improper channeling or inadequate release of norepinephrine.

A number of medical problems contribute to insomnia. Urinary obstruction or polyuria from other causes may cause frequent awakening. Pain from duodenal ulcers or heart disease may make sleep difficult. Sleep is disturbed by a number of brain disorders, including Alzheimer's disease or Parkinson's disease. Some patients have nocturnal delirium, hallucinations, or disorientation or wander during sleep. Anxious patients have difficulty getting to sleep, and depressed patients have difficulty staying asleep. While alcohol helps to initiate sleep, the sleep is often interrupted. Sleep apnea (interruption of breathing) is a common problem that may occur because of closing of the throat or simply because the respiratory center seems to "forget" to breathe. When the apnea results from throat problems, it usually involves loud snoring and is called obstructive apnea. Obstructive apnea reaches a peak in middle age and can sometimes be cured by surgery.

Caffeine, used to stay awake following a poor night's sleep, may interfere with the next night's sleep. Sleep-inducing drugs may produce a number of problems with daytime behavior and their effectiveness at night may diminish with time. A behavior producing night insomnia occurs when a patient is tired from a previous poor night's sleep and makes up for this by napping during the following afternoon. The next night's sleep is difficult because it is not needed, but the following afternoon the patient is tired again. The normal diurnal cycle becomes interrupted and the patient becomes neither adequately tired nor adequately rested.

CASE HISTORIES

❖ Multiple Sclerosis

M.P.'s illness began at the age of 35, 21 years before her death. Her initial symptom was a episode of blurred vision that spontaneously resolved. This was followed in a few months by the spontaneous onset of weakness in her right leg. Medical evaluation at that time showed horizontal nystagmus (rapid side-to-side eye movements), poor performance on repetitive finger to nose movements, and poor balance. All these findings related to dysfunction in the cerebellum, part of the brain responsible for coordination. A spinal tap showed normal pressure. The cerebrospinal fluid had a mildly elevated cell count (8 cells/mm), no visible bacteria, and normal glucose, but contained increased levels of protein. Cultures failed to grow bacteria. These findings led to a diagnosis of multiple sclerosis.

M.P. was treated with ACTH to stimulate production of her corticosteroids. She showed mild improvement in strength. However, her subsequent course was one of exacerbations and remissions. Each episode was associated with new symptoms and increased disability. Within 5 years she was wheelchair bound and 15 years later she was essentially paralyzed and bedridden, requiring nursing home care. She and her family had requested that no heroic measures be used should a threat to her life develop. This being a matter of record, when she developed pneumonia and respiratory failure, her death was unopposed.

The autopsy of M.P. showed muscle atrophy and fibrosis, with contraction deformity of the legs, arms, and back fixing her body permanently in a fetal position. The immediate cause of death was respiratory insufficiency caused by chronic and acute aspiration pneumonia. The spinal cord was not examined, but the brain had numerous foci of demyelinization, confirming the diagnosis of multiple sclerosis. Active and inactive disease areas were present.

Comment. M.P. had a classic case of progressive multiple sclerosis. This form of the disease is most often thought of, but MS can occur in both more mild and more severe forms. Even today there is no known cure or proven therapy. The adrenal stimulation by ACTH is similar to the corticosteroid therapy used today and is still as controversial as it was in the 1960s. Diagnostic tests have improved greatly. CT scans and especially MRI studies reveal remarkable detail inside the brain and spinal cord. CF protein studies can be diagnostic if oligoclonal banding is present and immunoglobulin indexes show increased production of antibody in the CNS. The protein studies demonstrate relatively pure antibodies that form discrete bands with protein electrophoresis, and suggest that the disease results from an active autoimmune process.

❖ Cerebrovascular Accident

M.F. is a 66-year-old retired auto worker who suddenly, quietly lost consciousness while eating dinner. He had not indicated pain and there were no epileptic-type movements. He had no history of recent head trauma. An ambulance was called and the paramedics found M.F. unresponsive but with essentially normal vital signs.

His medical history was positive for a tonsil and adenoid removal T and A and appendectomy as a child. For the past 6 years he had been taking oral hypoglycemia agents for diabetes ms and had been on hypertensive drugs since age 48. His physical examination revealed M.F. to be a comatose, obese male with a regular pulse at 98/min, a respiratory rate of 17, and blood pressure of 186/106 with no fever. The physical was otherwise unremarkable except for the neurologic findings. The patient showed no reaction to pin prick of the right arm, trunk, and leg. Withdrawal from deep pain was only noted on the left side; none was seen on the right. M.F.'s deep tendon reflexes DTRs were hyperreflexive on the right arm and leg and tested positive for Babinsky's sign (upturned toe response) on the right. These findings suggested loss of cortical motor and sensory function.

Laboratory evaluation revealed a glucose level of 180μg/dl, a cholesterol reading of 324 μg/dl, and a uric acid level of 9.6 μm/dl, but no other abnormalities. The ECG showed nonspecific ST-T changes without evidence of myocardial infarction. A chest x-ray and brain CT scan showed normal results. Conservative therapy with IV hydration, anticoagulation, and frequent turning was initiated. M.F. awoke within 48 hours. A repeat CT scan showed hypovascularity in the right internal capsule area with surrounding edema and no evidence of hemorrhage. Diagnosis of right-sided CVA, ischemic type (due to embolism) was made.

Although M.F. could follow simple commands and move his left side, he was confused and aphasic and his right arm and leg were paralyzed and without feeling. After 6 months of treatment, his right side continued to be paralyzed, but he could communicate and respond appropriately.

Comment. Cerebrovascular accidents (strokes) are a major cause of illness and death in the western world. They are usually associated with hypertension, which is also prevalent in the west. Most strokes are interruptions of blood flow within the brain. The vessels most often involved are the branches of the middle cerebral artery. They supply the internal capsule and basal ganglia and may be prone to disease because these vessels are relatively straight, allowing better transmission of blood and higher pressures. Full recovery from a stroke is slow; improvement can sometimes still be seen as late as 6 months after the event.

❖ Transient Ischemic Attack

A.K. is a 64-year-old grandmother who experienced transient blindness. Episodes had occurred three times in the past 2 weeks, each of which was sudden and without preceding symptoms, pain, or trauma. Each time the blindness involved the lower half of the vision in the left eye and was described as a "gray sharp curtain" that completely obscured that portion of her vision. Within 1 hour, normal vision slowly returned without leaving any visual defect.

A.K.'s medical history included hypercholesterolemia, weight-dependent diabetes Mus, and a recent hospitalization for management of angina. She drank rarely and smoked one pack of cigarettes per day for 30 years. Her only medication was sublingual nitroglycerine for anginal pain. She is on a low cholesterol, weight loss diet.

Her physical examination revealed bilateral neck bruits, harsh sounds heard with a stethoscope, indicating turbulence in the blood flow). Results of the remainder of the physical examination were within normal limits. Laboratory evaluation revealed a cholesterol level of 284 mg/dl. An ECG showed ischemia of the anterior portion of the left ventricle; an electroencephalogram (EEG) was normal, but Doppler ultrasonography showed bilateral carotid artery narrowings. Carotid angiograms confirmed significant atherosclerotic narrowings of both carotid bifurcations.

A.K. was acutely treated with antiplatelet agents and subsequently underwent a left carotid endarterectomy, followed by the same procedure on the right carotid artery. She has remained free of TIA symptoms for the last 2 years.

Comment. Reversible embolic strokes (TIAs) are often associated with ulcerating atherosclerosis, with defective heart valves, or with endocarditis. It is surprising how often an embolism, free in the bloodstream, will follow the same path as its predecessor and generate the same symptoms. Sometimes emboli in the eye can be seen in the retina as a sudden interruption of the vessel or as a hemorrhagic area. Therapy involves using antiplatelet drugs to prevent the formation of the small clots, and then removing their source. In this case A.K.'s doctors treated with aspirin first and then surgically removed the narrowed, ulcerated area of atherosclerosis.

❖ Atherosclerosis Senile Dementia Alzheimer's Type (SDAT)

M.K. is an 81-year-old grandmother who was observed by her children to be confused and apathetic. Her symptoms began slowly with recent memory loss and progressed with inappropriate responses to questions. The family dated the onset of her decline to a series of retinal detachments that left her virtually blind. Her one enjoyment in life had been reading but with loss of her sight she merely sat and was without interests. In time her inappropriate responses worsened until she became only rarely aware of what was occurring around her. Her medical history was remarkable for mild treated hypertension and congestive heart failure requiring digoxin. The family history revealed that her maternal mother died with senility. M.K. was a moderate drinker and smoker (two packs/day) prior to her decline.

Her physical examination showed her to be a thin, elderly apathetic female who was not aware of place, person, or time. Her head was covered with sparse, gray hair, and multiple seborrheic keratoses were found on the forehead and face. The eyes showed arcus senilis (yellow ring around the iris) and her vision seemed to be limited to large shapes. She was edentulous (without teeth). Her breasts were atrophic and without masses. The heart sounds were faint but normal with no extra sounds or murmurs. The chest was overinflated and the breath sounds had numerous rhonchi. The abdomen was unremarkable and the neurologic examination was normal with the exception of generalized weakness and decreased DTRs. No abnormal reflexes were noted. The family was advised that M.K. had senile dementia, Alzheimer's type (SDAT) and that her decline would continue.

During the next year she started to behave inappropriately by wandering away from the house, becoming more active at night, and becoming verbally abusive. When she became unable to control her bladder, the family could no longer care for her and placed her in a nursing home.

Comment. Some senile dementias and Alzheimer's disease have tended to merge into a single disease entity. In the recent past, the slower mental deterioration of the elderly was called senile dementia and the rapid but similar deterioration of a young (age 50s) person was called Alzheimer's disease. Because the pathologic condition of the brain seems virtually identical in the two processes, the later and usually somewhat slower progressing condition is now called SDAT. These etiologic speculations are important since genetics, therapies, and prognosis may be different in the two forms. It has been recently suggested that the pathologic amyloid of Alzheimer's is different from that found in the "normal" aged brain. If one turns out to be an accelerated process of the other, discovery of the critical factor may produce a cure for both.

❖ Parkinson's Disease

P.V.'s first symptom was shaking of his right hand, beginning when he was 62 years old. He did not remember when it started because it came on gradually and when he noticed it he assumed he was "getting old." There was no interference with function because the tremors disappeared while the hand was in use and reappeared when it was resting. About 1 year later when the shaking worsened and spread to his right arm and left hand, he went to his doctor.

His examination revealed a healthy but concerned 63-year-old man with a resting tremor in both arms but more severe on the right. The hand motion consisted of movement of the thumb over the first two fingers, a "pill rolling gesture." The tremor was not apparent while moving the affected limbs but returned when there was no intended motion. When the doctor passively moved P.V.'s arms he felt a rachet-like series of starts and stops, occurring with cog-wheel rigidity. P.V. reported that his wife said the tremor was not present when asleep. His face seemed not to change expression as he spoke and did not show the emotion expressed in his voice. His eyelids showed a slight tremor when lightly closed. His sensory examination and DTRs were normal. A diagnosis of paralysis agitans (Parkinson's disease) was made and P.V. was started on Levodopa.

After some initial improvement P.V's symptoms worsened over the next 4 years. The tremors had generalized, affecting his arms, legs, head, and face. He complained that it felt like he was in a body cast; conscious movement had become slow (bradykinesia) and so difficult that it was hard to rise from a chair or start walking. His first steps were small but would become larger and more rapid as if he were losing his balance. His voice was not monotonous and he had a stooped, ape-like posture. Although these symptoms developed while on medication, they rapidly worsened when he forgot to take it.

At age 76 R.V. was placed in a nursing home because his wife could no longer care for him. He was no longer capable of significant purposeful movement and drooled continually. He was incontinent (urinary) and was having difficulty with controlled defecation. During the next 2 years he became confused and finally demented. This was believed to be, at least in part, a side effect of his drug regimen. He died at the age of 79.

Comment. Parkinson's disease is a degenerative disorder affecting the substantia nigra and eventually the rest of the brain. Cases with nearly identical symptoms have occurred following viral encephalitis, repeated head blows (boxers), cerebral atherosclerosis, or following injections of a "designer" drug, MPTP. Although the tremor and rigidity are disabling, some patients with advanced disease are able to react in a perfectly normal way when under great emotional stress. Eventually one third of all people with this disease become demented. Current research has produced more effective drugs than were available to P.V. but these still only increase the quality of life during the disease. They do not appear to slow its progress. Experiments with transplanted substantia nigra tissue, while initially disappointing, are being continued.

❖ Astrocytoma

M.A. is a 53-year-old man who complained of severe headaches of 2 weeks duration. The present illness began with headaches that gradually increased in intensity over a 1 week period. They then abated and returned the day before admission. M.A. has no history of unconsciousness, trauma, seizures, blackouts, or sensory or motor losses. His past history reveals no prior headaches, no family history of migraines, and no medication use. He was hospitalized for a myocardial infarction at the age of 49, but otherwise has been in good health.

Physical examination showed a pulse of 92 with a blood pressure of 154/72 and no fever. The scalp was nontender and no bruits were heard over the cranium. The HEENT were significant for bilateral papilledema (optic neuritis and edema without inflammation). The cranial nerves were intact. The DTRs were brisk bilateral (somewhat hyper), and sensation was normal. The test for Babinski's sign showed normal responses. The chest, heart, abdomen, and peripheral vessels were unremarkable.

The patient was hospitalized after a CT scan showed a large right parietal-temporal necrotic mass within the cerebrum and extending into the basal ganglia. A right craniotomy produced a large amount of necrotic, yellow tumor classified as a glioblastoma multiform (highest grade astrocytoma). Radiation therapy was carried out.

During the next 9 weeks, the patient had only one symptom: an insatiable appetite. He gained 50 pounds and could not get enough to eat. This abnormal appetite continued even after his steroids were discontinued. The only other medication he was on was phenytoin (Dilantin). He was found unresponsive by his wife during the tenth week and died. At autopsy there was a large intraventricular hemorrhage and residual high grade astrocytoma present in the parietal-temporal area of the right cerebral cortex. Tumor had invaded the thalamus and hypothalamus. Two metastatic foci were noted in the left frontal lobe and the right cerebellum.

Comment. Glial brain tumors are almost always fatal because of the difficulty in removing all of the lesion without doing major damage to the vital functions. Another problem, as exemplified in this case, is the large size and extensive area of the brain involved without causing symptoms. The most common symptoms are headache, seizures, and loss of sensory or motor function. This case is interesting because the uncontrollable eating behavior suggests that either the tumor or the surgery damaged the hypothalamic satiety center.

CHAPTER 13

Birren JE, Woods AM: *Psychology of aging.* In Pathy MJS ed: *Principles and practice of geriatric medicine,* New York, 1985, John Wiley & Sons.

Carter AD: *The neurologic aspects of aging.* In Rossman I, ed: *Clinical geriatrics,* Philadelphia, 1986, J.B. Lippincott.

Cole G: *Neuropathology of aging.* In Pathy MJS, ed: *Principles and practice of geriatric medicine,* New York, 1985, John Wiley & Sons.

Dayan AD: *The central nervous system.* In Brocklehurst JC, ed 2: *Textbook of geriatric medicine and gerontology,* ed 2, New York, 1978, Churchill Livingstone.

Gorelick PB, Bozzola FG: Alzheimer's disease, clues to the cause, *Postgrad Med* 89(4):231-240, 1991.

Gottfries BG: The metabolism of some neurotransmitters in aging and dementia disorders, *Gerontology* 28:11-19, 1982.

McNeill TH: Differential changes of autonomic nervous system functions with age in men, *Am J Med* 75:249-257, 1983.

Nakra BRS, Grossberg GT, Peck B: Insomnia in the elderly, *Am Fam Physician* 3(2):477-463, 1991.

Samorajski T, Person K: *Neurochemistry of aging.* In Pathy MJS, ed: *Principles and practice of geriatric medicine,* New York, 1985, John Wiley & Sons.

14 *The Sense Organs*

❖ *The Senses and Their Functions*

Humans have traditionally thought of themselves as having five senses; sight, hearing, touch, smell and taste. The organs that convey these sensations enable us to coordinate our needs and actions with our environments. However, we do have some other major senses. The sense of equilibrium and the kinesthetic sense help us to deal with our external environment. The sense organ for equilibrium, located in part of the middle ear, maintains balance. The kinesthetic sense is our sense of limb position, also important in balance. The sense of temperature and the sense of pain are also important in dealing with both external and internal environments. Adding these would give us nine senses that enable us to deal with our external environment.

The sensation of hunger, thirst, fatigue, and vertigo as well as some sensations of pain, inform us about our internal environment. Because we depend on these senses, both externally and internally, it is easy to see that problems develop when they deteriorate with age.

❖ *Relevant Anatomy and Physiology*

SIGHT

Two of the most important senses enabling us to relate to our environment are sight and hearing. They are probably the most complex of our senses and when they begin to fail the impact can be devastating. The following paragraphs review the anatomic components of the eye that tend to fail with age.

The eyes are closed during sleep and are protected from injury by upper and lower eyelids (Figure 14-1). Both the inside of the eyelid and the surface of the eye are covered with a sensitive membrane, the conjunctiva. This membrane is lubricated and moistened by secretions of the lacrimal glands located in the outer margins of each eye.

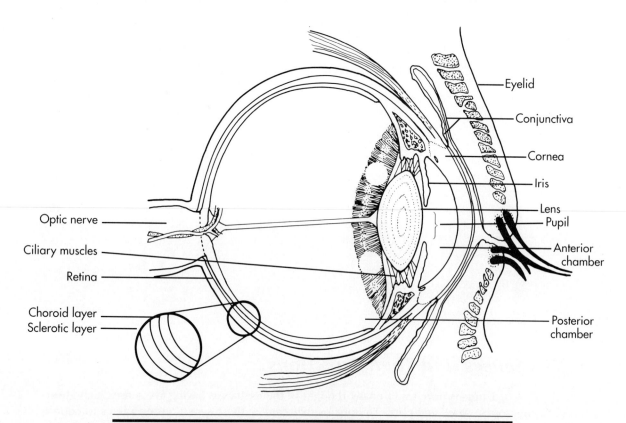

Figure 14-1   The anterior and posterior chambers of the eye are separated by the lens and ciliary muscles (found in the ciliary body). The anterior chamber contains watery aqueous humor; the posterior chamber contains thicker vitreous humor.

The tears secreted are distributed over the eye by blinking of the lids. Normal excesses are absorbed by the lacrimal sacs located at the corner of the eyes close to the nose. The absorbed secretions are deposited in the internal nares. Great excesses of tears may overflow onto the face under conditions of emotional stress or when the conjunctiva is irritated. Such irritation, called conjunctivitis, may be caused by infection, mechanical injury, irritants such as smoke, or because of allergy to otherwise nonirritating materials. Conjunctivitis and the damage it can cause the eye are greatly increased by inadequate secretion of tears. Dry eyes (a condition known as xerophthalmia) are more subject to infection than those that are normally lubricated.

The Structural Layers of the Eye

The fragile conjunctiva protects the thicker, tougher, transparent cornea. The cornea passes from the pupil to the "white" of the eye, where it becomes opaque and is known as the sclera, or the sclerotic layer. It is nonvascular and is composed entirely of connective tissue.

Beneath the sclera is a pigmented, vascular tissue called the choroid layer (see Figure 14-1). Under the cornea, an extension of this layer forms the iris, a colored ring of tissue that surrounds the pupil, a hole in its center. The iris expands or contracts the pupil, regulating the amount of light that is allowed to enter the eye. In dim light the iris expands the pupil, increasing the light that is allowed to pass through the lens to the retina, the light-sensitive layer that generates nerve impulses in response to light. The iris contracts the pupil in bright light to protect the retina. When the pupil size decreases it also sharpens focus, a "pin hole" effect.

Another extension of the choroid layer is the ciliary body (see Figure 14-1). This body overlies the circular ciliary muscle and continuously secretes aqueous humor into the anterior chamber of the eye. This fluid drains away through the Canal of Schlemm, and stable intraocular pressure is maintained. The lens is suspended by ligaments that are attached to the ciliary muscle. When the ciliary muscle contracts, the lens returns to its normal spherical shape, enabling the eye to focus on near objects. When the ciliary muscle relaxes, the elastic tissue stretches the lens to a flatter (less convex) shape, allowing focus on more distant objects. The process of focusing the lens is called accommodation.

Where the choroid layer lies beneath the sclera over the posterior five-sixths of the eyeball, it is blue-black. Here it has the dual function of absorbing light and providing nutrient to the sclera and the retina. If light were reflected instead of being absorbed after striking the retina, it would bounce around the inside of the eye, strike the retina repeatedly and distort the image.

Many people have either a misshapen eyeball or a misshapen or inelastic lens, making complete accommodation impossible. A lens made inelastic by age neither flattens adequately to see distant objects nor returns to its spherical shape to see things nearby. Some of the lost ability of the lens to stretch is the result of changes in proteins in the lens. Other loss occurs because the lens grows throughout life and becomes too large a mass for the elastic tissue to flatten.

The Retina

The innermost layer of the eyeball is the retina. It covers the inside of the posterior two-thirds of the eye and consists of three layers of cells (Figure 14-2). The anterior (pupillary) side of the retina consists of ganglionic nerves containing axons that become a part of the optic nerve and extend to the brain. These axons gather impulses from bipolar nerves that form the middle layer of the retina. The bipolar nerves receive impulses from the photoreceptor cells, which are rods and cones that lie on a pigmented base next to the choroid layer. This arrangement requires light to pass through the first two translucent layers before striking the rods and cones. Perhaps this provides additional protection for the rods and cones from extremely bright light.

The Rods. The most sensitive photoreceptors are rods. They are stimulated by very weak light and are concentrated more toward the edges of the retina. Away from the center of the retina, as many as 600 rods may ultimately feed into a single optic nerve fiber. This makes the retina much more sensitive to dim light near the edges.

The mechanism for generation of nerve impulses from light is quite simple. When light of the proper wavelength and enough intensity strikes a rod, it causes the pigment rhodopsin in the rod to split into two molecules, retinal and opsin. This reaction results in membrane changes and hyperpolarization of the rod. (Note that in most other recep-

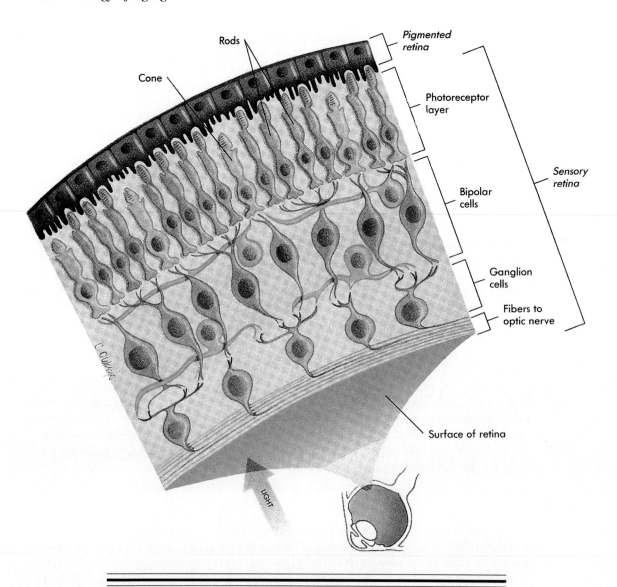

Rods

Cone

Pigmented retina

Photoreceptor layer

Sensory retina

Bipolar cells

Ganglion cells

Fibers to optic nerve

Surface of retina

LIGHT

Figure 14-2   Light must pass through the two outer layers of cells to reach the inner photoreceptive rods and cones in the retina.

tors, stimulation and transmission involves depolarization). This results in stimulation of the adjacent bipolar nerve, which in turn stimulates the optic nerve. Retinal and opsin recombine in a fraction of a second, readying the rod to fire again. Retinal is a derivative of vitamin A, some of which is lost from the retina and from the body in a variety of ways. For this reason, a person with inadequate vitamin A intake will lose the ability to see in dim light, where dependence is on rods rather than cones for vision.

The Cones. In the center of the retina is a small yellow circle, the macula lutea; in the center of that is an even smaller circle, the central fovea, or fovea centralis. (Fig-

Fovea centralis   Macula lutea

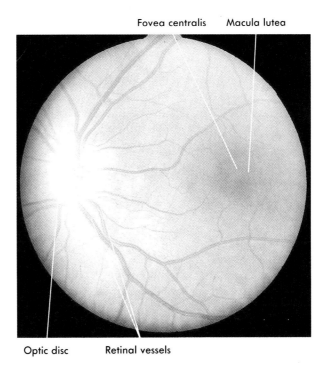

Optic disc   Retinal vessels

Figure 14-3   The retina as seen through an ophthalmoscope. This view, through the pupil, shows the posterior wall.

ure 14-3 and Color Plate 16, *C*). Both the central fovea and the macula lutea lack rods and consist of great concentrations of cones. Cones are receptive to light of different wavelengths (color). Our eyes focus the image on the central fovea and produce the image with the greatest detail and greatest color differentiation.

Cones need much more intense light to bring them to threshold than do rods. Because of this human eyes are only able to see color in fairly bright light. Each cone possesses one of three pigments sensitive to blue, red, or green wavelengths of light. If the wavelengths of the received light are so broad that they stimulate all three types of cone equally, the eye perceives white light. The stimulation of red, blue, and green in different proportions produces sensations in the brain that are interpreted as orange, yellow, and all of the other colors humans see. As with rods, the light splits the photoreceptive pigment into the light sensitive proteins, photopsins, and retinal. Vitamin A is necessary to regenerate lost retinal for the cones too but the lack of it will not be noticed until the deficiency is much more severe. People who lack cones altogether are totally color blind. They may, however, have a great concentration of rods in the central fovea and have excellent night vision.

Communication with the Brain

Once individual optic nerves are stimulated, the impulse is passed to the visual receptive areas in the occipital lobe of the brain. Some of the optic nerve fibers cross to the opposite side of the brain so that each eye stimulates both sides of the brain. Damage to the right occipital lobe (as may result from stroke or trauma) may therefore cause blindness in the left eye. The effect may also be loss of reception in the left field of both eyes. Vision problems are most often the result of eye changes or damage rather than brain damage.

HEARING
The Outer Ear

The pinna, auditory canal, and tympanic membrane make up the outer ear. Sound waves are collected by the pinna. They are concentrated in the auditory canal that penetrates the temporal bone of the cranium (Figure 14-4). The sound waves cause the tympanic membrane, or eardrum at the end of the auditory canal to vibrate. The auditory canal is lined with glands that produce cerumen, a modified sebaceous secretion generally called earwax. Cerumen seems to have some insect repellent characteristics. This quality, along with the fine hairs at the entrance of the canal, inhibits insects from entering the ear.

The Middle Ear

The middle ear consists of a small cavity containing three small bones: the malleus, the incus, and the stapes (see Figure 14-4). These bones are suspended from the sides and top of the middle ear and form a chain from the tympanic membrane to a membrane of the inner ear called the oval window. These bones magnify the force of vibrations received by the tympanic membrane and deliver the vibrations to the oval window. If they are removed, the intensity of sound required to stimulate the inner ear

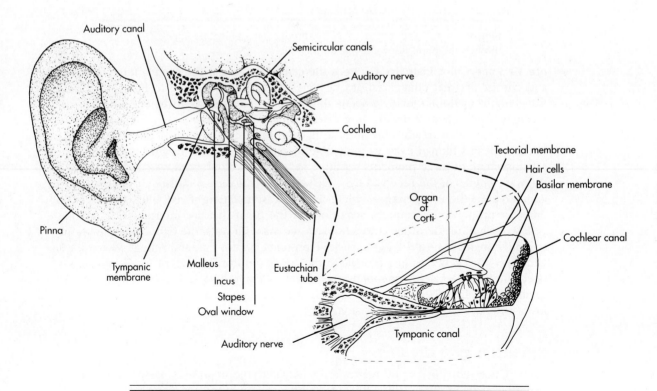

Figure 14-4   The outer ear includes the pinna, the auditory canal, and the tympanic membrane; the middle ear includes the three middle bones and the eustachian tube; the inner ear includes the cochlea, utricle, saccule, and semicircular canals.

increases by about 30 decibels, which is the approximate difference between a shout and a barely audible voice.

The middle ear is connected to the pharynx by a narrow opening, the eustachian tube. This tube permits the equalizing of pressure in the middle ear with the pressure of the air outside the body. When these pressures are not equal, the tympanic membrane does not vibrate freely and the sound is dampened. Most people experience this phenomenon when they have a cold and the eustachian tube is partially closed by inflammation. A change in altitude often causes ear pain that results from stretching of the tympanic membrane. Yawning or working the jaws may open the eustachian tube, equalizing the air pressure. This may cause the eardrums to pop into normal position and allow better transmission of vibrations. Infectious agents sometimes pass through the eustachian tube and become established in the middle ear. They may spread from there to the outer ear, temporal bone, inner ear, or even to the brain.

The Hearing Portion of the Inner Ear

The inner ear is composed of two sets of channels in the temporal bone that are connected to the vestibule, a small, central cavity (see Figure 14-4). The vestibule and three upper channels, (the semicircular canals), are concerned with balance. The lower channel is coiled like a a snail's shell and is called the cochlea. The cochlea is the part of the inner ear concerned with hearing, specifically the changing of vibrations to nerve impulses.

The cochlea is separated by two membranes into three canals. The upper vestibular canal and the lower tympanic canal of the cochlea are lined with periosteum and filled with a fluid called perilymph. They are united at the end farthest from the middle ear. A smaller canal, seemingly wedged between the tympanic and vestibular canals, is the cochlear canal. It contains Corti's organ, the structure that converts vibrations to nerve impulses. The cochlear canal contains a slightly different fluid called endolymph. This canal is separated from the vestibular canal by the thin vestibular membrane and from the tympanic canal by the much thicker basilar membrane.

The Reception of Sound. As discussed earlier, vibrations received by the tympanic membrane are passed on to the middle ear bones, which in turn deliver them to the oval window of the cochlear canal. The oval window eventually transfers the vibrations to the endolymph, where they are detected by hair cells.

Dampening Sound Waves. Waves tend to repeat with decreasing intensity the farther they are from the source. If this were to happen in the inner ear it would result in distortion and an inability to distinguish sounds. The flexibility of the vestibular membrane, the round window adjacent to the middle ear, and the basilar membrane absorb and dampen the sound waves, preventing repetition.

Conversion of Sound Waves to Nerve Impulses

The organ of Corti arises from the basilar membrane in the cochlear canal (see Figure 14-4). It consists of a series of hair cells, each attached to an auditory nerve fiber and surrounded by supporting tissue. The tips of the hair cells are in contact with the overlying tectorial membrane. The shock wave from each vibration of the oval window will crest at a point on the cochlear canal. At that point the wave causes the tectorial membrane to press down on the hair cells, stimulating the nerves. The impulses ultimately arrive at the auditory centers of the brain in a specific pattern that we interpret as a specific sound.

Distinguishing Pitch and Intensity of Sound. Intensity of sound varies with the amount of energy in each sound wave. Therefore a high amount of energy produces a sound wave with a high crest. More intense sound waves stimulate more hair cells at any given point on the organ of Corti. This is because some hair cells are more easily stimulated than others; only a very severe "slap" can stimulate the least sensitive ones. The brain interprets the increased number of impulses from the same point on the organ of Corti as more intense (louder) sound.

High frequency sound waves crest early in the cochlear canal and therefore stimulate hair cells close to the oval window. Low frequency sound waves do not crest until they are near the narrow end of organ of Corti far from the oval window. Nerves from different points along the organ of Corti go to different parts of the auditory center in the brain, creating different sensations that we interpret as different pitches. The number of hair cells stimulated at any point on the organ of Corti determine intensity or loudness of sound, but the location in the organ of Corti where the nerves are stimulated determines pitch.

The Auditory Nerves. A spiraling set of nerve fibers emanating from the hair cells of the organ of Corti form the cochlear branch of the auditory nerve. It joins the vestibular branch and passes through the thalamus to the auditory centers of the cerebral cortex in addition to passing on to the cerebellum and the reticular activating system of the brain stem. Unlike most other senses, it appears that while impulses can pass to the opposite side of the brain in at least three different places, most of the stimulation of the brain remains on the same side as the ear that received the sound.

Touch

The sense of touch involves stimulation of receptors located in the skin or deeper in the body. Most receptors are stimulated by pressure. The following list begins with easily stimulated receptors and ends with those most difficult to stimulate. As with many sense organs, those most easily stimulated are most easily ignored. Most of them are shown in Figure 14-5.

Root Hair Nerves

The dendrites of these extremely sensitive nerves are wrapped around the base of hair follicles. They are especially important in detecting the movement of air or other things over the surface of the skin and respond whenever the hair is moved. They may also provide early warning about the approach of an object to the skin.

Merkel's Disks

These disks are very sensitive touch (tactile) receptors located in the epidermis. They are disk-like endings of the dendrites of nerves and are attached to epidermal cells. They are found where a delicate sense of touch is especially important, such as the ends of the fingers or the lips.

Meissner's Corpuscles

These receptors are located in the papillary layer of the dermis, especially in the fingers, hands, soles of the feet, external genitalia, tongue, eyelids, and nipples. They detect light to moderate pressure. Meissner's corpuscles are important in recognition of exactly where the body is touched and in determining the texture of objects.

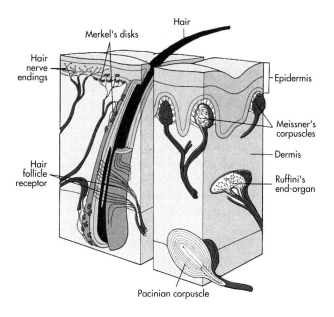

Figure 14-5 The sense of touch involves a number of different sense organs. Shown are root hair nerves at the base of the hair follicle; Merkel's disks and Meissner's corpuscles located just under the epidermis; and end organs of Ruffini and Pacinian corpuscles, situated deep within the dermis.

End Organs of Ruffini

Located deep in the dermis and in other deep tissues of the body, these nerves respond to continuous heavy pressure. They are multibranched, encapsulated endings of dendrites.

Pacinian Corpuscles

These receptors also are located deep within the dermis and subcutaneous tissue. They are found in the tissue of muscle, joints, and tendons as well and some occur in the lips and genitals. The Pacinian corpuscles sense pressure. They recover rapidly after being stimulated and are able to sense vibration and other frequently occurring stimuli.

SMELL

The sense of smell is technically known as the olfactory sense. Like taste, it involves chemical receptors, but these are located in the moist upper epithelium of each nostril. The dendrites of olfactory receptors actually project into the nasal cavity, and the axons extend through the ethmoid bone of the cranium where they synapse with the olfactory nerve. The olfactory receptors are surrounded with mucus secreting cells to keep them moist.

Olfactory nerves are extremely sensitive. Some substances have been reported to be detectable in concentrations of only 1/25,000,000,000 in air. Such sensitive nerves are said to have a very low threshold. Nerves so easily stimulated are easily exhausted. Limited thresholds can cause dangerous consequences. For example, methyl mercaptan, added to natural gas to make it easily detectable by smell, cannot be smelled by a person who has been exposed to it for a long period, even when it is in high concentration. A person who has become accustomed to a small gas leak therefore may not notice when the room fills with gas. The phenomenon of loss of smell after prolonged exposure is known as olfactory fatigue (or olfactory exhaustion).

Although color vision is the result of a mixture of three primary color receptors and taste is the result of four primary tastes, it has not been possible to identify a small number of primary smells. Fifty or more primary smell sensors have been postulated.

TASTE

The sense of taste is also called the gustatory sense. The sensory nerves are chemical receptors concentrated in the taste buds of the tongue but also found in the soft palate and pharynx. The sense of taste was discussed with the digestive system and will be briefly reviewed here.

What we taste is the result of the brain's interpretation of a combination of impulses from one or more chemical receptors responding to different compounds that they have contacted. For example, acids stimulate receptors that result in a brain sensation that we call sour. We interpret alkaline substances as bitter. Sugars and other compounds that stimulate a particular class of receptors are interpreted as sweet, and salts stimulate other receptors that result in salty sensations. Most foods give off airborne agents that we sense with the nose and interpret as smell. What we consider as taste actually results from a combination of different mixtures of the four basic tastes, and smell. For a discussion of the distribution of taste buds and taste senses on the tongue, see Chapter 7, The Digestive System.

EQUILIBRIUM

The sense of equilibrium is concerned not only with maintaining balance but also with the nerve centers governing the skeletal muscles for support and locomotion. This coordination is centered in the reticular formation extending from the top of the spinal cord through the medulla, thalamus, and hypothalamus. The sense of equilibrium can be roughly divided into two senses involving two different sense organs: the sense of static equilibrium and the sense of dynamic equilibrium.

Static Equilibrium

Static equilibrium is our sense of the position of the head relative to the pull of gravity. It involves two sense organs located above the cochlea, the utricle, and to a lesser degree, the saccule. The cells of each of these organs support a small (about 2 mm in diameter) sensory area called a macula (Figure 14-6). In each macula a series of hair cells arises from a bed of supporting tissue. Extensions called stereocilia extend from the hair cells into a gelatinous matrix. Lying above the gelatinous matrix are calcareous deposits called otoliths.

The pressure of the otoliths against different groups of hair cells informs the brain of the position of the head or of the extent of inertia if movement is begun. When the head is vertical, the otoliths press on certain hair cells, stimulating a branch of the vestibular nerve and creating a sensation that we have learned to interpret as "vertical." If the head is tilted, the otoliths press on different hair cells, stimulating different nerves and creating different sensations that we interpret accordingly. If we begin to move,

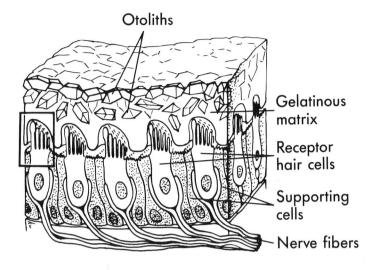

Otoliths

Gelatinous matrix

Receptor hair cells

Supporting cells

Nerve fibers

Figure 14-6 Changes in movement cause the otoliths to brush against the hair cells of the macula, alerting the brain to changes in the "status" of the body.

such as when we start to run, the combination of gravity and inertia will again cause the otoliths to press on certain hair cells, the associated nerves of which inform the brain.

Dynamic Equilibrium

The mechanism for sensing dynamic equilibrium is similar to the mechanism for static equilibrium. Above the utricle are three fluid filled semicircular canals. Each semicircular canal has an ampulla in its base. The ampullas are similar to maculas. They have hair cells that project into the fluid in the canal. When the head is moved or turned in any direction, the inertia of the fluid causes the canal to move around it. The friction of the fluid bends the stereocilia, thereby stimulating the hair cells and the vestibular nerves with which they are associated.

Bending the stereocilia in one direction stimulates the associated nerve; bending them in the opposite direction inhibits it. Because the canals are on different planes, any movement of the head will stimulate the nerves of at least one canal. The combination of impulses received by the brain through the vestibular nerves from the semicircular canals continually informs the brain of the direction (or dynamics) of the movement.

The movement of the endolymph in the semicircular canals is like that which results from turning a drinking glass with water in it. When the glass is turned the water remains in the same place, at least at first. However, friction between the glass and the water will eventually cause the water to turn along with the glass. When you stop turning the glass, the water may continue moving for a while because of its momentum.

Information from the vestibular apparatus is integrated with other information in the determination of body position and eye movements. This enables the eyes to predict where they will need to look next to "lock on to" and focus on an object. For this reason, if these nerves or the appropriate part of the brain are stimulated by some other means such as inflammation, tracking of the eyes may be disrupted, and dizziness and ataxia may result.

TEMPERATURE SENSE

Our sense of temperature is our sense of heat and cold. This sense of temperature gradations is poorly understood but it uses at least three different types of receptors: cold receptors, warmth receptors, and pain receptors. Cold receptors are scattered under the epidermis in differing concentrations in different parts of the body. Warmth receptors have not yet been identified but appear to be scattered under the skin in lower concentrations than are cold receptors. It appears that cold or warmth receptors are stimulated as the temperature rises or falls. As the temperature approaches a point at which damage may occur, the cold and warmth receptors fail to function but pain receptors react. The brain remembers that the pain was associated with increasing cold or heat and concludes that the pain is due to extreme cold or heat. The brain can be fooled, however, as in frostbite when damaged tissue gives a burning sensation.

PAIN

Pain receptors seem to be simple branching dendrites with little anatomic specialization. They can be stimulated by pressure, damaging temperatures, lack of oxygen, or a variety of other chemical or mechanical stimuli. Pain receptors are relatively hard to stimulate and do not easily exhaust. Therefore they continue to inform the body as long as a threatening situation exists. The stimulation of pain receptors produces a sensation of intense discomfort that in the brain causes us to take action limiting further damage to the body. This is clearly one of our most important senses, and when it is lacking (and it is with some rare individuals) there may be no warning of life-threatening damage and little motivation to prevent additional damage even if a problem is discovered. Diminished pain sense occurs with some elderly people and can be a serious problem.

Somatic Pain Receptors

Somatic pain receptors are found in the skin muscles, joints, and their associated fascia. They are considered organs of the sense of touch by many.

Visceral Pain Receptors

These receptors are located in the viscera such as the stomach, gall bladder, intestines, and other internal soft organs. We learn fairly rapidly where something hurts on the surface of the body, but it is often difficult for a person to tell where visceral pain is coming from. The misinterpretation of the origin of visceral pain is called *referred pain.*

Referred pain can occur in any part of the body (Figure 14-7). It results when the brain centers being stimulated are so close that the brain has difficulty in distinguishing the exact source of the pain. This is especially true if the brain is used to receiving pain impulses from one of the nerves in the complex but not the other. In this instance, the more commonly stimulated brain cells seem to be stimulated by "overflow" from the nerve tract carrying the pain impulses. For example, we are used to pain in the stomach but most of us have never felt pain in the appendix. Since nerves from the two areas utilize very close pathways in the brain, we usually interpret our first appendicitis pain as stomach pain. People with appendicitis often have vomiting as a symptom. Almost all of us have had esophageal pain referred to our forehead or eye after eating or drinking too much of something very cold. Serious consequences may result when one fails to

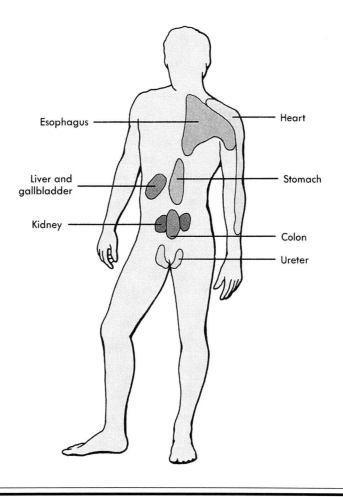

Figure 14-7   Areas of referred pain on the body surface.

recognize that the pain of a heart attack is usually referred to the left arm or throat. Sometimes upper gastric or lower esophageal pain is mistaken for a heart attack, or the reverse is true and the person having a heart attack thinks he or she has indigestion. Referred pain can seriously mislead those attempting to help someone with a medical problem.

Analgesia

The brain and spinal cord have a variety of areas specialized for the suppression of pain, or analgesia. In some instances a compound called enkephalin is released to suppress incoming pain impulses. In others, any of a variety of opiate compounds may be released. These find their way to opiate receptors, suppressing pain. One of the best known of these pain suppressors is a group of compounds called endorphins. A rapidly developing area of therapy involves controlling pain through the body's own pain-control mechanisms. People differ tremendously in their sensitivity and attitudes toward pain; because of this their ability to function while in pain differs greatly.

❖ *The Aging Sense Organs*

THE AGING EYE

Few changes in the aging eye can be observed externally. A squint may occur as a person, inhibited by decreased powers of accommodation, struggles to focus. Sometimes the orbicularis muscle migrates upward toward the the margin of the lower lid. When contracted, this muscle may cause the lower lid to turn inward, painfully rubbing the eyelashes against the cornea. This can become so severe that the lid is no longer held against the cornea, causing the cornea to become dry and subject to infection. Arcus senilis occurs when a circle of fat is deposited irregularly around the outside of the iris. Excessive watering of the eye can occur due to clogging of the lacrimal canal. Dryness can result from certain medications, vitamin deficiency, or diseases that prevent adequate secretion by the lacrimal gland. Scarring of the cornea may be visible, or the lens may appear cloudy because of cataract. The ability of the iris muscle to contract declines with age. This decline may be observed as the very small pupil seen in many elderly people. There also may be a conscious effort to decrease the pupil size in order to increase focus (by the pinhole effect) in an eye with decreasing lens elasticity and other factors that affect accommodation. However, such an effort decreases the amount of light entering the eye and may further reduce vision.

Sensitivity to dim light declines steadily with advancing age. The decline averages about 4% per year between ages 20 and 45 and is caused by a combination of reduction in maximum pupil size, changes in the ability of the lens to transmit dim light, and retinal changes involving decreased rod function. Also invisible to external examination is the loss of accommodation, presbyopia, that virtually all people will experience. Also invisible are a number of diseases and cellular changes that are discussed in the following section.

DISEASES AND CONDITIONS ASSOCIATED WITH THE AGING EYE
Presbyopia

Accommodation is the ability to change the focus of the eye to adjust for objects at any distance, near or far. With age, the lens loses elasticity, the ciliary muscles lose strength, and the focus becomes fixed. At least as importantly, the cornea becomes more flattened and loses much of its refractive power, requiring more lens adjustment. By age 45 or 50 the accommodation ability of most people has declined significantly. When this ability to accommodate is lost, the condition is known as presbyopia. While presbyopia is nearly universal among people, it is clearly one of the most annoying aspects of senescence.

Under normal conditions, light reflected from an object whether it is near or far is focused on the retina. Nearsightedness or farsightedness results when the eyeball is either too long or too short from front to back, or when a lens or cornea is improperly shaped. If the person is nearsighted, a condition known as myopia, the point of focus will be in front of the retina. If the point of focus is behind the retina, the condition is called hyperopia, or farsightedness. Figure 14-8 shows these conditions and their correction by the appropriate lens of the eye. People with presbyopia may be fitted with bifocal lenses that are divided into concave and convex halves. The eyes may require correction for distorting irregularities, or astigmatism, in addition to presbyopic corrections. This may require even the fixed point of focus to be corrected and necessitate trifocals.

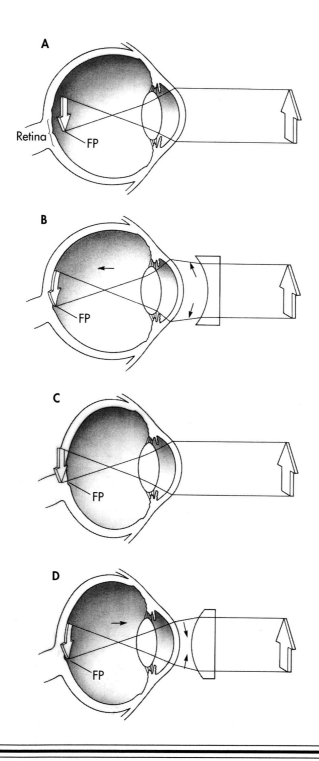

Figure 14-8 **A** In cases of nearsightedness (myopia) the focal point (FP) is in front of the retina. In cases of farsightedness (hyperopia) the focal point is behind the retina (**C**). **B** and **D** show lens corrections for these two conditions. Note that the image is inverted.

Cataract

Cataract is the most common correctable cause of blindness. There is a clear genetic predisposition for cataract but the cause is not fully understood. Opacities usually develop first around the periphery of the lens and do not cause problems until they have reached the central portion through which light passes when the pupil is constricted. Development of cataracts is usually associated with a loss in lens elasticity and development of presbyopia. With time cataracts will progress to opacity. In some instances the cortex of the lens may liquify, thus allowing irritating material to diffuse into the anterior chamber and cause general irritation. These cases require removal of the lens even when there is no likely improvement in vision because of other eye problems.

Cataracts are historically not removed until it is certain that vision dependent on external lens support is superior to vision with the cataract. Contact lenses have shown excellent success in compensating for the loss of the natural lens but many elderly can not tolerate even modern, extended wear soft contact lenses.

Modern surgery permits removal of the opaque, soft material in the lens, leaving the lens capsule in place. This is replaced with an intracapsular acrylic lens and can restore nearly normal vision. Acrylic lenses can sometimes be used even when the capsule is removed, and the surgery can now be done on an outpatient basis.

Glaucoma

Glaucoma may be a primary disease or the result of other problems that impair drainage from the anterior chamber of the eye. The disease usually develops slowly in the elderly, with no observable symptoms in the early stages. Under normal conditions, Canal of Schlemm drains aqueous humor from the anterior chamber and returns the fluid to the bloodstream. However, if it becomes blocked, glaucoma will result. Because aqueous humor continues to be produced, pressure builds, forcing the lens into the vitreous humor of the posterior chamber. This increases the pressure in the posterior chamber, causing damage to the retina. As the disease progresses, loss of peripheral vision results from the increased intraocular pressure, which damages the periphery of the retina first. This loss may develop to the point of leaving only "tunnel vision" (Figure 14−9). A slight increase in intraocular pressure causes retinal damage in some individuals. Others tolerate substantial increases in pressure with no observable damage.

Glaucoma is controlled by topically administered drugs that prevent muscle spasm around the canal of Schlemm that would normally return aqueous humor to the bloodstream. Severe cases require surgery and if not treated early enough, could result in destruction of the entire retina. An acute form of glaucoma exists, primarily in growing children. This form is of sudden onset and produces pain, especially when bright light suddenly enters a dilated pupil. Glaucoma is now the second most common cause of blindness after macular degeneration.

Macular Degeneration

This condition is the most common cause of blindness in the elderly. Badly affected people may lose the ability to read or see detail even with high magnification. The disease usually does not prevent seeing large shapes. Early in the disease, small yellowish spots appear on the retina. These are foci of degeneration between the retina and the choroid layer. No loss of vision occurs until the lesions accumulate fluid, forcing the retina away from its nutritive source, the choroid layer. Atrophy of these small spots of

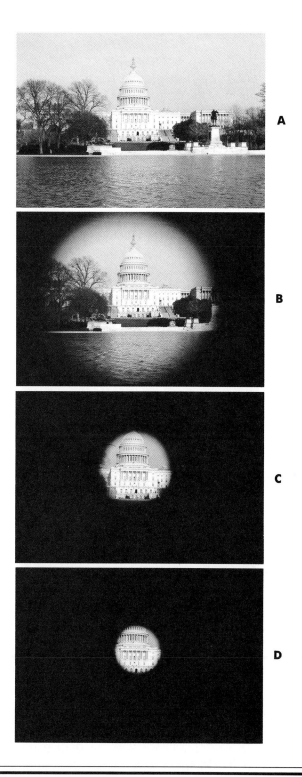

Figure 14-9   Progressive effects of glaucoma, from normal vision **(A)** to tunnel vision **(B)** and continuing vision loss **(C, D)**.

retina follows, resulting in loss of visual acuity. In some instances hemorrhage occurs either beneath the retina or through the retina into the posterior chamber of the eye. When this occurs, loss of vision is rapid.

To date macular degeneration is not treatable for most patients. Hemorrhage or severe detachment of the retina can be treated but this only slows the worsening of the condition. Often these patients also have cataracts, and removal of the cataract sometimes produces surprising improvement in vision. Macular degeneration affects only central vision; peripheral vision remains intact.

Detached Retina

The most common symptom of this condition is sudden deterioration of vision. It results from separation of a shrinking body of vitreous humor from the retina. When a person's head is suddenly moved, the shifting vitreous humor may stick to a part of the retina and pull it away from the choroid layer. Common symptoms include streaks or flashes of light in the peripheral visual field, along with a floating area of opacity.

Degenerative holes in the retina may occur and be followed by hemorrhages, further separating the retina from the nutritive choroid layer. Lasers and other new surgical tools are used to fix the retina to the choroid layer and to stop hemorrhage. Such procedures may prevent worsening of the condition but do not restore damaged retina to function.

Hemorrhage and other vascular problems seen in the retina are a common manifestation of diabetes. Diabetics often have early cataracts and may suffer primary nerve damage. Ophthalmologists and optometrists are often the first to diagnose diabetes, seeing the symptoms during routine eye examinations.

Ocular Atherosclerotic Embolism

Transient obstruction of blood vessels nourishing the retina may temporarily or permanently obscure vision. This is commonly caused by an embolism arising from an atheroma in a carotid artery, or from valvular heart disease. (See the discussion of transient ischemic attack in Chapter 13). Small platelet emboli may slowly pass through, producing transient loss of visual field. Treatment for complete occlusion of a retinal vessel rarely is possible before the tissue dies. In transient disease, the source of the emboli must be found and their formation prevented. Vascular spasm may sometimes cause transient complete obstruction from an otherwise incompletely blocked artery.

THE AGING EAR

Although there are exceptions, hearing loss is an expected consequence of advanced age. The loss often is manifested in inability to hear consonants while vowels are easily heard. This produces a distortion in the reception of speech that may not be apparent with tests of pure tone reception. This type of hearing loss is a subclinical problem until it interferes with one's job or communication with family members or friends. Even at that point a large number of affected people continue to ignore the problem and begin a path of progressive isolation from those around them. This is especially true of many older elderly who view it as only another irreversible consequence of age. One third of people over 65 suffer hearing loss to the point that it interferes with their lives.

Presbyacusis (also properly spelled presbycusis)

Age-related deafness is referred to as presbyacusis. Like so many aspects of senescence, it may result from many small, isolated insults or from specific diseases. It is difficult to separate deafness due to occupational noise from that due to presbyacusis. This difficulty has become a problem in a number of recent lawsuits.

The most important event that contributes to hearing loss is noise. This may be loud, sudden sounds that most of us recognize as damaging because they are followed by a short period of temporary deafness or more commonly, tinnitus, (ringing in the ears). It has been suggested that it is the moderately loud noise many people are exposed to in their every day lives that takes the greatest toll on the ears. Men suffer greater hearing loss than women, probably because of men's greater exposure to noise in both their work and play environment. Hearing loss is greatest regarding sounds of the highest frequency (pitch).

Contributors to deafness, other than noise, include damage from past infections, scarred or perforated tympanic membranes, excess ear wax, fluid in the middle ear, and chronic infection. Atherosclerosis may contribute to either deafness or tinnitus by restricting blood flow to the organ of Corti, auditory nerves, or the hearing centers of the brain. The following paragraphs examine hearing loss problems.

Conductive Hearing Loss. Age changes can occur in the conductive system of the outer and middle ear. A recent study reported that nearly one third of the patients complaining of hearing loss or deafness improved markedly when cerumen was removed from the external auditory meatus. The tympanic membrane may thicken with age and lose its elasticity.

The middle ear bones may become fixed to each other, or the stapes may become rigidly attached to the roof of the middle ear chamber. This fusion of middle bones is called otosclerosis. This conditions prevents adequate transfer of vibrations to the oval window. Fusion of any of the middle ear bones is treatable by removal of the fused bone or bones and replacement by a prosthesis. Evidence suggests that this is an inherited condition. Middle ear deafness can be distinguished from inner ear deafness because with the former, transfer of sound to the inner ear will still occur through the cranium but not through the air. This can be tested by holding a vibrating tuning fork to the cranium. If the middle ear bones are restricted, the sound will appear much louder when the tuning fork touches the head.

Cochlear Hearing Loss. With age, there is a loss of hair cells and nerves, especially in the portion of the Corti's organ near the oval window, where high-pitched sound is converted to nerve impulses. Hearing loss is usually greatest for sounds of the highest pitch. This is the most common form of neural atrophy, producing nerve deafness, and may be the mechanism for much reported presbyacusis. Hair cell degeneration usually precedes neuron loss but the two may occur together. Atrophy of the blood vessels supplying the organ of Corti can occur with age. This atrophy, as well as a number of drugs, may produce hearing loss or tinnitus. There is a loss of elasticity with thickening of the supporting tissues, and the vibrating ability of the basilar membrane declines.

Tinnitus. Ringing in the ears occurs in 37% of the population between 65 and 75 years old and in only 3% of people 18 to 24. Interestingly, it is slightly more common in women than in men and is not always associated with hearing loss. Many elderly find this "noise" to be a distraction and most believe that it is associated with *their* hearing loss. The cause is unknown although it has been associated with loud noise, inner ear damage, and brain damage such as that caused by atherosclerosis. So far there is no treatment. Instruments are available that mask the sound with another, more pleasant noise.

Loudness Recruitment. Many people find normal sound or slightly louder-than-normal sound unpleasant or even painful. Often such people have decreased hearing ability and respond to imperceptible speech by requesting that the speaker speak more loudly. When the speaker increases his volume, the listener flinches and comments "You don't have to shout." These people usually find most hearing aids unacceptable. The problem occurs when there is hair cell damage. Causes of such damage are unknown, and there is no treatment.

EQUILIBRIUM MALFUNCTION
Vertigo

Vertigo, or dizziness, is most often caused by problems with the balance apparatus, semicircular canals, saccule, or utricle. Just as hair cells are lost from the cochlea with age, they are lost from the ampullas of the semicircular canals. This loss may be as much as 40%. Atrophy also occurs in the maculas of the saccule and utricle with a loss of about 20% of the hair cells. Neural atrophy occurs here as well, with the degeneration of nerve cells of the labyrinthine portion of the auditory nerve. Ear infections may spread to the inner ear and cause vertigo. These factors contribute to a loss of balance stability and may result in an elderly person taking a more secure stance with the feet more widely spaced and the hands held out from the body.

Balance is affected by peripheral neuropathy. Diabetes and deficiency in vitamin B_{12} both contribute to peripheral neuropathy. Both of these problems are more common with age and produce cumulative damage. Therefore older people will have had such conditions longer and will have more peripheral neuropathy and greater problems with balance.

Menière's Disease

Menière's disease is a balance-related problem of unknown cause. It is believed to be the result of either overproduction or under absorption of endolymph resulting in an increase in endolymph pressure, although inadequate circulation has also been blamed. The symptoms include episodes of tinnitus and severe vertigo followed, at least initially, by long periods of remission. The disease often affects only one ear, and the ultimate result is usually deafness in the affected ear.

OTHER AGING SOMATIC SENSES

The sense of touch declines with age but only in hairless skin such as the palms of the hands. Here there is a decline in the number of Meissner's corpuscles and Pacinian corpuscles, but no decline has been observed in the three other skin receptors. A decrease in sensitivity to vibrations transmitted to skin (vibratory sensitivity) may be associated with loss of myelinated fibers in the dorsal roots of the spinal cord. This loss may amount to as much as 32% between ages 30 and 90. No change in hot and cold sensitivity has been reported.

A decline in the thirst reflex does appear in some elderly and can have serious consequences (see Chapter 8, The Excretory System.) The thirst reflex is part of the sense of thirst which also involves drying of the mouth and an increase in blood salinity. Response to low blood sugar as a part of the sense of hunger may decline too, but this is speculation. When mealtime approaches, a decreased gastric secretion due to atrophic gastritis may contribute to loss of the sensation of hunger. The result of this may appear as a lack of hunger response, causing an elderly person to fail to maintain an adequate intake of food. The extent to which this involves the gustatory center of the cerebral cortex is not understood. The sense of taste declines but this may be as much a result of environmental factors as it is of aging. There is no consistent evidence of loss of taste buds. Decline of sense of smell and any decrease in the brain's interpretive area may contribute to decreased perception of taste (see Chapter 7, The Digestive System). The sense of smell declines, probably as a result of loss of olfactory receptors. With age there is also a decrease in ability to identify previously known odors that can still be smelled.

CASE HISTORIES

❖ Cataract

E.H. is a 75-year-old recently retired machinist who entered a clinic for his second cataract surgery. His examination revealed that his left eye had an artificial intraocular lens with corrective soft contact lens. The right eye could detect light and large images under some conditions but was essentially blind and could not be corrected with external lenses. Funduscopy examination of the posterior of the eye, was unsuccessful due to a densely opaque cataract. A full medical consultation found him to be a healthy, elderly male suitable for cataract surgery.

E.H.'s decline in vision had been gradual and progressive over the past 12 years. He had worn glasses for myopia since an early age and at the age of 49 switched to bifocals with excellent results when his vision deteriorated to presbyopia. His first complaint regarding the cataracts occurred when he was driving at night and noticed halos of colored light around bright street lamps. This was followed by a 4-year period of painless, uncorrectable decline in vision in his left eye while the right eye seemed normal. Even though the vision in the left eye continued to deteriorate, E.H. experienced a period when the bifocals became unnecessary for that eye. This period was followed by "maturation" of the condition in the affected eye, resulting in blindness. Since E.H. was working as a machinist, cataract extraction with lens implantation was suggested. However, his fear of surgery or permanent blindness delayed this until deterioration of his right eye made it impossible for him to continue working. The left lens was then removed and replaced with a synthetic one, resulting in complete restoration of vision with the aid of glasses for that eye.

Three years after replacement of the left lens, the right lens was removed and replaced, resulting in restoration of vision correctable to 20/20 with glasses. While the surgery on the left eye had produced complete recovery in less than 3 months, the surgery on the right eye left E.H. with an annoying inflammation that persisted for over 9 months before gradually disappearing. E.H. has resumed working part-time as a machinist and is now without visual problems. His son, E.H. II, is also a machinist and at age 47 has just been told by his father's ophthalmologist that he is developing cataracts.

Comment. A cataract is any increase in opacity of the lens and occurs in 95% of people over 65 years of age. Fortunately, only a few of these are severe enough to cause significant reduction in vision. Clinically significant cataracts have a higher incidence in people with high serum lipids, glucose, or who have elevated blood pressure. The "second sight" phenomenon noted in this case with improvement in near vision as the cataract developed is the result of changes in the refractive index as the lens deteriorates. Fear or unavailability of surgery has resulted in a large number of cataract patients becoming totally blind.

❖ Glaucoma

F.P. is a 49-year-old woman known for her French cuisine who visited her ophthalmologist requesting new glasses. She had experienced poor acuity and blurred vision for 6 months. Lately she had become concerned that it was interfering with her tennis game. She had not noticed any eye pain, decreased tear formation, or excess watering. Her last glasses were prescribed 7 years prior to this visit.

Her past medical history included insulin-dependent diabetes mellitus which began at the age of 38 years and has been well controlled. She had an appendectomy at an early age and a cervical fusion for a spinal nerve problem at the age of 45.

Physical examination found F.P. to be a thin, middle-aged healthy female in no distress. Her eyelids were normal with intact extraocular movements. The sclera, cornea, and iris were unremarkable. Her pupils were equal and reacted to both light and accommodation. Funduscopy showed early diabetic retinopathy with a rare fluffy infiltrate. The left optic cup was mildly enlarged. Tonometry revealed increased intraocular pressure bilaterally (28 mm Hg in the right eye and 32 mm Hg in the left; Normal values are less than 22.) Examination of the anterior chambers revealed normal angles between the iris and the trabecular meshwork. Refractive examination indicated 20/30 vision in the right eye and 20/40 vision in the left eye without correction of the blurring by any lens combination. Repetition of the examination 2 weeks later continued to show increased intraocular pressure and was consistent with a diagnosis of glaucoma.

F.P. was placed on pilocarpine eye drops and the intraocular pressure returned to normal. The patient reported a moderate improvement in vision but continued to experience some blurring uncorrectable by lenses.

Comment. Glaucoma is the second leading cause of blindness in the United States. Approximately two percent (1 out of 50) of the population over the age of 45 have the condition, and many are not aware of it. This is because the vast majority of the cases are without symptoms except for a gradual and slow decrease in visual acuity. The most common symptom, loss of peripheral vision (tunnel vision), usually does not appear until substantial permanent damage to the retina has occurred. As we age there is a "natural" decrease in vision and any event that accelerates that process, even if it does not lead to blindness, should be prevented if possible. Without periodic measurements of intraocular pressure (tonometry), glaucoma will be discovered only after permanent damage has been done.

❖ Retinal Detachment

J.A. is a retired 81-year-old bacteriologist who suddenly developed severely reduced vision. When he awoke one morning he was immediately aware of a gray area reducing the vision of the outermost half of his right eye. He told his ophthalmologist that he had experienced no pain, flashes of light, trauma, or other eye symptoms prior to this event.

His medical history revealed coronary artery bypass surgery for unstable angina 4 years earlier. He has been taking dipyridamole (Persantine) and digoxin since the bypass. There were no other significant problems except a functional depression that began 18 months ago following the death of his wife. He has been wearing glasses for severe myopia since childhood.

Examination revealed an elderly, overweight, short male who appeared depressed. Both lower eyelids showed mild ectropion, or turning out of the lid with accumulation of tears, while the upper lids showed mild ptosis, or dropping. Eye function and movement were normal. Funduscopy showed a retinal detachment on the nasal margin of the right eye affecting 40% of the retina and a smaller one in the upper temporal portion of the left eye.

J.A. was referred to a retinal surgeon, and retinal reattachment by laser coagulation was performed within the week. Following a recovery period of a few days, his vision returned to normal. He has had no further eye problems and his depression seems to have disappeared.

Comment. In the elderly, retinal detachment often occurs without evidence of trauma. It occurs as a sudden loss of part of the visual field and sometimes produces symptoms of spontaneous flashes of light. The condition is more common in men and in people with myopia. It is also a common complication of cataract surgery. Reattachment of the retina usually restores vision to the condition before detachment unless the retina has suffered permanent damage.

❖ Menière's Disease

V.N., a 58-year-old diabetic insurance executive, went to his physician because of an episode of acute vertigo associated with nausea and vomiting. The attack occurred spontaneously with a sudden onset and was completely disabling. It was not associated with fever, infection, trauma, or pain. The episode lasted several hours and then ended suddenly. During the attack, the patient experienced a loud, roaring tinnitus that was more prominent in his left ear. For a short period following the attack he was dizzy whenever he moved his head. His medical history revealed well controlled diabetes mellitus and two similar but slightly less severe episodes the previous year. He had consulted a physician following the second episode and was treated for a viral infection.

By the time of his current examination, his symptoms had all abated except for a persistent but less intense ringing in his left ear. Physical examination at that time showed a thin, fit elderly male with normal vital signs. His pupils were equal and reacted to light and accommodation with normal extraocular movements. There was no nystagmus. His hearing was grossly normal bilaterally with a mild deficit on the left. The neurologic examination (including balance and eye-hand coordination) was normal, as was the remainder of the physical. Following ear, nose, and throat (ENT) consultation, a diagnosis of Menière's disease was made.

Follow up showed that similar episodes occurred during the next few years, alternating with periods of remission. By 7 years following diagnosis, V.N.'s hearing in the left ear had progressively deteriorated to deafness. Various therapies including histamines, antihistamines, and belladonna were tried all without significant results.

Comment. Menière's disease involves an accumulation of excessive amounts of fluid in the canals of the inner ear. As the fluid accumulates the pressure first stimulates and eventually kills the nerves of the organ of Corti and the ampulae. The stimulation causes the severe dizziness and perception of sound (tinnitis). Over time the nerve endings atrophy and function is lost. There is no known therapy and the cause of the disease is unknown.

BIBLIOGRAPHY

Corso CF: *Aging sensory systems and perception,* New York, 1981, Prager.

Doty RL and others:, Small identification abilit: changes with age, *Science* 226:1441-1443, 1984.

Fisch L: *Special senses — the aging auditory system.* In Brocklehurst, ed: *Textbook of geriatric medicine and gerontology,* New York, 1978, Churchill Livingstone.

Graham A: *The eye.* In Pathy MJS ed: *Principles and practice of geriatric medicine,* New York, 1985, John Wiley & Sons.

Kasper RL: *Eye problems of the aged.* In Reichel W, ed: *Clinical aspects of aging,* Baltimore, 1983, Williams & Wilkins.

Mills R: *The auditory system.* In Pathy MJS, ed: *Principles and practice of geriatric medicine,* John Wiley & Sons. New York, 1985.

Yoder MG: *Geriatric ear, nose, and throat problems.* In Reichel W, ed: *Clinical aspects of aging,* Baltimore, 1983, Williams & Wilkins.

15 Lifestyles: Environmental Factors and Aging

David Spreadbury PhD
Associate Professor of Nutrition,
University of Osteopathic Medicine
and Health Sciences,
Des Moines, Iowa

The first 14 chapters of this book have described how body systems degenerate with age, and the diseases in some of those systems that hasten loss of function. Genetic background affects the age-related rate of change in organ systems but, it is also influenced by environmental factors and choices such as diet, use of drugs, exercise, and stress. The range of capabilities seen among persons between 60 and 85 years of age demonstrates varied rates of aging. Centenarians frequently give a variety of "recipes" for their longevity.

The cumulative effects of environmental factors, or lifestyle, can be profound. Seven characteristics of a prudent lifestyle (listed in Table 15-1) have been associated with improved health and a gain in longevity. A 45-year-old man who followed all 7 practices lived an average of 11 years longer than a man who followed 3 or less.

In another example of the effect of lifestyle on longevity, the nonsmoking population of Mormons in California has been intensively studied. The risk of mortality from all cancers and cardiovascular diseases among members of this group is only half that of

---

Table 15-1 Lifestyle Practices Associated with Improved Health and Increased Longevity

1. Sleeping 7-8 hours per night
2. Eating breakfast daily
3. Not eating between meals
4. Maintaining weight (+19% to −5% of ideal weight [men]; +9.9% to −5% of ideal weight [women])
5. Regular exercise
6. Moderate or no use of alcohol
7. Not smoking

Adapted from Belloc (1973).

the general population. Among the 25,000 high priests of the Mormon church, the risk of cancer mortality is around 30% of that of the general population; their risk of cardio-vascular disease is only 14% of that of the general population. These men routinely avoid tobacco, alcohol, and caffeine. They exercise regularly and get sufficient sleep. A group of the general white population selected for a similar lifestyle showed similar results.

This chapter will consider how lifestyle factors, especially those that can be modified, affect the aging process and the various pathologic conditions that occur more frequently as we age.

NUTRITION AND AGING

The elderly are a heterogeneous group, comprised of individuals within a forty-year age range. Inter-individual variation among members of this group exceeds that among younger individuals. Great variation in biologic age, the effects of chronic disease, and the effects of medication all contribute to this heterogeneity and probably to the nutritional requirements of many individuals. There are, however, some basic requirements for optimum nutrition.

Basic Requirements

Over forty nutrients must be ingested daily to maintain good health. The Recommended Dietary Allowances (RDA) of the National Academy of Sciences Food and Nutrition Board contains suggested daily allowances for many of these (Table 15-2). These allowances are not the minimum needed to prevent deficiency disease, but rather an intake that will assure most healthy individuals a margin of safety.

The suggested allowance for each nutrient varies according to age, sex, and unique physiologic states, such as growth or pregnancy. Many of the recommendations are still controversial. The final age division used is 51 years and above, which implies that the requirement for specific nutrients remains the same from the fifth through the eighth or ninth decades of life. This may not be the case, for little is known regarding any change in requirements that might take place from late adulthood through old age.

Table 15-2   Recommended Dietary Allowances for Protein, Vitamins and Minerals Per Day for Adults 51 years and above—Revised 1989*

	PROTEIN g	VITAMIN A μgRE	VITAMIN D μug	VITAMIN E mgαate	VITAMIN K μug	VITAMIN C mg	THIAMIN mg	RIBOFLAVIN mg	NIACIN mg NE
Males	63	1,000	5	10	80	60	1.2	1.4	15
Females	50	800	5	8	65	60	1.0	1.2	13

	VITAMIN B6 mg	FOLATE μug	VITAMIN B12 μug	CALCIUM mg	PHOSPHORUS mg	MAGNESIUM mg	IRON mg	ZINC mg	IODINE μug	SELENIUM μug
Males	2.0	200	2.0	800	800	350	10	15	150	70
Females	1.6	180	2.0	800	800	280	10	12	150	55

*Food and Nutrition Board, National Academy of Sciences National Research Council 1989

Energy. The age-related loss of lean tissue from the body reduces the basal metabolic rate, which together with a decline in physical activity lowers the daily caloric requirement and reduces food intake (Figure 15-1).

Other factors such as a decreased sense of taste, loss of teeth, lack of appetite, depression, chronic systemic disease, prescribed drugs, alcohol abuse, poverty, or social isolation may also reduce food intake and lead to what is termed *the anorexia of aging.*

There are wide variations, even among young adults, in the daily calorie intake required to maintain appropriate body weight. These differences can be as great among older adults; therefore no specific daily calorie requirement can be given. Intake should be sufficient to maintain body weight, or low enough to ensure some weight loss if obesity is a problem.

There is no evidence to suggest that the dietary source of energy should be altered as we age. The current recommendation for adults of no more than 30% of total calories from fat is applicable to older adults. Fifty-five per cent of total calories should be provided by carbohydrate, mostly as complex carbohydrate containing fiber, with not too great a reliance on energy derived from sucrose.

Protein. The recommended dietary protein allowance for a young adult is 0.8 g of protein per kg of body weight per day. There is little evidence that this requirement is markedly different for the elderly adult. This allowance is easily achieved and usually exceeded by the habitual intake of large quantities of a protein-rich diet by the young, but it is harder to achieve with the lower volume of food consumed by the elderly, especially if food preference and economic pressures have increased the proportion of carbohydrate-rich foods in the diet. It is important to ensure that the elderly maintain an appropriate protein intake by maintaining or slightly increasing the quantity of lean meat or low-fat dairy products eaten each day.

Micronutrients. The specific requirements of the elderly for vitamins and trace metals and minerals are largely unknown, therefore the standards for young adults are usually used. A number of surveys of the elderly have revealed inadequate intakes of vitamin D, folate, some of the B vitamins, calcium, magnesium. and zinc. Some of these inadequacies occurred while elderly individuals were eating diets supplied by hospitals and nursing homes.

Total calorie intakes of 1000 calories or less per day are commonly recorded among the elderly, especially among the estimated 15% of Americans over age 65 who are suffering from dementia. At least 1200 calories per day provided by nutrient-rich foods are required for an adequate micronutrient intake. If this cannot be achieved, a complete vitamin and mineral supplement can be used, provided that it supplies no more than about 100% of the RDA per day for each nutrient.

Water. The typical adult requires about 2.5 liters of water per day under normal conditions. High air temperatures can greatly increase the amount required. About 1.5 liters of this requirement are derived from food. The remaining liter (approximately 5 cups) must be obtained from various liquids. Inadequate water intake usually causes thirst, but this mechanism is not always effective in the elderly; it is necessary to encourage the regular intake of water and other beverages. Fluid intake and loss in urine in the elderly should be closely monitored in hospitals and nursing homes.

Research in Nutrition and Aging

As explained in Chapter 2, the free radical theory of aging attributes physiologic deterioration to accumulated cellular damage caused by free radicals. These compounds

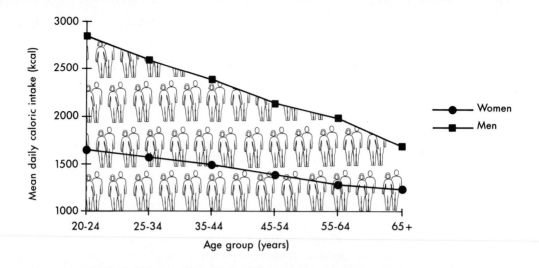

Figure 15-1 Mean daily calorie intake for white men and women from age 20 to above 65 years. Drawn from data of the National Center for Health Statistics.

contain atoms with unpaired electrons, which makes them very reactive. Free radicals are normally neutralized by enzymes that need adequate amounts of dietary copper and selenium for optimum activity. Vitamins E and C, and vitamin A and the carotenoids may also reduce damage from free radicals. A diet rich in these nutrients may therefore increase the removal of free radicals and slow the rate of cell deterioration.

Data from experiments with laboratory animals clearly demonstrate that dietary manipulation can increase or decrease longevity in many species by slowing the basic process of aging. Severe underfeeding during the first half of life doubles the life expectancy of laboratory rats. This technique, known as chronic energy intake restriction (CEIR), provides adequate intakes of all nutrients except energy, which is restricted to 60% or 70% of voluntary intake. Such restriction has resulted in some rats surviving for 1800 days, which is more than double normal, and roughly equivalent to a human life span of 180 years. The protein content of the diet also influences the prevalence of some typical diseases of aging in rats. Glomerulonephrosis, myocardial fibrosis, and prostatitis all appear more frequently in animals fed high protein diets. The effect of diet on malignant neoplasms is more complex but generally calorie restriction reduces the number of tumors of most sites. When high-fat diets are fed to genetically susceptible pigs, accelerated atherosclerosis results, and the animals die from premature cardiovascular disease.

It is tempting to extend these results to the human population, where obesity, a symptom of too high an energy intake, is associated with decreased life expectancy. Renal, myocardial, and prostate diseases are also common in human populations that eat a protein-rich diet. Current research is investigating the effects of both excessive dietary protein and fat and chronic underfeeding on nonhuman primates.

The Skeletal System

The age-related loss of body cell mass is usually viewed as a benign phenomenon, whereas bone loss is treated more as an age-related pathologic condition probably because the latter frequently results in bone fractures with severe clinical complications. In practice, the excessive loss of either muscle or bone can have severe consequences (see Chapters 4 and 5). Gender, race, and individual genotype all influence peak bone mass, which is usually achieved around 30 years of age. Men tend to be heavier boned than women, black races tend to be more heavily boned than whites, and some families have a heavier bone structure than others. The daughters of white women who have osteoporosis typically are at high risk for the disease.

Throughout the period of growth, the diet must contain adequate levels of all essential nutrients. Tables 15-3, 15-4, and 15-5 contain daily allowances for nutrients that are better understood. Calcium, phosphorus, and magnesium are especially important for bone growth, as all are major components of bone. Fluoride and possibly boron may also be required for complete mineralization.

The importance of calcium in bone formation is described in Chapter 4. During maximum growth an intake of 1200 mg of calcium per day is recommended. Currently for an adult, 800 mg per day is the recommended intake, equivalent to the calcium in 24 oz. of milk. Actual intakes are often below this level, especially if milk and milk products are excluded from the diet, or among young women restricting food intake for weight control. Lactose intolerant women, who have avoided milk, have a higher than average rate of osteoporosis. Calcium supplements of up to 2,000 mg per day are often suggested, but there is little clinical evidence that this appreciably slows bone loss. The use of a fluoride supplement results in increased bone mass but of even greater fragility.

Other factors can also affect the rate of bone loss. High intakes of both animal protein and sodium, typical of the American diet, increase calciuria. Bone loss is also increased by cigarette smoking and the heavy use of alcohol. In contrast, obesity appears to reduce bone loss, possibly the only health benefit of this condition. For women, the loss of estrogen following menopause or total hysterectomy increases bone loss significantly. This can be reduced by estrogen replacement therapy.

Osteomalacia is a decalcification of bone secondary to a Vitamin D deficiency. Adequate quantities of this vitamin are synthesized in the skin by exposure to ultraviolet light; only 10 minutes exposure of the head and arms to midday summer sunlight is required. A conversion to the active form then takes place in the liver and kidney. Additional vitamin D is derived from the diet, especially from milk and margarine, which are fortified with this vitamin.

Vitamin D synthesis in the skin declines in the elderly, as does its conversion in the kidney. Age, infirmity, or living conditions may also prevent even the limited sun

Table 15-3 Recommended Dietary Allowances of Minerals Per Day for Growth-Revised 1989*

	AGE YRS	CALCIUM (mg)	PHOSPHORUS (mg)	MAGNESIUM (mg)	IRON (mg)	ZINC (mg)	IODINE (µg)	SELENIUM (µg)
Children	1-3	800	800	80	10	10	70	20
	4-6	800	800	120	10	10	90	20
	7-10	800	800	170	10	10	120	30
Males	11-14	1200	1200	270	12	15	150	40
	15-18	1200	1200	400	12	15	150	50
	19-24	1200	1200	350	10	15	150	70
Females	11-14	1200	1200	280	15	12	150	45
	15-18	1200	1200	300	15	12	150	50
	19-24	1200	1200	280	15	12	150	55

*These are for nutrients for which less information is available and do not appear in the Recommended Dietary Allowances. Food and Nutrition Board, National Academy of Sciences—National Research Council, 1989.

Table 15-4 Recommended Dietary Allowances of Protein and Vitamins Per Day for Growth—Revised 1989*

	AGE YRS	PROTEIN (g)	VIT. A (µg/RE)	VIT. D (µg)	VIT. E (mgαTE)	VIT. K (µg)	VIT. C (mg)	THIAMIN (mg)	RIBOFLAVIN (mg)	NIACIN (mgNE)	B_6 (mg)	FOLATE (µg)	B_{12} µg
Children	1-3	16	400	10	6	15	40	0.7	0.8	9	1.0	50	0.7
	4-6	24	500	10	7	20	45	0.9	1.1	12	1.1	75	1.0
	7-10	28	700	10	7	30	45	1.0	1.2	13	1.4	100	1.4
Males	11-14	45	1000	10	10	45	50	1.3	1.5	17	1.7	150	2.0
	15-18	59	1000	10	10	65	60	1.5	1.8	20	2.0	200	2.0
	19-24	58	1000	10	10	70	60	1.5	1.7	19	2.0	200	2.0
Females	11-14	46	800	10	8	45	50	1.1	1.3	15	1.4	150	2.0
	15-18	44	800	10	8	55	60	1.1	1.3	15	1.5	180	2.0
	19-24	46	800	10	8	60	60	1.1	1.3	15	1.6	180	2.0

*These are for nutrients for which less information is available and do not appear in the Recommended Dietary Allowances. Food and Nutrition Board, National Academy of Sciences—National Research Council, 1989.

Table 15-5 Estimated Safe and Adequate Daily Dietary Intakes of Selected Vitamins and Minerals for Growth*—Revised 1989

AGE YRS	BIOTIN (mg)	PANTOTHENIC ACID (mg)	COPPER (mg)	MANGANESE (mg)	FLUORIDE (mg)	CHROMIUM (mg)	MOLYBDENUM (mg)
1-3	20	3	0.7-1.0	1.0-1.5	0.5-1.5	20-80	25-50
4-6	25	3-4	1.0-1.5	1.5-2.0	1.0-2.5	30-120	30-75
7-10	30	4-5	1.0-2.0	2.0-3.0	1.5-2.5	50-200	50-150
11+	30-100	4-7	1.5-2.5	2.0-5.0	1.5-2.5	50-200	75-250
Adults	30-100	4-7	1.5-3.0	2.0-5.0	1.5-4.0	50-200	75-250

*These are for nutrients for which less information is available and do not appear in the Recommended Dietary Allowances. Food and Nutrition Board, National Academy of Sciences—National Research Council, 1989.

exposure required for synthesis of this vitamin. The problem becomes more acute if foods fortified with vitamin D are not eaten regularly. Surveys indicate that the elderly commonly achieve only half the daily recommended intake for dietary vitamin D.

The elderly should be encouraged to eat vitamin D-rich foods such as fortified milk and margarine, and naturally oily ocean fish like mackerel, sardines, or salmon. Regular short periods of exposure to sunshine should be arranged whenever possible. Vitamin D supplements may be recommended for individuals at high risk of bone loss.

The Muscular System

The body can be divided into metabolically active body cell mass, extracellular water, body fat, and bone and connective tissue. The body cell mass begins to decrease early in adult life. The bulk of the loss is in skeletal muscle, which decreases in weight by 40% between young adulthood and age 70. There are also reductions in the mass of internal organs. The liver decreases in weight by 20%, and the kidneys are reduced by 9% during the same period. At the same time, body fat increases; therefore total body weight does not always show a corresponding fall until age 70 to 75, when body fat also decreases.

A healthy adult requires around 35g of dietary protein per day to replace normal losses. If intake falls below this amount, body protein is lost and muscle and organ weight will decrease. Typical protein intakes in the U.S. are generally above this minimum, so they are not usually a factor in the loss of lean tissue. Even so, age brings about a steady decline in the weight of organs such as the liver.

The Digestive System

Age-related changes and pathologic conditions of the digestive system were reviewed in chapter 6. Physiologic changes in the gastrointestinal tract and associated organs may be influenced by the nature of the diet and then in turn may affect how dietary components are digested and absorbed.

Approximately 30% of Americans have lost their teeth by age 65. Inadequate dental care and sucrose added to the typical western diet contribute to this process. An inadequate intake of fluoride during tooth development also contributes to tooth loss. After most teeth are lost, a modification of diet often takes place.

A decrease in taste perception occurs as old age is reached. It is possible that in some cases an inadequate intake of zinc is responsible, for this nutrient is known to play a role in the maintenance of taste acuity. Reduced taste perception may lead to less enjoyable eating and a decrease in food intake.

Reflux esophagitis and hiatal hernia are common conditions in the obese. When such conditions exist, specific foods or beverages such as chocolate, nuts, or coffee may exacerbate these conditions.

The gastric atrophy and achlorhydria of aging can impair the absorption of vitamin B_{12}. Malabsorption of this vitamin often exists for several years before it results in pernicious anemia. Injections of the vitamin are then required. Diminished secretion of acid from the stomach can also reduce iron absorption, leading to an iron deficiency anemia. Decreased gastric function will also impair overall digestive efficiency.

Enzymes for the digestion and absorption of the various components of the diet are contained in the secretions of the pancreas and the mucosa of the small intestine. The volume of enzymes secreted decreases with age but remains sufficient for adequate absorption of nutrients until extreme old age. Carbohydrate digestion and absorption are not significantly reduced until the eighth decade of life.

Constipation is very common in the elderly, and can lead to discomfort and diminished food intake, affecting overall nutrition. Low intakes of dietary fiber and fluids, inadequate physical activity, and the effects of medications can all be contributory factors. Persistent constipation can lead to overflow incontinence.

As many as 50% of all individuals who reach 90 years of age have developed diverticula in the colon (see Chapter 6). This condition is less prevalent in countries where the diet is high in fiber. Raising dietary fiber levels from the typical 15 g per day to 40 g per day reduces pain and other symptoms in patients suffering from diverticulitis.

The Excretory System

Evidence from the Baltimore Longitudinal Study on Aging shows that an age-related decline in kidney function is not inevitable. One third of 254 subjects followed for 24 years had no absolute decrease in renal function. Both human and animal data suggest that external factors throughout life can influence the rate of renal deterioration in the latter stages of life. Non-insulin dependent diabetes mellitus, atherosclerosis, and hypertension are all widespread in the older American population, and are known to cause renal damage. The prevalence of these three diseases can be reduced by dietary and other lifestyle changes and will be discussed in later sections of this chapter.

It has been suggested that interactions among the above diseases, medications, and a high protein intake may account for much of the renal deterioration attributed to aging. The typical American protein intake is around 100 g per day, most of which is derived from meat. This is approximately double the suggested allowance and can induce a high renal blood flow and the eventual loss of glomeruli.

The Cardiovascular System

Atherosclerosis. The cause of atherosclerosis is complex. As indicated in Chapter 9, family history, high serum cholesterol, cigarette smoking, high blood pressure, obesity, diabetes mellitus, and a lack of physical activity all increase the likelihood of the disease. Symptoms of coronary artery disease are likely to occur when 60% of the internal surface of the affected blood vessel is covered in atheromatous plaques. The presence of several of the adverse risk factors described above can bring down the age when this point is reached from 80 to 40 years.

Cholesterol is transported through the blood in several lipoproteins. The two main components of total serum cholesterol are found in low density lipoprotein (LDL), and high density lipoprotein (HDL). Cholesterol associated with the LDL component promotes arterial damage, whereas cholesterol found in the HDL component has a protective effect. For this reason laboratories routinely estimate the fraction of the total serum cholesterol carried in each of the lipoproteins. Cholesterol in the low density lipoprotein is the larger component of the total, so total serum cholesterol is often used as an indicator of the risk of arterial disease. A value of less than 5.2 mmol/L (200 mg/dL) is now advocated for adults, and 3.9 mmol/L (150 mg/dL) is thought to be an optimum level. The significance of higher values in those over 65 years of age and free of symptomatic coronary heart disease is less clear. At present it is still prudent to encourage persons in this age group with hypercholesterolemia to modify diets that contribute to the condition.

Serum cholesterol is controlled by the liver and is genetically determined, but it is also affected by diet. The typical western diet, high in saturated fat and cholesterol and low in fiber, raises LDL cholesterol. Diet affects HDL cholesterol less, but this lipoprotein is lower in the obese and those who avoid aerobic activity. In societies where fat and calories are less plentiful and physical activity is more strenuous, average cholesterol values are lower and rise less with age: therefore atherosclerosis and its clinical consequences are much less common.

Cigarette smoking is a major cause of occlusive arterial disease of peripheral blood vessels (in the leg for example) but it is not known what role diet plays in this condition. Long-term vitamin E supplementation has improved arterial flow and lessened the symptoms of intermittent claudication resulting from the condition.

Essential Hypertension. Elevated blood pressure correlates with increased atherosclerosis and with the incidence of strokes (CVAs) Around 30% of adults in the United States have blood pressure above 140/90 mm Hg and are therefore classified as hypertensive. Two million children also have blood pressure that exceeds the desirable level for young individuals. In the U.S., blood pressure usually rises with age. In the Framingham study, representative of the American population, 45.5% of women and 61.1% of men aged 75 to 84 were classified as hypertensive. In less developed societies, average blood pressures are lower and rise less with age. Genetic factors modify individual response to diet, body weight, level of physical activity, alcohol intake, and level of stress, but all of these elements are known to affect blood pressure regardless of genetic constitution.

Humans are equipped with a very efficient mechanism for conserving sodium but have difficulty excreting an excessive intake. For some individuals, habitually high sodium intake is associated with an increase in blood pressure. The physiologic requirement for sodium is around 8.5 mEq (500 mg) per day, whereas intakes are typically 85 mEq to 130 mEq (5000 to 7000 mg) per day, mostly in the form of sodium chloride added at various stages of food preparation. The modern diet also contains a low ratio of potassium to sodium, which exerts a hypertensive effect.

High levels of dietary fat, especially saturated fat, increase blood pressure, possibly by modifying prostaglandin synthesis in a way that has an adverse effect on the peripheral vasculature. The consumption of more than two alcoholic drinks per day increases blood pressure. When an alcoholic enters a treatment program, blood pressure can fall by as much as 20 mm Hg. When obesity and hypertension coexist, a reduction in weight will usually reduce blood pressure. Diet can therefore favorably influence serum lipoproteins and help control blood pressure, thereby minimizing the age-related pathologic changes that occur in the aging cardiovascular system.

Blood. As described in Chapter 9, blood is a complex mixture of cellular and subcellular structures suspended in plasma. Inappropriate diets can adversely affect both the cellular and plasma components of blood. Deficiencies in the availability of some trace elements (such as iron) and vitamins (such as B_{12} or folate) disrupt normal erythrocyte synthesis, resulting in anemia. Failure to absorb vitamin B_{12} or a dietary deficiency of vitamin B_{12} or folate eventually results in megaloblastic anemia. Vitamin B_{12} is readily obtained from all meat, and a reserve is stored in the liver. Green leafy vegetables are the major source of dietary folate, and these are frequently omitted from the diet of the elderly. An inadequate intake of folate is therefore the most likely cause of a megaloblastic anemia of dietary origin, and can be reversed by a supplement of this vitamin.

Plasma albumin values are normally around 40 g/L to 60 g/L but can be reduced by stress such as an infection. Low values can also indicate depletion of body protein following long periods of inadequate dietary protein intake. This can be encountered in elderly individuals who are existing on a "tea and toast" diet.

Immunity. Aging is accompanied by an increase in the prevalence of autoimmune and immune complex disease, infections, and cancer, all of which are associated with a decline in immunocompetence. Both humoral and cell-mediated immune responses display age-related changes, and although all cells of the immune system are affected, deterioration is most pronounced in the T cells (see Chapter 9). Experimentally, activity of the immune system responds to changes in diet, ambient temperature, psychologic stress and exposure to toxic compounds. Therefore it is not clear what proportion of the loss in immunocompetence seen in the elderly is the result of external factors.

Protein-calorie malnutrition leads to some of the same decline in measures of immune function as aging, but the effect of a dietary protein deficiency on the immune system is not straightforward. When a deficiency occurs, some parameters of immune function are adversely affected, whereas others appear to be unaffected or enhanced. Chronic energy intake restriction in laboratory rats maintains immunologic competence in aged animals and prevents much of the disease that usually occurs with a deteriorating immune system.

Zinc is an essential dietary trace element involved in multiple cellular enzyme systems. A deficiency state leads to major immunodeficiencies, which can be reversed by a dietary supplement of zinc. Daily intakes of zinc among the population of the United States are frequently found to be less than the 12 mg and 15 mg recommended for females and males respectively. Intakes of zinc tend to fall as food intake declines in old age. There is speculation that marginal chronic zinc deficiencies develop, accounting for some of the loss of immunocompetence in the elderly. Zinc supplements improve cellular immunity in some elderly subjects.

Low serum levels of vitamin E have been associated with susceptibility to infections and delayed hypersensitivity reactions, a measure of T-cell mediated immunity. Short-term vitamin E supplementation has improved immune responsiveness in healthy elderly individuals who were eating a diet assumed to be adequate in all nutrients. A supplement containing vitamins A, C, and E improved cell-mediated immunity in a group of elderly long-stay hospitalized patients.

The Endocrine System

Diabetes Mellitus. The most common endocrine defect encountered in middle and late adulthood involves the insulin-secreting cells of the pancreas, and the role of insulin in controlling blood glucose. Aging Americans commonly exhibit glucose intolerance, which is indicated by higher than expected blood glucose values after a glucose challenge. This may result from a decline in insulin receptor activity and hence sensitivity of target tissues to either the action of insulin that occurs with age, or to the production of defective insulin (see Chapter 12). Elevated blood glucose, or hyperglycemia, accelerates the degeneration of proteins throughout the body, leading to the nephropathies, neuropathies, and blood vessel damage commonly seen in diabetics.

The elevated blood glucose levels of approximately 10 million Americans progress beyond glucose intolerance to type II or non insulin-dependent diabetes mellitus (NIDDM), a disease that usually becomes symptomatic in middle or late adulthood. The disease is intimately connected with obesity, and most type II diabetics are overweight. A genetic predisposition to this disease is seen within families and within specific racial groups. For example the Pima Indians of the S.W. United States experience a very high rate of type II diabetes. NIDDM begins with a loss of sensitivity to insulin at the insulin receptors in target tissues, provoking an increased secretion of insulin from the pancreas and hyperinsulinemia while maintaining hyperglycemia.

Lifestyle factors are influential in the development of type II diabetes. It is believed that high fat intakes and a low expenditure of energy in physical activity contribute to development of the disease. Both factors also predispose to obesity. Epidemiologic evidence confirms that the disease is uncommon in areas of the world where fat and protein intakes are low and starchy foods are the main source of calories.

Chromium, a dietary trace element, forms part of an organic complex called the glucose tolerance factor. This complex is involved in the binding of insulin to its receptors by an unknown mechanism. Surveys have indicated marginal intakes of chromium, especially among the elderly, with evidence of a reduction in blood glucose among a group of elderly given a combined chromium nicotinic acid supplement.

The Nervous System

Some of the most debilitating changes associated with aging are those that affect the central nervous system and lead to dementia, or generalized mental deterioration.

Diet and physical activity can modify the course of hypertension and atherosclerosis in many individuals. Both diseases can result in clinical and subclinical cerebrovascular disease, which impairs intellectual function by reducing blood flow to areas of the brain. Existing clinical cardiovascular disease has been found to be a significant predictor of decreased performance on several cognitive function tests, although the pattern of effects was not consistent. Non insulin-dependent diabetics, who usually have advanced vascular deterioration, are especially likely to have suffered some decline in cognitive function. These changes result from specific pathologic conditions and therefore are not a necessary sequel to disease-free aging.

The decreased caloric requirement of old age often causes a substantial drop in food intake, which can reduce nutrient intakes below adequate levels. Low protein intake, a problem sometimes encountered with the elderly, reduces the availability of the amino acids tyrosine and tryptophan, which could restrict synthesis of the neurotransmitters dopamine and norepinephrine, or serotonin.

In addition to being essential for the synthesis of red blood cells, vitamin B_{12} is required for maintaining the myelin sheaths around nerves. A deficiency of B_{12}, generally brought about by an impaired absorption, can lead to nerve degeneration and paralysis. The typical age when this breakdown occurs is around 70 years. Deterioration of gastric function is the most common cause.

An adequate intake of vitamins and minerals depends on a total intake of at least 1200 kcalories per day derived from a wide range of fresh foods. Serum levels of vitamin C and the B vitamins are frequently low in the elderly; it is generally assumed that the low values are indicative of poor intakes rather than an age-related decrease in absorption. In a group of independent living elderly, poorer performance in tests of nonverbal abstract thinking and memory were associated with lower serum levels of the vitamins mentioned above. Severe deficiencies of some vitamins provoke dementia that responds to supplements, but in general, attempts at supplementation to improve cognitive function where mild deficiencies were suspected have been unsuccessful.

Up to 35% of all individuals who survive to age 80 suffer mild to severe dementia. Multiple small infarcts in the brain, alcohol, or a major vitamin deficiency may all cause dementia, but more than 50% of cases are said to be a result of Alzheimer's disease. Sometimes there is a genetic predisposition to the disease. The ingestion of aluminum has been suggested as a contributory factor, but little evidence of a cause and effect relationship has been found.

DRUGS AND AGING
Polypharmacy

Physiologic changes of aging affect the rates of drug absorption, metabolism, and excretion, as discussed in Chapters 7 and 8. This means that for any given drug dose, blood levels in the elderly are often higher and remain so for longer periods than in younger individuals, increasing the likelihood of an adverse reaction. In addition, prescription medications taken over long periods can cause renal damage. Popular over-the-counter medications such as the nonsteroidal antiinflammatory drugs and acetaminophen, an analgesic, can also cause loss of kidney function.

Drugs may be prescribed for several conditions, sometimes by different physicians, without the realization that they have similar pharmacologic effects, or that there might be interactions between drugs. For example, a number of drugs possess anticholinergic activity with side effects that include blurred vision, urine retention, dry mouth, and increased intraocular pressure. Mild side effects from one such medication might be tolerable, whereas the combined effects of two or three could be unacceptable. In other cases, the effects of a cholinergic drug can be altered by the subsequent prescription of one or more anticholinergic drugs.

In addition to these potential interactions, the elderly are prone to modifying drug intake themselves. If a prescribed dose of a drug has failed to alleviate symptoms, then taking more is seen as a logical step. In other cases if symptoms occur, a drug prescribed for a spouse or friend with similar symptoms is used, or to save money medication left over from treating a previous illness is used.

Diet-Drug Interactions

The absorption of a drug may be reduced or delayed by the timing and composition of a meal. For example, the absorption of some antibiotics can be reduced if they are taken with milk. The drug L-dopa is an amino acid and if taken at the same time as a high protein meal, will have to compete with the amino acids from protein for absorption. Consequently drug absorption is reduced.

If diet composition changes radically during a course of medication, blood levels of the drug can be substantially changed. High carbohydrate diets tend to reduce the rate of drug metabolism, whereas a high protein diet has the reverse effect.

Drugs can also interfere with the absorption of nutrients. Cimetidine, used for decreasing acid secretion in peptic ulcer disease, can disrupt the complex process of vitamin B_{12} absorption, which begins in the stomach. A B_{12} deficiency anemia can result.

Over-the-Counter Drugs

Many analgesics, antiinflammatory drugs, and laxatives are widely used by the elderly. These can interact with prescribed medications. For example, the prolonged use of acetaminophen by an individual taking phenytoin can lead to an accelerated rate of acetaminophen metabolism and an increased level of metabolites that is potentially damaging to the liver. The excessive use of aspirin or indomethacin to treat arthritis can lead to gastrointestinal bleeding and iron deficiency anemia.

Laxatives are frequently abused by the elderly. A variety of agents are used for their laxative effect. Drugs that alter intestinal motility and electrolyte transport, such as those contained in senna, cascara, and phenolphthalein, can interfere with the absorption of nutrients and increase plasma protein losses via the intestine. If the user has a low potassium intake, prolonged use of these substances can precipitate potassium deficiency. Mineral oil is another frequently used laxative that interferes with the absorption of fat soluble vitamins, including vitamin D. This interference can be a precipitating factor in osteomalacia.

The use of antacids containing aluminum and magnesium hydroxide can give rise to a phosphate deficiency because these substances combine with dietary phosphorus to form insoluble phosphates.

The elderly are thus very susceptible to problems associated with drug use. Many of the chronic effects of long-term drug use are unknown. Potential renal damage has been discussed, and it is likely that liver damage also occurs. Mental confusion is frequent among the elderly; in many cases drug use can be a contributing factor. Nevertheless, we are likely to see an increase in the use of both prescribed and over-the-counter medications used to treat the variety of maladies that occur with aging.

Vitamin and Mineral Supplements

Up to 50% of elderly individuals in some communities admit to self-dosing with high levels of several vitamins and minerals. This has not been shown to be beneficial, and a number of toxic effects have been recorded.

High doses of vitamin E are sometimes taken in the belief that they combat heart disease. If this is done by a patient receiving anticoagulants the effect of these drugs on vitamin K can be increased with serious consequences. High daily intakes of vitamin A or D from supplements can be toxic. Regarding vitamin A, symptoms include loss of appetite, bone pain, loss of hair, and liver damage, and can occur at intakes only 10 times the RDA. Vitamin D taken at doses greater than 5 times the RDA can lead to overabsorption of calcium, which is deposited in organs such as the kidney. Symptoms can include loss of appetite, diarrhea, and mental confusion.

Supplements of water soluble vitamins have been considered less of a hazard but intakes of vitamin B_6 around 20 times the RDA have given rise to peripheral neuropathies affecting speech and gait.

Zinc supplements of up to 5 times the RDA are commonly taken, an action which reduces the level of HDL cholesterol in serum. Low HDL cholesterol concentrations are associated with an increased risk of heart disease. Higher intakes of zinc also interfere with the absorption of iron and copper, and can suppress the immune system.

ALCOHOL AND THE AGING BODY
The Digestive System

Liver size and hepatic blood flow normally decrease with age, but both can increase with concomitant alcohol use. Drug metabolism and excretion frequently involve enzymes found in the liver. The metabolism of drugs by this system is often slowed by normal aging and the use of alcohol can significantly alter this liver function, leading to unanticipated effects from prescribed drugs.

Seventy percent of young adults use alcohol but only 43% of persons aged 65 to 74 years are users. Of this elderly group 8% have more than two drinks per day, compared with 15% of young adults. Alcohol use often increases in the elderly during the stresses of illness or after the loss of a spouse.

Alcohol absorption and elimination are relatively unaffected by age, but for any given dose blood alcohol concentration is higher in an older individual than in a younger individual. This effect is probably a function of body water volume, which decreases with age and could result in less dilution of any alcohol present.

Alcoholism

The current rates for alcohol and drug dependence were determined by the Epidemiologic Catchment Area (ECA) program. This study differed from those published previously by using only diagnoses made according to the DSM-III-R criteria for alcohol dependence. The lifetime prevalence of alcohol dependence rates for men was found to be 27% for 18 to 29-year-olds, 28% for 30 to 44-year-olds, 21% for 45 to 64-year-olds, and 14% for those over 65 years old. The corresponding rates for women were 7%, 6%, 3% and 1.5%. Dependence on prescription and over-the-counter drugs was also found to be a significant problem.

Despite the lower prevalence among the elderly compared with younger age groups, alcohol dependence is still regarded as a serious problem. As the proportion of elderly increases, the total number of elderly alcoholics will rise. A high percentage of elderly individuals admitted to psychiatric units or seen as outpatients use alcohol heavily.

Risk factors for alcoholism among the elderly are similar to those predisposing earlier in life and include maleness, low income, a lower level of education, and a history of depression. The effects of heavy alcohol consumption by the elderly can be mistaken for psychosocial deficits attributed to growing old. As a result they may easily be overlooked and are frequently undiagnosed.

The social consequences for the elderly heavy drinker may be far different than for the younger alcoholic. Because elderly drinkers are often retired, housebound, and living alone, there is no potential for intervention as a consequence of social failures at home or at work. The adverse effects are most likely to be psychiatric and medical.

A youth-oriented society often turns a blind eye to drinking by the elderly; it may be condoned even within a family. Alcohol may be viewed as a comfort to a group that has little else and that imposes minimal harm at that stage of life. Physicians sometimes suggest alcoholic drinks to older patients who have difficulty sleeping. Even though increased availability raises the risk of use and dependence, some retirement communities have a happy hour before the evening meal.

Heavy alcohol intake may cause irreversible damage to the gastrointestinal, cardiovascular, and central nervous systems of an elderly person. The risk of atrophic gastritis is increased, leading to upper G.I. bleeding and anemia, and impaired vitamin B_{12} absorption and anemia. Liver damage can reduce the production of active vitamin D and precipitate osteomalacia. Cardiomyopathy is a common problem among elderly alcoholics. Alcohol has an acute dose-related negative effect on the brain that is reflected in the results of several neurobiologic tests. The deficit increases with the age of the subjects, which suggests that the aging brain is more sensitive to the effects of alcohol than the younger brain. The chronic effect of regular alcohol ingestion on brain function is more controversial. Alcohol intake may accelerate some of the cognitive changes associated with normal aging, but the difficulty in defining disease-free aging in the central nervous system, together with the problem of identifying effects specific to alcohol, complicate the situation. Nevertheless, end stage alcoholism is associated with severe neurologic disease resulting from the toxicity of alcohol and accompanying vitamin deficiencies. The older the alcoholic, the greater the degree of cerebral atrophy that has occurred.

According to Zimberg, the elderly alcoholic may be continuing a behavior established earlier in life, albeit with a different pattern. In these cases personality factors are important in the development and maintenance of the behavior. Occasionally spontaneous remission occurs when perhaps the effects of alcohol become less pleasant or side effects begin to be more severe. A second group of elderly alcoholics is composed of individuals with no previous history of the condition. In these cases it is more likely that the intake of alcohol began in response to stressful external factors such as isolation or loneliness, physical illness, bereavement, or retirement. This second group may partly explain the late age of onset detected in the elderly subjects in the ECA program.

For the younger alcoholic, treatment is focused on the drinking behavior itself, and utilizes group therapy based on the Alcoholics Anonymous model. This may also be an effective approach with the elderly alcoholic of many years standing. For alcoholism that begins in the later years, an intervention that seeks to correct some of the social deficits referred to above may also be successful. In these cases reestablishing the individual's sense of usefulness and contribution to society is very important, and the formation of a strong social network is a powerful preventive factor.

EXERCISE AND AGING

The ability to perform physical activity usually declines with age. What is not clear however, is how much of this deterioration is an inescapable part of aging rather than a result of disuse. Much of the data documenting these changes was obtained from a largely sedentary population. It may therefore be a measure only of lack of physical activity and not aging per se.

Basic Requirements

Exercise appears to offer protection against several of the diseases of aging and is a factor in maintaining good health in the elderly. However 73% of individuals above age 65 do not exercise regularly, compared with 56% of the 30-year to 44-year age group. Older women appear to be especially sedentary. Ideally a strength and fitness program should start early in life and be maintained with necessary modifications into old age. It is, however, beneficial to begin even late in life (Figure 15-2). Some very old individuals have achieved impressive increases in muscle strength through progressive resistance training, and there is ample evidence that major increases in cardiovascular fitness can be achieved by the elderly.

The greatest benefits are derived from an aerobic cardiovascular conditioning program. Most authorities suggest that a complete medical examination be performed before an older person begins a conditioning program, especially if there are any known risk factors for coronary artery disease. Part of this examination should be a treadmill test, a measure of the cardiac response to exercise. Such a test identifies those with asymptomatic circulatory disease and will provide information useful in formulating an exercise program for a specific individual.

An exercise prescription includes the type of activity, the intensity with which it is to be performed, and the duration and frequency of each session. Musculoskeletal problems that limit exercise will also be taken into account. The physical activities that can be incorporated in a conditioning program are those that employ the large muscle groups of the body (Table 15-6). Any one or a combination of these activities can be used to fill the exercise prescription.

The intensity of exercise should be sufficient to exert a training effect but not high enough to overtax the aged musculoskeletal system. Level of exertion is usually expressed as target heart rate (THR). A target heart rate of 60% of maximum heart rate (MHR) is given as the minimum required to improve cardiovascular endurance. However, in the very elderly sedentary individual, 40% of maximum heart rate or no more than 20 beats per minute above resting heart rate may be a more prudent starting point. Maximum heart rate is approximately 220 minus age in years.

Figure 15-2   Appropriate physical activity is always beneficial. **A**, Passive exercise for a wheelchair-bound nursing home resident. **B**, An elderly jogger. **C**, Elderly participants in a dance class. Sponsored by The Dance Exchange.

The Dance Exchange, founded in 1976 by dancer/choreographer Liz Lerman, is a nonprofit arts organization whose work is based on the inseperable concerns of artistic excellence and community involvement.

Dance Exchange programs include:

*Dancers of the Third Age, a senior adult modern dance company, presents informal participatory performances for people of all ages.

*Community Crossover, offers dance classes and composition workshops for senior adults, people in hospitals, children, the mentally and physically challenged, people living with Aids and other particular populations.

*Liz Lerman Dance Exchange, an intergenerational modern dance company, presents formal concert work and community based residencies nationally and abroad.

The power of dance rests in a process of personal expression that has the potential to actively engage participants of any age on many levels-mental, emotional, spiritual as well as physical. In all of its work, The Dance Exchange demonstrates its philosophy that dance is for everyone.

A more precise method for calculating THR is based on a graded exercise test and should be used whenever possible. Individuals should be encouraged to monitor their own pulse rates during a pause in exercise. Activity at this intensity should be performed continuously for 15 to 45 minutes three to five times per week.

A 60-year-old adult who has been totally sedentary might begin an exercise program at three periods of 15 minutes per week, keeping exercise heart rate to 96 beats per minute, or approximately 60 percent of the calculated maximum. The duration of each session would then be increased by no more than 10% per week, provided the participant was free from undue tiredness or any muscle or joint pain.

When the desired duration for each session is reached, healthy individuals can gradually increase THR to 70% of MHR over a period of months. Frequency can also be increased to 4 to 5 times per week. Some individuals may reach the optimum exercise

Table 15-6   Physical Activities Suitable for Inclusion in a Cardiovascular Conditioning Program for the Elderly

Walking	Table Tennis
Running	Squash
Aerobics	Racquetball
Dancing	Badminton
Calisthenics	Volleyball
Fencing	Roller skating
Swimming	Rowing (stationary or on water)
Cross country skiing	Bicycling (stationary or mobile)
Golf (carrying clubs)	

prescription in less than 6 months, whereas others may spend a year or more working up to a suitable level. Many of the activities listed in Table 15-6 utilize only the large muscles of the legs, allowing upper body atrophy to proceed unchecked. This can be improved by an exercise schedule with light weights, or by using some of the resistance machinery now available.

The Skeletal System

Physical activity plays a major role in the development and retention of bone. Groups of highly trained young athletes have significantly greater bone mass than their counterparts who do not exercise, and middle aged and elderly joggers have spine and leg bones that are 40% more dense than those of sedentary contemporaries. However, there is no clear evidence regarding the effect of particular types or levels of exercise on bone mass. But so many other benefits accrue from exercise that it is prudent to advocate a moderate level of weight-bearing activity for all. Even in cases of clinical osteoporosis a carefully designed program of weight-bearing physical activity can slow bone loss, and in some cases promote a slight accumulation of bone. A secondary benefit is an increase in muscular strength and the reduced likelihood of falls.

Osteoarthritis. Interest in the relationship between osteoarthritis and physical activity has centered on the fear that weight-bearing activities such as jogging will induce degenerative arthritic changes in the weight-bearing joints by the time old age is reached. It has been stated that elderly joggers may end up with wonderful cardiovascular systems but in wheelchairs. However, there is no evidence to support this. The elderly recreational joggers with dense bones referred to above have normal joint spaces as often as their sedentary peers. The extreme stress of very high mileages at a fast pace for many years endured by elite athletes can lead to osteoarthritis of the hip. Moderate exercise combined with an appropriate energy intake will help combat obesity and thereby reduce the risk of joint damage resulting from excess body weight. Non–weight-bearing exercise in water maintains maximum joint mobility and is encouraged whenever possible in both osteoarthritis and rheumatoid arthritis patients.

The Muscular System

The habitual level of physical activity is a major determinant of the rate of muscle loss. A 1981 cross-sectional study by Heath and others compared young endurance athletes age 22 ± 2 years with master athletes age 59 ± 6 years and found that lean body weights did not differ significantly and that both groups had almost identical low percentages of body fat. Such studies do not settle the issue of an inevitable age-related loss of lean tissue versus atrophy induced by inactivity, for it can be argued that those genetically predisposed to maintain lean mass throughout life are more likely to become master athletes.

Observations made on elderly individuals subjected to a physical training program are more persuasive. Persons in their seventies who begin an endurance program based on vigorous walking, then progressing to jogging, increased their lean body mass over a 12-month period. Persons in their nineties given high intensity strength training for 8 weeks increased total midthigh muscle area by 9 ± 4.7% and average strength by 174 ± 31%. These changes significantly improved functional mobility. When the elderly engage in physical activity they gain lean body mass. Continued physical activity throughout life may well slow down the loss of lean tissue that usually accompanies aging.

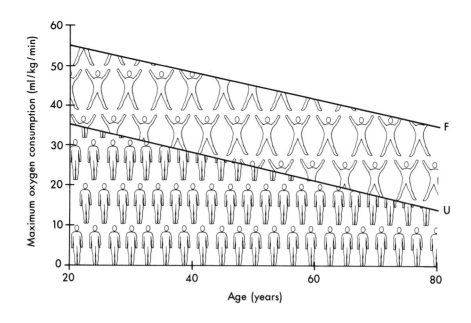

Figure 15-3 Decline in VO_{2max} with age for men, comparing aerobically fit (F) with aerobically unfit (U) individuals. The estimated rate of decline is approximately 9% per decade for the unfit and 5% per decade for fit (Heath et al, 1981). Note that the elderly fit can have a higher VO_{2max} than unfit 20-year olds.

The Respiratory System

Maximum oxygen consumption (VO_{2max}) indicates how much oxygen can be processed by the body and is a measure of the individual's capacity for aerobic exercise. This usually declines steadily with age, primarily because of a loss of oxygen-consuming muscle. The rate of decline is slower in aerobically fit individuals, and their absolute levels remain higher therefore a fit 70-year-old can have a higher value than an unfit 40-year-old (Figure 15-3). A 30% rise in VO_{2max} was demonstrated in a group of sedentary older individuals enrolled in an endurance training program.

The Cardiovascular System

Major physiologic changes take place in the heart during aging (see Chapter 10), but these are accompanied by compensatory adjustments that maintain cardiac output. Very little decline has been found in the pumping efficiency of the heart between the ages of 25 and 79 years in healthy subjects.

Physical activity, even without weight loss, reduces blood pressure and also beneficially raises the amount of cholesterol carried by high density lipoprotein, thus reducing atherosclerosis. In addition, appropriate exercise strengthens the heart, maintains the elasticity of blood vessels, and encourages the proliferation of collateral circulation (additional small blood vessels that can help prevent a heart attack). Those who exercise regularly also have a better chance of surviving a heart attack.

Immunity

Regular exercise increases several indices of immune function in older subjects, and the marked increase in natural killer cell activity in response to exercise is as great in elderly subjects as in much younger individuals. Physical activity may therefore be required to maintain maximum efficiency of the immune system.

The Endocrine System

Diabetes Mellitus.. Exercise is important in maintaining normal glucose metabolism. Comparisons between groups of master athletes and young athletes have shown little difference in glucose tolerance or insulin sensitivity, and a direct relationship has been found between aerobic fitness and the efficiency of insulin action in older male subjects. In diabetics the efficiency of insulin action is improved by exercise. Physical activity is strongly protective against NIDDM in older men at high risk of this disease and among women aged 34-59 years. The deterioration in endocrine control of blood glucose that frequently appears in the older adult may therefore often be a function of diet, increased body fat, and lack of physical activity rather than an unavoidable result of aging.

The Nervous System

Very fit older subjects have been found to have higher scores on fluid intelligence tests than less fit individuals. Aerobic training raised scores on several neuropsychological tests in sedentary subjects when compared with untrained control group subjects. Reaction time is a combined function of the peripheral and central nervous systems, and prolonged reaction times have always been associated with aging, mainly because of a slowing of the processing of sensory input in the central nervous system. The reaction times of older exercisers however, are similar to those of younger individuals, and considerably faster than those of the sedentary elderly. Regular exercise may stimulate perfusion of the brain by blood and nutrients and thus promote better functioning of the central nervous system.

STRESS AND AGING

Physiologic changes take place in the body in response to stress. Although these changes predispose an individual to some of the diseases of aging, the effect of stress on general age-related changes is less clear.

Immunity

Several animal studies have demonstrated the immunosuppressive effects of stress, but there is limited data from studies with humans. Among medical students, high scores for stressful life events and loneliness are associated with lower levels of natural killer cell activity. These cells are a major bodily defense against malignant cells and viral infection. Regardless of life stress scores, natural killer cell activity was reduced by the stress of taking examinations. Among a group of elderly women, severe threats, depressed mood, and dissatisfaction with social support were all related to some reduction in immune status.

Stressful life events such as divorce or death of a spouse are associated with increased rates of morbidity and mortality, but at present there is little evidence in these situations that establishes cause and effect between depressed immunity and increased rates of disease.

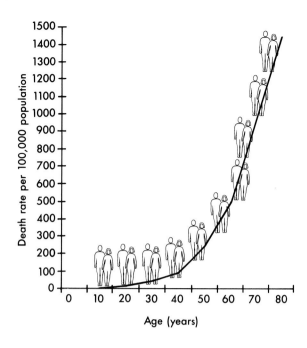

Figure 15-4   Rise in death rate from cancer with age. Drawn from data published by the SEER program of the National Cancer Institute, 1985.

LIFESTYLES AND CANCER
Epidemiology

Malignant neoplasms occurring at sites throughout the body are initiated by a variety of agents, but their incidence rises sharply with age (Figure 15-4).

It is not certain what proportion of these are an unavoidable consequence of aging and what proportion are a result of exposure to modifiable environmental factors. Such known factors include tobacco use, diet, exposure to toxic chemicals or ionizing radiation, and viruses. Tobacco use and dietary factors are thought to account for up to 60% of all cancers. The rates for specific cancers are 30 or 40 times more in some countries than in others and can vary fivefold from one industrialized country to another, which strongly suggests that environmental factors contribute to the disease. The complexity of the carcinogenic process and the exposure of individuals to a variety of environmental agents, which change over a lifetime, often make it very difficult to link specific factors with cancers at particular sites. Thirty or forty years may elapse between the first stage of carcinogenesis, when the cellular genetic material is damaged, and the appearance of a diagnosable tumor. One environmental agent may be involved in the initiating event and another in the promotional stage required to progress to malignant cells.

The international trends in cancer mortality rates are not reassuring. Approximately half of all cancer in the world occurs among one fifth of its population in the developed countries. In the developed countries studied between 1968 and 1987, all forms of cancer, except that of the lung and stomach, have increased in persons over age 54. These substantial and rapid increases (for example, the approximate doubling of brain and nervous system cancer rates in persons aged 68 to 84) during this period are very disturbing and, demand intensive investigation.

Within the United States many cancers are increasing in incidence, especially among the elderly. Figures 15-5 and 15-6 show the trends in total cancers at all sites and for the lung and bronchus for three older groups of the population during a recent 15-

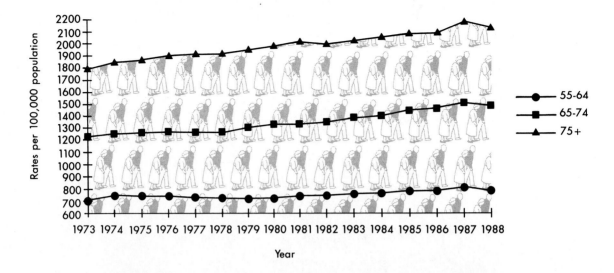

Figure 15-5   Age specific cancer incidence rates for both sexes and all races. Totals for all primary cancer sites excluding lung and bronchus. Rates age-adjusted to the 1970 U.S. standard population. Drawn from data published by the National Cancer Institute, 1991.

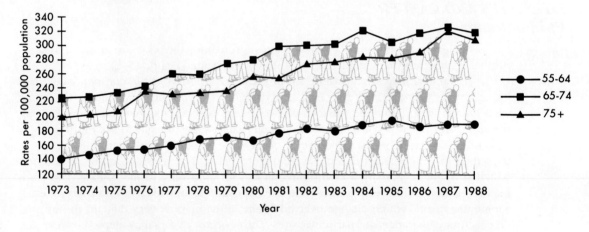

Figure 15-6   Age-specific incidence rates for cancer of the lung and bronchus for both sexes and races. Rates are age-adjusted to the 1970 U.S. standard population. Drawn from data published by the National Cancer Institute, 1991.

year period. The rates for each age group are age-adjusted, which removes the effect of an increasing number of older individuals surviving within each group between 1973 and 1988.

Gastric cancer rates in the United States have fallen steadily over the past 50 years, whereas in Japan they remain among the highest in the world. The reasons for this decline in the United States are not certain, but a reduced intake of foods preserved with salt and nitrite together with an increased intake in vitamin C may be involved.

Table 15-7 Environmental agents that have been linked with an increased risk of cancer at certain sites.

SITE	SUGGESTED ENVIRONMENTAL FACTOR INCREASING RISK
Oral cavity	Tobacco—Chewed or smoked
Esophagus	Smoking—effect amplified by alcohol
Stomach	Salt preserved and smoked foods
Liver	Mycotoxins, alcohol, hepatitis B, vinyl chloride, anabolic steroids
Distal colon	High intake of dietary fat and meat, low intake of fiber and calcium. Northerly latitude
Rectum	Alcohol
Kidney	Obesity, smoking
Bladder	Smoking, industrial chemicals
Pancreas	Smoking
Lung	Smoking
Prostate and ovary	Diet high in saturated or omega 6 polyunsaturated fat
Breast	Same as for prostate and ovary, plus having no children or late childbearing
Uterine cervix	High-fat diet, obesity, genital warts, multiple sex partners

The Environment

There are known associations between environmental agents and tumors at certain sites (Table 15-7)

Nutrition

Epidemiologic evidence suggests that the consumption of a diet high in fat and calories is the dietary factor most closely associated with an increased risk of cancer of the colon and endocrine-related cancers of the breast, ovary, and prostate. Experiments with several species of laboratory animals support this suggestion for spontaneously appearing or chemically induced tumors at these sites are more frequent when the animals are fed a high-fat diet. This promoting effect exists for both saturated fats, usually of animal origin, and polyunsaturated fats of the ω-6 type, such as corn, soy, and sunflower oil. On the other hand, highly monounsaturated fats such as olive oil appear not to have a cancer enhancing effect, whereas ω-3 polyunsaturated fats, as found in fish oil, exert a protective effect.

The consumption of green and yellow vegetables appears to have a protective effect against cancer at a number of sites. Several dietary components that can interfere with carcinogenesis have been identified. The best studied of these is beta carotene, a plant compound found in many orange or yellow fruits and vegetables and also in green leafy vegetables. It is converted to vitamin A in the body and has been associated with preventing cancer in epithelial tissues. Other plant substances not regarded as human nutrients also offer protection. Sulforaphane, an isothiocyanate, found in members of the cruciferae, such as the cabbage family, may act in a similar way, as do the indoles and phenols found widely among plants. Consumption of fruits and vegetables is recommended as part of an adequate diet for the elderly (Figure 15-7).

It has been suggested that high intakes of vitamins E, C, and A have a protective effect, but the results of many studies are contradictory.

Cancers of the colon, breast, uterus, and prostate occur more frequently among the obese. When tumors of the breast or colon are induced in laboratory rats, the incidence of tumors is much lower if the animals are fed a calory-restricted diet. This effect occurs regardless of the fat content of the diet.

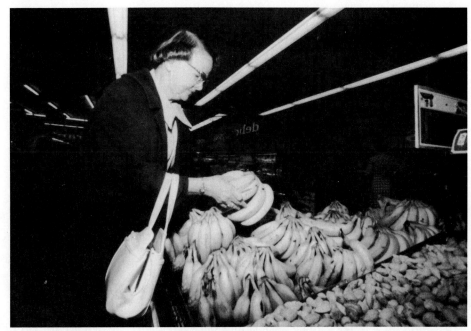

© NATIONAL INSTITUTE ON AGING

Figure 15-7   It is important for elderly persons to consume an adequate amount of fruits and vegetables.

Exercise

Treadmill exercise reduces the incidence of chemically induced colon tumors in rats by 108%.

Among humans, regular exercise reduces the risk of breast cancer, and a job that involves physical activity lowers the risk of colon cancer. A recent prospective study of 13,344 men and women for approximately 8 years found that deaths from cancer declined significantly as the participants level of fitness rose. These findings make it clear that the risk of cancer as an individual ages can be influenced by a pattern of diet and physical activity over many years. There is a genetic predisposition to some cancers, but environmental factors play a major role: an adopted child faces five times the risk of cancer if the adoptive parents have died of the disease.

CONCLUSION

The general view of old age is that it is a time of declining physical and mental capacity with an increasing likelihood of serious disease. This is not a universal prospect, however, for an appreciable number of humans reach advanced old age with their physical and mental capabilities only marginally reduced. In fact the great functional variation between individuals in the seventh and eighth decades of life is in itself remarkable. This has led to the concept of biologic rather than chronologic age.

Many of the tissue changes described in previous chapters are inevitable; differences in the rate at which they take place are to a large extent genetically determined. Nevertheless, there is evidence that lifestyle factors may influence the rate of change (Figure 15-8). It also appears that if our physical and mental capacities are continually

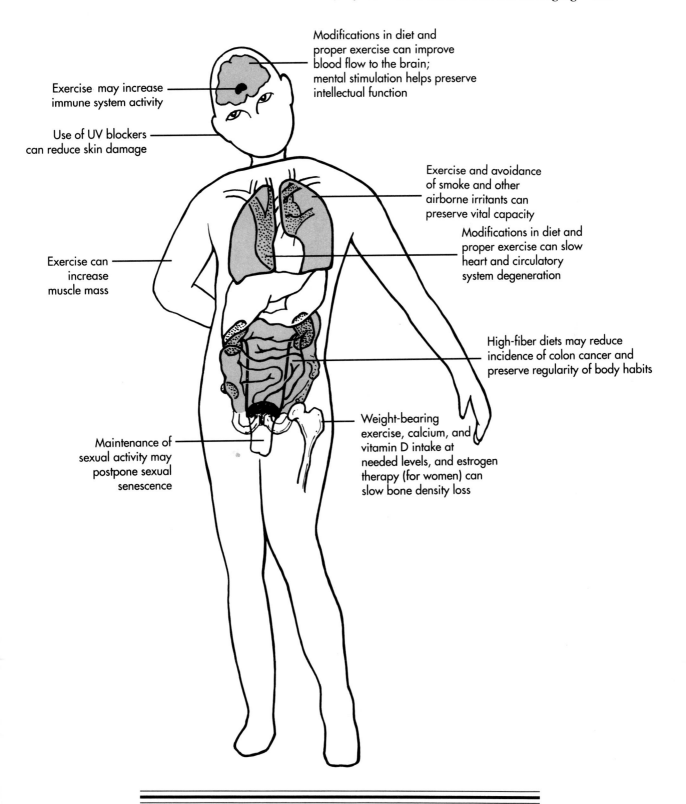

Exercise may increase immune system activity

Use of UV blockers can reduce skin damage

Exercise can increase muscle mass

Maintenance of sexual activity may postpone sexual senescence

Modifications in diet and proper exercise can improve blood flow to the brain; mental stimulation helps preserve intellectual function

Exercise and avoidance of smoke and other airborne irritants can preserve vital capacity

Modifications in diet and proper exercise can slow heart and circulatory system degeneration

High-fiber diets may reduce incidence of colon cancer and preserve regularity of body habits

Weight-bearing exercise, calcium, and vitamin D intake at needed levels, and estrogen therapy (for women) can slow bone density loss

Figure 15-8   Lifestyle factors can affect the aging process.

challenged, compensating adjustments occur to maintain overall performance at a high level despite age-related decrements in specific characteristics. To put it simply, we are much more likely to rust out than wear out.

In many cases the actual aging process is overlaid with the effects of disease. There is now abundant evidence that modification of the typical lifestyle of the western world can greatly reduce the incidence of or delay the diseases of aging. The compelling evidence from the Mormon studies cited at the beginning of this chapter indicates that a modified lifestyle that is still within the American mainstream is adequate to dramatically reduce deaths from cancer and coronary artery disease. An increase in physical activity from zero to a very modest level was sufficient to reduce all causes of mortality substantially. It appears that relatively small changes, if practiced over a long period, can make great differences to our prospects as elderly citizens: the continuing decline in the incidence of death from gastric cancer, stroke and coronary heart disease confirms that a population can beneficially change its habits.

Despite many unavoidable changes that occur as we age there are many choices that we make concerning the way in which we live. Much can be done to reduce the risk of premature disease and maintain optimum function of most organ systems. It has been said that we can choose to "die young as old as possible."

BIBLIOGRAPHY

Belloc NB: Relationship of health practices and mortality, *Prev Med* 2:67-81, 1973.

Benfante R, Reed, D: Is elevated serum cholesterol level a risk factor for coronary heart disease in the elderly?, *JAMA* 263:393-396, 1990.

Blair, SN, and others: Physical fitness and all-cause mortality. Prospective study of healthy men and women, *JAMA* 262:2395-2401, 1989.

Blair SN, Brill PA, Kohl, HW: Physical activity patterns in older individuals. In Spirduso WW, Eckert HM, (eds): *Physical activity and aging,* Champaign, 1989, Human Kinetics.

Block JE, Genant HK, Black DM: Greater vertebral bone mineral mass in exercising young men, *West J Med* 145:39-42, 1986.

Block JE and others: Does exercise prevent osteoporosis?, *JAMA* 257:3115-3117, 1987.

Brenner BM, Meyer TW, Hostetter TH: Dietary protein intake and the progressive nature of kidney disease: the role of hemodynamically mediated glomerular injury in the pathogenesis of progressive glomerular sclerosis in aging, renal ablation, and intrinsic renal disease, *N Engl J Med* 307:652-659, 1982.

Chandra, RK: Nutritional regulation of immunity and risk of infection in old age, *Immunology* 67: 141-147, 1989.

Consensus Conference: Osteoporosis, *JAMA* 252:799-802, 1984.

Davis DL and others: International trends in cancer mortality in France, West Germany, Italy, Japan, England and Wales and the U.S.A., *Lancet* 336:474-481, 1990.

Delvin EE, Imbach A, Copri, M: Vitamin D nutritional status and related biochemical indices in an autonomous elderly population, *J Clin Nutr* 48: 373-378, 1988.

Duchateau J and others: Beneficial effects of oral zinc supplementation on the immune response of old people, *Am J Med* 70:1001-1004, 1981.

Elsayed M, Ismail AH, Young RJ: Intellectual differences of adult men related to age and physical fitness before and after an exercise program, *J Gerontol* 383-387, 1980.

Engel GL: A lifesetting conducive to illness. the giving up-given up complex, *Ann Intern Med* 69:293-299, 1968.

Enstrom JE: Health practices and cancer mortality among active California Mormons, *J Nat Cancer Inst,* 81:1807-1814, 1989.

Fiatarone MA and others: High-intensity strength training in nonogenarians, *JAMA,* 263:3029-3034, 1990.

Fiatarone HW and others: The effect of exercise on natural killer cell activity in young and old subjects, *J Gerontol* 44:M37-M45, 1989.

Food and Nutrition Board, Subcommittee on the Tenth Edition of the RDAs, *Recommended Dietary Allowances, Ed 10, Washington, DC, 1989, National Academy Press.*

Goodwin JS, Goodwin, JM, Garry, PJ: Association between nutritional status and cognitive functioning in a healthy elderly population, *JAMA* 249(21), 1983.

Gorman KM, Posner JD: Benefits of exercise in old age, *Clin Geriatr Med* 4(1): 181-192, 1988.

Grundy SM: Cholesterol and coronary heart disease: A new era, *JAMA* 256:2849-2859, 1986.

Guth PH: Physiologic alterations in small bowel function with age and the absorption of d-xylose, *Am J Dig Disease* 13:565-571, 1968.

Haeger K: Long time treatment of intermittent claudication with vitamin E, *Am J Clin Nutr* 27:1179-1181, 1974.

Heath GW and others: A physiological comparison of young and older endurance athletes, *J Appl Physiol.* 51:634-640, 1981.

Helmrich SP and others: Physical activity and reduced occurrence of non-insulin-dependent diabetes mellitus, *N Engl J Med* 325:147-152, 1991.

Hertzog C, WarnerSchaie K, Gribbin K: Cardiovascular disease and changes in intellectual functioning from middle to old age, *J Gerontol* 33:872-883, 1978.

Hollenbeck CB, Haskel W, Rosenthal M: Effects of habitual physical activity on regulation of insulin-stimulated glucose disposal in older males, *J Am Geriactr Soc* 33:273-277, 1985.

Kannel WB: Nutritional contributors to cardiovascular disease in the elderly, *J Am Geriactr Soc* 34:27-36, 1986.

Kiecolt-Glaser J and others: Psychosocial modifiers of immunocompetence in medical students, *Psychosom Med* 46:7-14, 1984.

Korenchevsky V: *Physiological and pathological* aging. In Bourne GH, ed. Basel, 1961, Karger.

Kritchevsky D: Influence of caloric restriction and exercise on tumorigenesis in rats, *Proc Soc Exp Biol Med* 193(1):31-34, 1990.

Lane EN and others: Long distance running bone density and osteoarthritis, *JAMA* 255:1147-1151, 1986.

Lindeman RD, Tobin J, Shock N: Longitudinal studies on the rate of decline in renal function with age, *J Am Geriatr Soc,* 278-285, 1985.

Manson JE, Rimm EB, Stampfer MJ, et al: Physical activity and incidence of non-insulin-dependent diabetes mellitus in women: *Lancet* 338:774-778, 1991.

Marti B and others: Is excessive running predictive of degenerative hip disease? Controlled study of former elite athletes. *BMJ* 299:91-93, 1989.

Masoro EJ: Assessment of nutritional components in prolongation of life and health by diet, *Proc Soc Exp Biol Med* 193(1):31-34, 1990.

Mazess RB, Forman SH: Longevity and age exaggeration in Vilcabamber, Ecuador, *J Gerontol* 34:94-98, 1979.

McNaughton ME and others: Stress, social support, coping resources and immune status in elderly women, *J Nerv Ment Dis* 178:460-461, 1990.

Meydani SN and others: Vitamin E supplementation enhances cell-mediated immunity in healthy elderly subjects, *Am J Clin Nut* 52:557-563, 1990.

Miller NS, Belkin BM, Gold MS: Alcohol and drug dependence among the elderly: epidemiology, diagnosis, and treatment, *Compr Psychiatry* 32(2):153-165, 1991.

Mishara BL, Kastenbaum R: *Alcohol and old age.* In *Seminars in psychiatry,* New York, 1980, Grune & Stratton.

National Cancer Institute, Cancer Statistics Review, 1973-88, U.S. Department of Health and Human Services, National Institutes of Health Publication No. 91-2789, 1991.

National Cancer Institute, Cancer Incidence and Mortality in the United States SEER, 1973-81, U.S. Department of Health and Human Services, National Institutes of Health Publication No. 85-1837, 1985.

Novak LP: Aging, total body potassium, fat free mass and cell mass in males and females between ages 18 and 85 years, *J Gerontol* 27:438-443, 1972.

Penn ND and others: The effect of dietary supplementation with vitamins A, C, and E on cell-mediated immune function in elderly long-stay patients: A randomized controlled trial, *Age Ageing* 20:169-174, 1991.

Resnick NM, Greenspan SL: 'Senile' osteoporosis reconsidered, *JAMA* 261: 1025-1029, 1989.

Riggs BL and others: Effect of fluoride treatment on the fracture rate in postmenopausal women with osteoporosis, *N Engl J Med* 322:802-809, 1990.

Rile V: Mouse mammary tumors, alteration of incidence as apparent function of stress, *Science* 189:465-467, 1975.

Rodeheffer RJ and others: Exercise cardiac output is maintained with advancing age in healthy human subjects: cardiac dilatation and increased stroke volume compensate for a diminished heart rate, *Circulation* 69:203-213, 1984.

Ross MH: In Winick M, Ed: *Nutrition and Aging,* New York, 1976, Wiley & sons.

Rudman D, Cohen ME: Nutritional causes of renal impairment in old age, *Am J Kidney Dis* 16:289-295, 1990.

Scott RB, Mitchell MC: Aging, alcohol, and the liver, *J Am Geriatr Soc* 36:255-265, 1988.

Seals DR and others: Endurance training in older men and women, *J Appl Physiol* 57:1024-1029, 1984.

Selye H: Stress, aging and retirement, *J Mind Behavior* 1:93-110, 1980.

Shamburek RD, Farrar JT: Disorders of the digestive system in the elderly, *N Engl J Med* 322:438-443, 1990.

Shock NW: Physiological aspects of aging in man, *Annu Rev Physiol* 23:97-116, 1961.

Sidney KH, Sheperd RJ, Harrison JE: Endurance training and body composition of the elderly, *Am J Clin Nut* 30:326-333, 1977.

Sorensen T and others: Genetic and environmental influences on premature death in adult adoptees, *N Engl J Med* 318:727-32, 1988.

Steen B: Body composition and aging, *Nutr Rev* 46:45-51, 1988.

Urberg M, Zemel MB: Evidence of synergism between chromium and nicotinic acid in the control of glucose tolerance in elderly humans, *Metabolism* 36:896-899, 1987.

Walford RL: *The immunologic theory of aging.* Baltimore, 1969, Munksgaard.

Weinsier RL, Norris D: Recent developments and treatment of hypertension: dietary calcium, fat, magnesium, *Am J Clin Nutr* 42:1331-1338, 1985.

Weisburger JH: Nutritional approach to cancer prevention with emphasis on vitamins, antioxidants, and carotenoids, *Am J Clin Nutr* 42:2265-2375, 1991.

Zang Y, Talay P, Cho C et al: A major inducer of anticarcinogenic protective enzymes from broccoli: isolation and elucidation of structure, *Proc Natl Acad Sci USA* 89:2399-2403, 1992.

Zimberg S: *Alcohol abuse among the elderly.* In Carstensen LL, Edelstein BA, eds: *Handbook of clinical gerontology,* New York, 1987, Pergamon.

A Side Effects in the Elderly of Some Commonly-used Medications

Integumentary System

Sulfonamides (antibiotics used to treat infection) may cause rashes or purpura. This effect may be intensified if diuretics (used to increase urine flow) are also being used to treat hypertension.

Respiratory System

Beta-adrenergic blockers, used in eyedrops to treat glaucoma, may cause bronchospasm (asthma-like symptoms) in persons with a history of bronchial asthma.

Digestive System

Antiarrhythmic drugs, used to control cardiac arrhythmias, and anticholinergic drugs, used to treat Parkinson's disease, may cause xerostomia.

Cardiac glycosides, used to treat congestive heart failure, may cause loss of appetite, nausea, vomiting, or diarrhea.

Potassium-sparing diuretics may cause hyperkalemia.

Excretory System

Anticholinergic drugs, used to treat Parkinson's disease, amantadine, an antiviral used to treat influenza Type A, and antiarrhythmic drugs, may cause pain and/or difficulty in urination.

Cardiovascular System

Certain antiarrhythmics and diuretics may cause excessive hypotension.

Cardiac glycosides, used to treat congestive heart failure, may cause abnormal heart beats or arrhythmia. Beta blockers, used in eyedrops to treat glaucoma, may decrease heart rate and cause blood pressure to fall.

Blood

Sulfonamides, used to treat infection, may cause bleeding.

Aspirin and phenylbutazone, an antiinflammatory, may inhibit blood clotting.

Nervous System

Amantadine, used to treat influenza Type A, may cause ataxia, confusion, lethargy, or slurred speech.

Antiarrhythmics may cause ataxia or dizziness.

Antidepressants may cause confusion or temporary speech blockage.

Antihistamines, used to control allergic reactions, may cause confusion or psychotic symptoms.

Indomethacin, an analgesic and antiinflammatory, may cause confusion or drowsiness.

Cardiac glycosides, used to treat congestive heart failure, may cause confusion, dizziness, headaches, or nervousness.

Sense Organs

Selected antibiotics (aminoglycosides) may cause hearing loss.

Beta-adrenergic blockers, used to treat cardiac arrhythmia, may enhance susceptibility to hypothermia.

Cardiac glycosides, used to treat congestive heart failure, may cause visual disturbances.

B *Sample Case History*

The case histories presented in this text have been edited for clarity and brevity. This sample shows the amount of detail and the technical terminology used in actual medical case histories. (See "Multiple Myeloma" on p. 211.)

SAMPLE CASE HISTORY

This 63 Y/O M presented complaining of generalized weakness and back pain. His past history included numerous episodes of prior back or flank pain. Four attacks of flank pain occurred over a 7 year period 30 years PTA. All started abruptly during the night and cleared with several weeks of bedrest. Two years PTA a similar episode of back pain occurred and required hospitalization. This episode resolved without specific therapy and all clinical studies were negative. One year PTA he was seen in another hospital with flank pain and was treated wtih antibiotics for prostatitis. Two months PTA he was again hospitalized for two weeks with back pain. All of his consultations (General Surgical, Orthopedic and Medical) were negative. His creatinine was 1.0 mg/dl and his BUN was 21 mg/dl. AN IVP was normal.

He was admitted to ILH for evaluation of flank pain. It varies with trunk motion, does not radiate and is not affected by laughing, coughing or sneezing. He denies urinary symptoms (urgency, frequency or dysuria) but feels the pain is related to his kidneys.

His past history is positive for a herniorrhaphy, unknown date; appendectomy, at 16 years of age; gastroenterostomy for ulcer disease, 20 years PTA; and a partial thyroidectomy for an oxyphilic adenoma, 1 year PTA.

The ROS revealed easy fatigability for 6 months, a normal urine output, stable weight and the only medication was an occasional aspirin.

On physical examination his vital signs were: weight 72 kgm, height 180 cm, BP 170/90, P 80/min and regular, RR 18/min, and T 36.6. His HEENT were WNL. The eye exam showed PERLA with benign fundoscopy. His neck contained no palpable lymph nodes and there was a well healed scar on the right. The chest was clear to P and A. The heart tones included a normal S1 and S2 without murmurs or additional sounds. The abdomen was soft wtih normal BS and no tenderness. There was a healed appendectomy scar and no organomegaly or masses. The extremities were unremarkable as was the back. Specifically no tenderness could be elicited and there was no pain on knee flexion or with straight leg raising. The neurological exam was completely normal without sensory deficits or weakness. The DTRs were symmetrical and 2+.

Numerous radiologic studies were performed. The lumbosacral spine showed minor degenerative changes. The thoracic spine showed a compression fracture of the 11th vertebra of undetermined age with a radiolucency in the anterior and right lateral aspect of the compressed bone. In addition there was osteoporosis with mild degenerative changes. The gall bladder and chest (PA and lateral) were normal. A lower GI study showed multiple diverticula. On IVP the kidneys were large with poor dye concentration.

Laboratory examination showed: Hgb 8.9 gms/dl, MCV 90, WBC 2,400 with 2 eos, 1 band, 58 PMNs, 32 lymphs and 6 monos. The platelets were normal. A UA found clear yellow urine with a sp. gr. of 1.008, pH 6 and all tested components negative. The serum studies found the calcium to be 9.3 mg/dl with a phosphorous of 5.8 mg/dl (normal to 4.5). The BUN and creatinine were elevated at 68 and 9.1 mg/dl and the uric acid was 12.0 mg/dl. All of the enzymes, LDH, SGOT and AP, were WNL. Serum protein electrophoresis found a TP of 6.6 gm/dl with 2.9 gm of albumin (normal over 3.5), alpha 1 of 0.49, alpha 2, 1.18, beta 0.88 and gamma 1.12. The comment was low albumin, increased alpha 1 and a diffuse gamma increase suggestive of chronic inflammation.

The hospital course demonstrated continued back pain and a marked decrease in urine output after the IVP. He remained afebrile. Urine studies demonstrated 4.8 gms of protein in his 24 hour urine collection and electrophoresis demonstrated most of this was a monoclonal band that typed as kappa light chain. A sternal bone marrow aspirate revealed numerous plasma cells with a monotonous pattern and a diagnosis of multiple myeloma was made. A renal biopsy demonstrated the tubules to be plugged with a dense, acellular proteinaceous material compatible with myeloma protein precipitate. He was treated with chemotherapy, renal dialysis and plasmaphoresis. Death was due to complications of renal failure.

C Table of Normal Physiological Values With Abbreviations

Normal values depend on many variables. These include the patient population used in the study that serves as a baseline for "normal," the patients' sex and age, their preparation (diet, state of health), time of day or year study is done, the method used to determine the values, the units of measure used to report them, and the reagents used in the testing. That is why normal values for most tests change from laboratory to laboratory and even within laboratories if they change methods, instruments, reagents, or reporting units. For these reasons, the normal values differ in some of our case presentations. The following table is based on standard methods for a 50- to 60-year-old adult and is adjusted for sex and age only when such adjustments are pertinent.

TEST	NORMAL VALUES	SIGNIFICANCE
adrenocorticotropic hormone (ACTH)	< 60 µg/ml; Diurnal variation	↑ Hyperpituitary ↑ Hypocortisol ↓ Hypopituitary ↓ Hypercortisol
Acid phosphatase	0.01-1.8 IU/L	↑ Prostate cancer ↑ Prostate hyperplasia
Albumin	3.4-5.5 g/dl, blood	↓ In malnutrition ↓ Liver disease ↓ Inflammation
Aldoase	0-7.4 IU/L	↑ Muscle disease
Aldosterone	0-7.4 mg/dl blood	↓ Hypoadrenal ↑ Hyperadrenal ↑ Some hypertension
Alkaline phosphatase	50-125 IU/L	↑ Bone disorders ↑ Liver disease ↓ Hypothyroidism
Bilirubin	0-1.4 mg/dl	↑ Red cell destruction ↑ Liver disease

Continued.

TEST	NORMAL VALUES	SIGNIFICANCE
Blood gases	$P_{O_2} > 85$ mm Hg	↓ Lung insufficiency
	P_{CO_2} 35-48 mm Hg	↑ Lung insufficiency
Blood sugar, fasting	60-105 mg/dl	↑ Diabetes mellitus
		↑ Pregnancy
		↓ Hyperinsulinism
		↑ Cushing's disease
Blood pH	7.35-7.45	↑ Hyperventilation
		↑ Cushing's disease
		↓ Hypoventilation
		↓ Addison's disease
Blood urea nitrogen (BUN)	7-23 mg/dl	↓ In malnutrition
		↑ Renal failure
Calcium (Ca)	8.4-10.1 mg/dl	↑ Osteoclastic activity
		↑ Bone cancer
		↑ Hyperparathyroidism
		↓ Hypoparathyroidism
		↓ Vitamin D deficiency
Complete blood count (CBC)		
White blood count (WBC)	3.9-10.6 cells/mm^3	↑ Total in leukemia
		↑ Total in infection
		↓ Total in some anemias
		↓ Total in chemotherapy
	42%-75% Neutrophils	↑ Infection
	2%-4% Eosinophils	↑ Allergies and inflammation
	0.5-1% Basophils	↓ Allergies
	3-8% Monocytes	↑ Mononucleosis chronic infections
	20-52% Lymphocytes	↑ Some leukemias
Hematocrit	Male 39%-54%	↑ Dehydration
	Female 36%-45%	↑ High altitude
		↑ Some bone marrow hyperfunction
		↑ Renal anoxia
		↑ Pulmonary disease
		↓ Most leukemias
		↓ Anemia
		↓ Chronic disease
Hemoglobin	Men 13.3-17.5	All above (hematocrit)
	Women 11.7-15.7	All above (hematocrit)
Mean corpuscular volume (MCV)	81-100 μg/L	↑ Hemorrhage
		↑ Some anemias
		↓ Iron deficiency
Platelets	100,000-500,000/mm3	↑ Some leukemias
		↑ Bleeding
		↑ Stress
		↓ Drug reaction
		↓ Most leukemias
Red cell count (RBC)	Men 4.5-6.2 ml/mm^3	see Hematocrit
	Women 4.2-5.4 ml/mm^3	see Hematocrit

TEST	NORMAL VALUES	SIGNIFICANCE
Carcinoembryonic antigen (CEA)	0-5.0 mg/ml	↑ GI malignancy
		↑ GI inflammatory disease
Cortisol	AM 5.5-20.0 μcg/dl	↑ Hyperadrenal
	PM 2.0-10.0 μg/dl	↓ Hypoadrenal
Creatinine	Male, 0.5-1.2 mg/dl	↑ Renal disease
	Female, 0.5-0.9 mg/dl	↓ Malnutrition
		↑ Muscle mass increase
Creatinine kinase (CK)	Male 10-160 IU/L	↑ Hypothyroidism
	Female 10-135 IU/L	↑ Heart attack
		↑ Other muscle disease
MB fraction of CK		
		↑ Heart attack
Erythrocyte sedimentation rate (ESR)	Male 0-13 mm/h	↑ Multiple myeloma
	Female 0-20 mm/h	↑ Inflammatory disease
Iron	25-200 μg/dl	↓ Iron deficiency anemia
Lactic dehydrogenase (LDH)	90-180 IU/L	↑ During myocardial infarction
		↑ Tissue necrosis often associated with cancer
		↑ Liver and lung disease
Lipids	See specifics below	↑ Diabetes mellitis
		↑ Hypothyroidism
		↓ Hyperthyroidism
		↓ Malnutrition
Cholesterol	< 200 mg/dl desired	↑ Hypercholesterolemia
		↓ Acute hepatitis
		↓ Malnutrition
High density lipoproteins (HDL)	> 40 mg/100 ml	↓ Hypercholesterolemia
		↑ Regular exercise
Low density lipoproteins (LDL)	< 180 mg/100 ml	↑ Hypercholesterolemia
Triglycerides	40-150 mg/100 ml	
Phospholipids	145-200 mg/100 ml	
Fatty acids	190-420 mg/100 ml	
Norepinephrine	70-750 pg/ml	↑ Adrenal tumor
		↑ Stress
Potassium	3.5-5.0 mEq/L	↑ Hypoaldosteronism
		↑ Acute kidney failure
		↓ Vomiting, diarrhea
		↓ Severe malnutrition
Prostate specific antigen (PSA)	0-4.0 ng/ml	↑ Prostate hyperplasia
		↑ Prostate cancer
Protein, total	6.0-8.0 gm/dl	↑ Severe dehydration
		↓ Hemorrhage
		↓ Severe malnutrition
		↓ Liver disease
Prothrombin time	11.5-13.5 seconds	↑ Liver disease
		↑ Congenital clotting factor deficiency
Serum glutamic oxaloacetic transaminase (SGOT) (AST)	12-45 IU/L	↑ Cell necrosis in muscle, lung, liver, heart

Continued.

TEST	NORMAL VALUES	SIGNIFICANCE
Serum glutamic-pyruvic transaminase (SGPT) (ALT)	8-55 IU/L	↑ Liver damage as with SGOT but more specific for liver
Sodium	135-145 mEq/L	↑ Starvation ↑ Dehydration ↓ Kidney failure ↓ Cushing's disease
Total iron binding capacity (TIBC)	250-450 μg/dl	↑ Iron deficiency ↑ Blood loss
Tetraiodothyroxine (T4)	4.5-11.0 μg/dl	↑ Hyperthyroid ↓ Hypothyroid
Thyroid stimulating hormone (TSH)	0.38-6.10 mIU/ml	↑ Thyroid deficiency ↓ Excessive thyroid
Vitamin B_{12}	100-700 pg/ml	↓ Deficiency ↑ Liver disease
Vanillylmandelic acid (urine) VMA	0.0-9.0 mg/24	↑ Adrenal tumors ↑ Stress
Urine		
Acetone	0	↑ Fasting ↑ Diabetic acidosis
Albumin	0-trace	↑ Hypertension ↑ Kidney disease Temporary strenuous exercise
Aldosterone	2-22 μg/24 h	↑ Hyperadrenal function ↓ Hypoadrenal function
Ammonia	20-70 mEq/L	↑ Liver disease
Calcium	< 150 mg/day	↑ Hyperparathyroidism ↓ Hypoparathyroidism ↑ Bone cancer
Creatinine clearance	100-140 ml/min	↑ Some kidney disease ↓ Advanced age ↑ Pregnancy
Glucose	0	↑ Diabetes mellitis ↑ Hyperthyroidism ↑ Hyperadrenal cortex function
pH	4.6-8.0	↑ Alkalosis ↑ Urinary infection ↓ Acidosis ↓ Dehydration ↓ Emphysema
Specific gravity	1,003-1,030	↑ or ↓ Fluid imbalance

Glossary

The words used in this glossary were selected from the text because students might not be familiar with their scientific meaning. The meaning of the word roots is provided to help in learning the word and sometimes to explain word usage. Derivation of word roots are indicated by abbreviations; L., Latin; G., Greek; Fr., French; Ar., Arabic; ME., Middle English. Definitions given are specific to the way in which the word is used in this text and often exclude other uses.

Pronunciations are keyed to sounds as noted in the following list. We have tried to avoid extensive use of diacritical marks: the exceptions are the long "i" sound (T), the dipthong d̈, and the distressed ə.

a as in *able*	ī as in *ice*
ä as in m*a*gic	i as in s*i*t
ah as in f*a*ther	o as in *o*at
aw as in j*aw*	oo as in f*oo*d
e as in *ea*t	ow as in flo*w*er
ə as in m*e*t	u as in m*u*scle

Primary stresses are indicated by capital letters; secondary stresses by accent marks.

Abductor (äb-DUK-tor) (L. one who takes away) Muscle that moves a limb away from the axis of the body.

Abscess (ÄB-səs) (L. away, to go) The disintegration of tissue and formation of a cavity in which pus accumulates.

Acetabulum (as′-i-TAB-yu-lum) (L. vinegar cup) The hip socket, composed of the ilium, pubis and ischium.

Acetylcholine (as-e′-til-KO-len) (L. vinegar + G. intestines) An acetic acid ester of choline; in humans, the most widely used transmitter of impulses between nerves and across junctions of nerves and muscles.

Achlorhydria (a′-klor-HI-dre-ah) (L. without + G. green water) Stomach condition characterized by lack of production of hydrochloric acid; typical of atrophic gastritis.

Acinus (a-SEN-us) (L. grape) A small lobule of a compound gland. The pancreatic acini are the digestive glands of the pancreas.

Acromegaly (äk′-ro-MəG-ah-le) (G. extremity + great) Condition resulting from excessive secretion of growth hormone in an adult, producing enlargement of the extremities and facial bones and usually resulting in diabetes.

Actinic (äk-TIN-ik) (G. ray) keratosis (kər′ ah-TO-sis) (G. horny + excessive) Horny skin lesion varying from flat and thickened to raised and warty, resulting from excessive exposure to the sun's rays. They sometimes develop into squamous cell cancers.

Addison's disease (English physician 1860) A condition involving deficiency of adrenocortical hormones, especially cortisol and aldosterone. Inadequate sodium levels and acidosis develop.

Adductor (ä-DUK-tor) (L. one who brings or adds to) Muscle that moves a limb toward the axis of the body.

Adenocarcinoma (äd′-e-no-kar′-si-NO-mah) (G. glandular cancer) Malignant tumor that usually develops from glandular tissue and retains glandular form.

Adenoma (ad′-en-O-mah) (G. glandular tumor) Neoplasm of glandular origin. The term implies that the tumor is benign, as compared to adenocarcinoma.

Adrenal (ahd-RE-nal) (L. near kidney) Endocrine glands located above the kidneys that secrete adrenalin (epinephrine) and a variety of steroid hormones, e.g. cortisone.

Adrenalin (ah-DRəN-ah-lin) Trade name commonly used for epinephrine, the stimulating hormone released by the adrenal medulla.

Adrenocorticotropic hormone (ah-dren′-o-kor′-ti-ko-TRO-pik) (L. near, kidney, rind + G. a turning) Protein hormone produced by the anterior pituitary. It causes the release of steroid hormones from the cortex of the adrenal glands, especially cortisone.

Adenohypophysis (ah-DEN-o-hi-pof′-i-sis) (G. gland, grows under) Anterior lobe of the pituitary. See *pituitary*.

Albumin (äl-BYU-min) (L. white) Water soluble protein that is the primary lubricant of blood and helps to keep the blood at the same osmotic pressure as the tissues.

Aldosterone (al-DOS-ter-ON) Steroid hormone released by the adrenal cortex that increases absorption of salt from the loop of Henle in the nephron, resulting in increased absorption of water from the urine and increased blood pressure.

Allergy (ÄL-ur-je) (G. other + work) Hypersensitive state resulting from exposure to a foreign material (antigen). The condition is the result of byproducts of an antigen/antibody reaction.

Alveolus (äl-VE-ol-us) (L. small belly) Irregular pouches at the end of respiratory bronchioles in the lungs. Also the secretory pouch of a gland. Most gas exchange takes place in the alveoli of the lungs.

Alzheimer's disease (ÄLTZ-hi-merz) (German neurologist) Rapidly progressing dementia beginning before age 80 and characterized by atrophy with formation of neurofibrillary tangles, especially in the hippocampus of the brain. A similar condition after age 80 is now called senile dementia of the Alzheimer's type (SDAT).

Amygdaloid nucleus (ah-MIG-dah-loyd) (G. almond) Complex of nuclei located in the temporal lobes of the cerebral cortex; a major part of the limbic system; involved with olfactory associations and with intense emotions.

Amyloidosis (äm′-i- loy′-DO-sis) (G. starch, abnormal condition) Deposition of an insoluble starch-like protein in tissues. Amyloid deposits are frequently found in some senile tissues.

Amyotrophic lateral sclerosis (a′-mi-o-TROF-ik) (G. against muscle growth + L. on one side + G. hardening) Progressive degeneration of the neurons of the corticospinal tract and of motor cells of the brain and spinal cord; often fatal in two or three years.

Analgesia (än′-al-JEZ-e-ah) (G. without pain) Elimination of pain while retaining consciousness.

Anemia (ah-NE-me-ah) (G. without blood) Abnormally low number of red blood cells, reducing in oxygen carrying capacity. A hematocrit of 41 or less (hemoglobin 13.7) for men or 36 or less (hemoglobin 12) for women.

Aneurysm (ÄN-yur-iz-um) (G. widening) Localized dilation of an artery or bulge in an arterial wall, often one weakened by a disease process.

Angiotensin (än′-je-o-TEN-sin) (G. vessel + L. to stretch) Angiotensin I, an inactive enzyme in the blood, is converted to angiotensin II by renin released from the kidney. Angiotensin II causes the release of aldosterone from the adrenal cortex resulting in increased salt and water reabsorption from the urine. Angiotensin II also causes vasoconstriction. Both of these factors increase blood pressure.

Anorexia of aging (än′-o-REK-se-ah) (G. lack of appetite) Reduced food intake of the elderly caused by age-related factors such as loss of taste, teeth, or other physiological factors.

Antidiuretic hormone (än′-ti-di′-yu-RəT-ik) (G. against urination) Also known as ADH or vasopressin; secreted by the posterior lobe of the pituitary it increases permeability of the collecting tubules of the nephrons to water, thereby increasing blood volume and decreasing urine volume.

Antigen (ÄN-ti-gən) (G. against + creation) Substance, usually foreign to the body, that causes B lymphocytes (plasma cells) to produce antibodies that attach to it.

Antihistamine (än′-tī-HIS-tah-men) (G. against, tissue, + amine) Compound that counteracts the effects of histamine.

Aplastic anemia (a-PLÄS-tik) (G. to fail to form) Failure of the bone marrow to produce adequate numbers of red and white blood cells and platelets.

Appendicitis (ah-pen′-di-SĪT-us) (L. to add or hang on + inflammation) Inflammation of the appendix, a vestigial extension of the cecum.

Apraxia (ə-PRÄX-e-ah) (G. without action) Inability to perform learned skills such as the manipulation of objects; may involve the inability to recognize the function of an object.

Arachnoid layer (ah-RÄK-noyd) (G. spider + form) Middle layer of the meninges, the membrane between the dura mater and the pia mater surrounding the brain that is associated with a weblike network of blood vessels.

Arachnodactyly (ah-räk′-no-DÄK-til-e) (G. spider + finger) Abnormally long fingers and toes; an inherited feature that can be associated with Marfan's Syndrome.

Arcus senilis (AHR-kus se-NIL-us) (L. bow + old) Yellowish ring around the margins of the cornea, produced by a fatty deposit, that appears in the eyes of some elderly persons.

Areola (ə-re-O-luh) (L. small space) Distinct circular area surrounding a central point; the areola of the breast is darker than the surrounding skin and encircles the nipple.

Arrhythmia (ə-RITH-me-ah) (G. without rhythm) Cardiac arrhythmia; any deviation from the normal heartbeat.

Arteriole (r-TĒR-e-ol) (L. little artery) Smallest arteries in the arterial tree. Blood passes from arterioles into other arterioles or capillaries.

Arthropathy (r-THROP-ah-the) (G. joint disease) Any disease of the joints.

Aspiration (äs′-per-A-shun) (L. to breathe) Pulmonary aspiration refers to the inhaling of foreign matter such as water or food particles into the lungs; also the removal of fluids from a body cavity by use of an aspirator.

Asteatosis (ä-stee-ah-TO-sis) (G. without tallow or oil) Skin condition resulting from inadequate sebaceous secretions; the skin becomes dry and prone to infections.

Asthma (ÄZ-mah) (G. panting) Disease involving difficulty in breathing; often an allergic response; involves spasmodic constriction of the bronchi resulting in wheezing.

Astigmatism (uh-STIG-mah-tiz′-um) (G. without point) Inability to focus an entire picture on the retina. It results from a defect in the surface of the cornea, lens or retina and is corrected with a lens that compensates for the defect.

Astrocyte (ÄS-tro-sīt) (G. star cell) One of the supporting cells (neuroglia) of the central nervous system. Astrocytes are derived from neural ectoderm and have many processes that hold axons, dendrites, and blood vessels in position.

Ataxia (ä-TAKS-e-ah) (G. without order) Inability to coordinate muscles enough to move effectively.

Atelectasis (ät′-ə-LəK-ta-sis) (G. imperfect expansion) Collapse of a lung, resulting in loss of pulmonary function.

Atherosclerosis (äh′-thər-o-sklə-RO-sis) (G. mealy + hardening) Development of fatty, cholesterol-laden deposits (atheromas) in the wall of an artery. These deposits may project into the arterial lumen and erode the arterial wall. They may calcify and be referred to as hardening of the artery.

Atheroma (äth′-ər-O-mah) (G. meal-y + tumor) Fatty, cholesterol-laden deposit that forms in arteries.

Atrium (A-tre-um) (L. vestibule) Chamber of the heart that receives blood from the lungs or the rest of the body. Upon contraction it passes the blood on to the ventricles. Also a structure at the end of the respiratory bronchioles that passes air on to the alveoli.

Atrophy (ÄT-ro-fe) (G. Atropos, goddess of death) Wasting away; decrease in the size of a cell, tissue, or organ.

Autonomic (aw′-to-NAHM-ik) (G. self + law) Division of the nervous system that operates without conscious control. It is divided into two sub-systems, the sympathetic (originating from cranial and sacral spinal nerves) and the parasympathetic (originating from thoracic and lumbar spinal nerves).

Axilla (äk-SIL-ah) (L. wing) Angle between the upper arm and the body, the armpit.

Axon (ÄKS-ohn) (G. axle) Process of a neuron that carries an impulse away from a nerve cell body and is capable or releasing a neurotransmitter such as acetylcholine.

Azotemia (a′-zo-TE-me-ah) (Fr. nitrogen + G. blood) Excessive urea in the blood, usually as a result of inadequate kidney excretion.

Baroreceptor (BAH-ro-re-səp′-tor) (G. weight + L. that which receives) Pressure receptive nerve endings located in large arteries, especially the internal carotid. Stimulation of the baroreceptors decreases heart rate and results in vasodilation, lowering the blood pressure.

Basophil (BA-so-fil) (G. base + to love) Polymorphonucleated granulocytes whose granules take a purple basic stain. These cells release histamine and heparin and make up about 1% of the white blood cells.

Benign (be-NĪN) (L. friendly) Nonmalignant; favorable for recovery.

Bilirubin (bil′-e-ROO-bin) (L. bile + red) Orange-yellow pigment produced from the breakdown of hemoglobin and excreted by the liver in the bile.

Bradycardia (bra′-de-KAHR-de-ah) (G. slow + heart) Abnormally slow heart rate, less than 60 beats per minute.

Bronchiogenic (brong′-ke-o-JEN-ik) (L. bronchi + G. creation) Originating from or in the bronchi. Bronchiogenic carcinoma includes the most common forms of lung cancer and originates in the bronchi.

Bronchiole (BRONG-ke-ol) (L. the smallest branches of the respiratory tree) The small branches, extending from the tertiary bronchi to the respiratory bronchioles and alveoli.

Bronchus (BRONG-kus) (L. intermediate lung tube) The two main branches of the trachea are called primary bronchi; these divide into secondary bronchi which further divide to form tertiary bronchi. All bronchi are protected by cartilaginous rings. Inflammation of the bronchi is called bronchitis.

Bruit (BROO-e) (Fr. noise) Abnormal noise produced by blood flowing around an obstruction; also, any abnormal organ sound.

Bulbourethral gland (bul′-bo-yu-RE-thral) (L. swollen + root + urethra) Paired mucus-secreting glands located at the base of the urethra, also called Cowper's glands.

Bursitis (bur-SĪ-tis) (G. pouch + inflammation) Inflammation of a lubricating pouch or bursa. Bursae exist between frictional surfaces in the body and may accumulate excessive fluid and become painful if injured.

Calcific tendinitis (kal-SIF-ik ten′-din-ĪT-is) (L. limestone + tendon inflammation) Inflammation and calcification of the tendons.

Calcitonin (kal′-si-TON-in) (L. limestone + G. tone) Hormone that stimulates production of osteoblasts and increases their deposition of bone; produced by the parafollicular cells of the thyroid gland.

Calculus (KĂL-kyu-lus) (L. little stone) Hard, often crystalline deposits found in the kidney (renal calculi or kidney stones) or urinary bladder (bladder stones). They may form in several organs and may be inorganic, usually calcium, or organic compounds such as uric acid.

Carbidopa (KĂHR-bi-do'-puh) Synthetic compound used to reduce the formation of dopamine outside the brain, thereby reducing the side effects of treatment of Parkinson's disease with L-dopa.

Carcinoma (kahr'-si-NO-muh) (G. crab or cancer) Malignant growth arising from epithelial (lining) cells.

Carotene (KĂR-o-ten) (L. carrot) Orange pigment found in body fat and many vegetables. Some forms may be converted to vitamin A in the body.

Catabolite (kah-TĂB-o-līt) (G. throwing down, shattering) Any product of catabolism (the enzymatic breakdown of a molecule into its components).

Cataract (KĂT-ə-räkt) (G. to break down) Any increase in opacity of the lens of the eye. It occurs to some degree in 95% of people over age 65.

Catecholamine (KĂT-ə-kol'-ə-men') Compounds released by the adrenal medulla, epinephrine for example, and some neurons, that stimulate and work in parallel with the sympathetic nervous system. Another example, norepinephrine, is a neurotransmitter. They cause vasoconstriction and stimulate heart rate and force.

Cecum (SE-kum) (L. blind pouch) Dilated portion of the colon into which the small intestine empties. The appendix is attached to and is a vestigial part of the cecum.

Cerebellum (ser-ə-BəL-um) (L. small brain) Part of the brain located behind the cerebrum and dorsal to the brain stem. Like the cerebrum, it is divided into two hemispheres. It is concerned with coordinating motor impulses.

Cerebral cortex (sər-E-bral KOR-təx) (L. brain + bark) Layer of gray matter on the surface of the cerebral hemispheres that integrates high mental functions and is therefore the center for thought.

Cerebrospinal fluid (sə-RE-bro-spi'-nal) (L. brain + spine) Ionically balanced liquid that bathes the brain and spinal cord and protects them from physical shock. It is secreted by the choroid plexes in the ventricles of the brain and is absorbed by the arachnoid granulations located under the arachnoid layer of the meninges.

Cerebrum (sə-RE-brum) (L. brain) Largest and most anterior part of the brain; the center for thought, reasoning, sensation, and motion, divided into two four-lobed hemispheres.

Cerumen (sə-ROO-min) (L. wax) Waxy secretion produced by the modified apocrine sweat glands of the external ear canal.

Cervix (SəR-vix) (L. neck) Neck of any structure or organism; examples include the cervical vertebrae of the neck or the narrow portion (cervix) of the uterus.

Cherry angioma (ăń-je-O-mah) (G. vascular) Raised red, benign skin tumor composed of blood vessels.

Cholecystokinin (ko'-le-SIS-to-kīn'-in) (G. bile sack + to move) Hormone secreted by the mucosa of the upper intestines and carried by the bloodstream that causes the gall bladder to release bile and the pancreatic acini to release digestive enzymes.

Chordae tendineae (KOR-dah TəN-din-e) (L. cord + tendon) Tendons that anchor the cusps of the mitral and tricuspid valves to the papillary muscles of the ventricles of the heart.

Choroid (KO-royd) (G. skin + form) Black, highly vascular membrane on the inner side of the sclera that provides nutrient to the retina and sclera and absorbs light.

Chromosome (KRO-mo-som) (G. colored body) Coiled double strand of DNA (deoxyribonucleic acid) containing many genes and found in the nucleus of each body cell. Humans have 46 chromosomes made up of 23 similar pairs.

Chyme (KĪM) (G. juice) Semifluid slurry produced by the digestion of food.

Ciliary (SIL-e-är'-e) (L. fringe of short hairs) Fringe of hairs or hair-like processes. The ciliary muscle attaches the lens to the sclera of the eye and its contraction allows the lens to gain a more spherical shape.

Cirrhosis (si-RO-sis) (G. orange or yellow + condition) Disease of the liver in which the liver cells are destroyed and replaced with scar tissue; associated with hepatitis and toxins such as alcohol.

Cisterna chyli (sis-TəRN-ah KĪ-le) (L. small container + juice) Series of abdominal lymph channels and their associated nodes that form a reservoir for lymph.

Claudication (KLAW-di-ka′-shun) (L. lameness) Appearance of pain especially in the muscles of the legs or arms during exercise. The pain disappears when with rest. Results from a inadequate blood supply during exercise.

Coccyx (KAHK-siks) (G. beak of a cuckoo) Bone composed usually of four fused vertebrae that begin at the base of the sacrum and terminate in a point. This curved bone is shaped like the bill of a cuckoo.

Cochlea (KO-kle-ah) (L. snail shell) Part of the inner ear concerned with receiving vibrations, dampening them, converting them to nerve impulses, and sending the impulses to the brain through the auditory nerve. It is snail shaped and lies in a spiral cavity, the labyrinth, in the temporal bone.

Collagen (KOL-ah-jən) (G. glue + to make) Protein molecule that combines with other similar molecules to form scar tissue, tendons, ligaments, cartilage, and the protein matrix of bone. Collagen is scattered through all organs.

Collateral (ko-LĂT-r-al) (L. together + side) Usually used in association with circulation. Collateral blood vessels are accessory or nearby vessels that can enlarge and partially or completely supply a tissue deprived of its normal blood supply.

Colon (KOL-un) (L. large intestine) Large intestine, extending from the cecum to the rectum. It functions to absorb water and vitamins and to compact the feces.

Colostomy (ko-LOS-to-me) (L. large intestine + mouth) Severing of the large intestine and creation of an exit for the feces through the abdominal wall.

Conjunctiva (KUN′-junk-TI-vah) (L. the membrane of the eye) Membrane that covers the inner surface of the eyelids and the outer surface of the eyeball.

Conjunctivitis (KUN-junk′-tiv-Ī-tis) (L. membrane of the eye + inflammation) Inflammation of the membrane of the eye.

Constipation (kahn-sti-PA-shun) (L. crowding together) Difficult or infrequent evacuation of sometimes hardened feces from the rectum.

Cornea (KORN-e-ah) (L. horny) Tough, transparent anterior covering of the pupil of the eye, continuous with the sclera, the "white" of the eye that continues around the eyeball.

Corpus albicans (KOR-pus ĂL-bi-känz) (L. body + white) Degenerated corpus luteum; the whitish scar tissue left after a corpus luteum ceases to function.

Corpus luteum (KOR-pus LOOT-e-yum) (L. body + yellow) Temporary endocrine gland that forms from the torn tissue of the ovary after ovulation. The luteal cells secrete progesterone, which causes production of the secretory endometrium of the uterus.

Cortex (KOR-təks) (L. bark) Outer part of a structure, as in the cortex of the kidney (renal cortex) or cortical bone, the dense outer part of a bone.

Cortisol (KOR-ti-sol) (L. rind, + solution) Antiinflammatory steroid hormone produced by the adrenal cortex; also known as hydrocortisone or cortisone.

Corticosteroid (kor′-ti-ko-STəR-oyd) (L bark + G. solid) Steroid hormones produced by the adrenal cortex, e.g., cortisone.

Costochondritis (kos′-to-kon-DRĪ-tis) (L. rib + G. cartilage + inflammation) Inflammation of the cartilages of the ribcage. It causes pain when breathing or sometimes when moving.

Creatinine (kre-AH-tin-īn) (G. flesh) Waste product of the metabolism of phosphocreatine and used to measure glomerular filtration (kidney function).

Crescentic glomerulonephritis (kre-SEN-tik glo-mer′-yoo-lo-ne-FRĪT-is) (L. shape of new moon + G. small ball + inflammation) Inflammatory disease of the kidney involving crescent-like deposits in Bowman's space around the glomeruli; the result of deposition of an antigen-antibody complex but is not known to follow any particular bacterial infection

Crypts of Lieberkuhn (LE-ber-ku′-un) Intestinal glands, that secrete the enzyme enterokinase as well as some digestive enzymes.

Cushing's syndrome (Boston surgeon) Condition involving hypersecretion of the adrenal cortex with excessive cortisone, aldosterone, and androgen production. The condition may result from adrenal disease or tumor or excessive secretion of ACTH.

Cutaneous (kyu-TA-ne-us) (L. skin) Pertaining to skin.

Cuticle (KYU-ti-kl) (L. skin) Epidermis or outer layer of skin; also the free outer surface of an epithelial cell.

Cyanotic (sī′-ə-NO-tik) (G. blue) Bluish color of the skin and lips when the blood is not adequately oxygenated; usually results from inadequate lung or heart function.

Cystocele (SIS-to-sel) (G. sac + hernia) Prolapse (hernial protrusion) of the urinary bladder into the vagina resulting from atrophy or other damage to the pelvic floor, often from estrogen deprivation.

Cytopenia (sīt′-o-PEN-e-ah) (G. cell + poverty) Abnormally low numbers of any of the cellular components of blood. Pancytopenia refers to deficiency of all the cellular elements of blood.

Decubitus (de-KYOO-bi-tus) (L. to lie down) As in decubitus ulcers (bedsores) acquired as a result of lying on a part of the body long enough to kill tissue because of lack of circulation.

Deglutition (de′-gloo-TISH-un) (L. swallowing) Act of swallowing.

Dementia (de-MəN-she-ah) (L. without mind) General term for mental deterioration or malfunction.

Dendrite (DəN-drīt) (G. tree) Branching process of a neuron that has neurotransmitter receptors or specialized endings for the reception of stimuli. Dendrites conduct impulses toward the neuron cell body.

Deoxyribonucleic (de-ox′-e-RĪ-bo-nu-kle′-ik) **acid** DNA, a large coiled molecule that provides the code for making the proteins of the body. The code for each protein is called a gene. A single chromosome contains a double DNA strand that may contain thousands of genes.

Depressor (de-PRES-or) (L. one who presses down) Muscle that pulls a body part down.

Dermatome (DUR-mah-tom) (G. skin + to cut) Skin segments supplied by individual spinal or cranial nerves. They are remnants of early embryological segments.

Dermis (DUR-mis) (L. skin) Fibrous, nutrient-bearing portion of skin. It lies between the epidermis and the subcutaneous fat.

Detrusor muscle (de-TROO-sər) (L. to thrust) Muscle contained within the wall of the urinary bladder that expels urine from the body.

Diabetes mellitus (dī′-ah-BE-tez MəL-īt-us) (G. to go through + honey) Condition where glucose accumulates in the bloodstream and is not moved into the body's tissues. Some passes through the kidneys into the urine. It may result from inadequate insulin, abnormal insulin, or loss of insulin receptor sites.

Diabetic nephropathy (dī-ah-BəT-ik nəf-ROP-ah-the) (G. to go through + kidney disease) Kidney disease resulting primarily from the vascular and abnormal metabolic effects of diabetes; usually involves diffuse glomerulosclerosis.

Diaphysis (dī-AF-i-sis) (G. away from + growth) Shaft of a long bone.

Diastole (dī-ÄS-to-le) (G. expand + set) Period between heart contractions. At this time the ventricles are relaxing and filling with blood. The diastolic blood pressure is the approximate minimum pressure in the arteries between heartbeats.

Diencephalon (dī′-ən-SəF-ah-lahn) (G. second + brain) Portion of the brain anterior to the mesencephalon and consisting of the walls and floor of the third ventricle. It includes the thalamus and hypothalamus.

Diopter (DĪ-op-tur) (G. optical instrument for measuring angles) Unit of measurement of the refractive power of the eye, the combined effect of the lens and the cornea.

Distal (DIS-tal) (L. distant) Remote; farthest from the point of origin of a structure. Opposite of proximal.

Diuresis (dī′-yoo-RE-sis) (G. through + urine) Increased production of urine.

Diverticulitis (di′-vur-tik′-yoo-LI-tis) (L. to turn aside) Inflammation of a herniated portion of the colon, a diverticulum.

Dopamine (DO-pah-men) (acronym of dioxyphenylethylamine) Neurotransmitter produced by the substantia nigra in the brain. Deficiency in this compound produces the condition known as Parkinson's disease.

Duodenum (doo-o-DEN-um) (L. twelve) Portion of the small intestine adjacent to the stomach; the most important digestive center in the body.

Dura mater (DOO-rah MAH-tur) (L. tough + mother) Outermost of the meninges. The tough dura mater protects the brain from penetration by bone fragments in the event of severe skull fracture.

Dysfunction (dis-FUNK-shun) (G. difficult + function) Partial failure or impairment of an organ in performing its purpose.

Dysphasia (dis-FA-ze-ah) (G. disordered + speaking) Failure to coordinate words or put them in proper order; often results from cerebrovascular accident.

Dyspnea (DISP-ne-ah) (G. difficult + breathing) Difficulty in getting enough air; gasping for breath.

Ectoderm (əK-to-dərm) (G. outer + skin) Outermost of the three embryonic germ layers; forms the epidermal layer of skin, its products, and the nervous system.

Ectropion (ək-TRO-pe-on) (G. to turn out) Turning out of the eyelid as a result of muscular atrophy or damage to the facial nerve.

Ectopic (ək-TOP-ik) (G. out of place) Any structure or event that is out of place. Ectopic heartbeats arise outside of the normal pacemaker; ectopic pregnancies implant outside the uterus.

Edema (ə-DE-mah) (G. swollen) Swelling as a result of increase in fluid in the intracellular spaces in a tissue. Lymphadema, for example, involves swelling as a result of blockage of the lymphatic drainage of a tissue.

Elastin (e-LÄS-tin) (L. capable of returning to its original shape after distortion) Yellow scleroprotein found in elastic tissue that is flexible when wet and tends to return to its original shape after being stretched.

Embolism (EM-bol-izm) (G. to throw in) Blocking of a blood vessel by a thrombus (clot), thereby depriving the downstream tissue of oxygen; a cerebral embolism is referred to as a form of stroke.

Emphysema (əm-fi-SEM-ah) (G. inflated) Lung condition involving loss of alveolar walls resulting in increased amounts of nonfunctional lung area (increased dead air space).

End organs of Ruffini (RUF-in-e) (Italian anatomist) Sensory nerve endings located deep within the dermis and in other tissues of the body. They respond to heavy continuous pressure and are branching, encapsulated ends of dendrites.

Endarterectomy (ənd′-art-tər-əK-to-me) (G. within + artery + to cut out) Surgical removal of the tunica intima of an artery; done when an atheroma has enlarged or replaced the arterial inner lining.

Endocardium (ən′-do-KAR-de-yum) (G. inside + heart) Tunica intima of the heart; the inner lining of the atria and ventricles.

Endolymph (əN-do-limf) (G. within + L. water) Fluid of the cochlear canal, the inner canal of the cochlea containing the organ of Corti.

Endometriosis (ən′-do-me′-tre-O-sis) (G. within + womb + abnormal increase) A condition in which endometrial cells exist outside the uterus and cause pain or discomfort when they respond to estrogenic hormone fluctuations.

Endometrium (ən'-do-ME-tre-yum) (G. within + womb) Epithelium of the inner surface of the uterus. A nonsecretory layer is produced during the first half of the uterine cycle under the influence of estrogens and a secretory layer is produced in the second half of the cycle under the influence of progesterone. This tissue provides nutrients and an anchoring place for the early embryo.

Endorphins (ən-DOR-finz) (G. in + god of dreams) Opiate-like compounds found in the brain that block the sensation of pain, natural analgesics.

Endosteum (ən-DAWS-te-yum) (G. inner + bone) Membrane lining the marrow cavity of bone.

Endothelium (ən'-do-THE-le-yum) (G. inner + nipple) Cells lining the inner surface of the blood vessels; vascular epithelium.

Enzyme (əN-zīm) (G. in + leaven) Protein catalyst; compound that increases the rate of a chemical reaction.

Eosinophil (e'-o-SIN-o-fil) (G. red of dawn + love) Polymorphonucleate granulocyte whose granules stain red with eosin. Eosinophils are somewhat phagocytic and release antihistamines. They make up 2% to 4% of the white blood cells.

Ependymal cell (əp-ən-DĪ-mahl) (G. outer garment) Layer of cuboidal epithelial cells that lines the ventricles of the brain and the central canal of the spinal cord. The outer part of the brain is similarly protected by the pia mater.

Epidermis (əp-i-DER-mis) (G. outer skin) Ectodermal portion of skin lying above the dermis.

Epidermophytosis (əp'-e-der'-mo-fi-TO-sis) (G. upon + skin + plants + excessive) Skin infection by fungi, especially the genus *Epidermophyton*. Ringworm.

Epinephrine (əp'-e-NEF-rin) (G. above + kidney) Primary stimulatory hormone released by the adrenal glands located above the kidney. The "fight or flight" hormone.

Epiglottis (əp'-e-GLOT-is) (G. above + entrance to the larynx) Flap of tissue that covers the entrance to the larynx (glottis) during swallowing.

Epiphysis (e'-PIF-i-sis) (G. end + growing) Head of a long bone, separated from the shaft or diaphysis by the epiphyseal cartilage plate, from which growth in bone length occurs.

Erythematous (ər'-i-THEM-ah-tus) (G. red) Similar to the red inflammation of skin resulting from dilation of superficial capillaries, as would occur with a mild burn.

Erythrocyte (e-RITH-ro-sit) (G. red cell) Red blood cell, which develops from an erythroblast and is the oxygen-carrying cell of the blood. The oxygen is attached to hemoglobin in the cytoplasm of the cell.

Erythropoetin (e-rith'-ro-po-E-tin) (G. red + to make) Hormone made by the kidneys in response to inadequate oxygen. It stimulates production of red blood cells, thereby increasing the oxygen-carrying capacity of the blood.

Esophagitis (e'-sof-a-JĪT-is) (G. to carry food + inflammation) Inflammation of the esophagus, the tube connecting the mouth with the stomach.

Esterase (əS-tər-ase) Any compound formed by the combination of an alcohol and an acid by the removal of water. The "ase" ending signifies an enzyme. Acetylcholine esterase aids the splitting of the acetate from the choline components by adding water to the molecule.

Estradiol (əs'-tra-DĪ-ahl) (G. gadfly + Ar. essence, oil) Potent natural steroid responsible for female secondary sexual characteristics.

Estrogen (əS-tro-jen) (G. that which drives men mad, gadfly) General term for the steroid hormones responsible for female secondary sexual characteristics. Examples: estradiol, estrone.

Eustachian (yu-STA-shun) (L. tube) Air passage between the middle ear cavity and the pharynx that equalizes air pressure on the two sides of the tympanic membrane.

Extensor muscle (əks-TEN-sor) (L. one that lengthens) Muscle that straightens a limb. If contracted excessively, the limb may bend an abnormal direction (hyperextension).

Exocrine (əKS-o-krin) (G. outside + to separate) Gland that excretes its product through a duct; the salivary glands and pancreatic acini are exocrine glands.

Exophthalmos (eks'-of-THAL-mos) (G. outward + eye) Abnormal protrusion of the eye. It may be due to enlargement of the tissues behind the eye and is often associated with excessive thyroxine production (exophthalmic goiter).

Fascia (FASH-e-ah) (L. band) Sheet or band of fibrous connective tissue beneath the skin, investing a muscle or other organ of the body.

Fecal impaction (FE-kl im-PÄK-shun) (L. refuse + forced in) Large, firm, immovable mass of stool in the rectum or colon.

Fibrillation (fi'-bril-A-shun) (L. small fiber + event) Abnormal uncoordinated contraction of individual fibers in the heart. A fibrillating heart fails to move blood and is therefore normally a terminal event.

Fibrinogen (fi-BRIN-o-gen) (L. fiber producer) Protein produced by the liver that makes up 0.3% of blood plasma. In the presence of clotting factors it precipitates and forms the fibers of blood clots.

Fibroblast (FĪ-bro-blast) (L. fiber + G. producing or germinal) Connective tissue cell that lays down collagen fibers. Undifferentiated fibroblasts may form osteoblasts or chondroblasts.

Fibroepithelial polyp (fi'-bro-əp'-i-THE-le-al POL-ip) (G. fibrous outer tissue + nipple-like) Benign, soft, nonpigmented, fleshy tumors of skin. A skin tag.

Fibrosis (fi-BRO-sis) (L. fiber + G. disease or excessive) Abnormal increase in the collagenous fiber in a tissue, usually as a result of degenerative disease or trauma.

Fimbriated (FIM-bre-a-ted) (L. fringe) Structure with many finger-like projections; a fringe. For example, the fimbriated infundibulum guides ovulated ova into the lumen of the oviduct (Fallopian tube).

Flexor (FLəK-sor) (L. one that bends) Muscle that bends a part of the body.

Fovea centralis (Fo-ve-ah sen-TRAL-is) (L. a pit) Depression in the center of the macula lutea of the retina. It is the point of normal focus and has a high concentration of cones but no rods.

Funduscopy (fun-DUS-ko-pe) (L. a blind pouch + G. to look) Examination of the interior of the eye.

Ganglion (GANG-gle-ahn) (G. knot) Any concentration of neurons outside the central nervous system that functions as a nerve distribution center.

Gastrin (GÄS-trin) (G. stomach) Polypeptide hormone released by the mucosa of the stomach. It increases the release of pepsinogen and hydrochloric acid.

Genome (JE-nom) (G. to produce) Complete set of genes for an organism.

Gingivitis (jin'-je-VI-tis) (L. the gum + inflammation) Inflammation of the fibrous tissue covered by the periodontal membrane and fixed to the mandible and maxillary bone.

Glaucoma (glaw-KO-mah) (G. dull eye) Eye disease involving increased intraocular pressure and subsequent hardening and degeneration of various eye structures, especially degeneration of the retina.

Globulin (GLOB-yu-lin) (L. small ball) Proteins that make up 2.3% of the plasma of blood. They are divided into three forms, alpha, beta, and gamma. Gamma globulins are composed of antibodies.

Glomerular nephritis (glo-MER-yu-lar ne-FRĪ-tis) (L. small ball + G. kidney, inflammation) Inflammation of the glomeruli of the kidney, usually as a result of deposition of a streptococcal immunocomplex in the glomeruli following a strep throat.

Glomerulus (glo-MER-yoo-lus) (L. small ball) Ball composed of a specialized arteriole constructed to leak water and certain small molecules such as urea under pressure, but to retain larger molecules such as proteins. It is located within the Bowman's capsule in each nephron of the kidney. The glomeruli are the filtering structures of the kidneys.

Glottis (GLAHT-is) (G. entrance to the larynx) Entrance to the larynx, including the vocal cords.

Glucose (GLOO-kos) (G. sweetness) Blood sugar; dextrose in solution; the primary form of sugar found in blood.

Glucagon (GLOO-kə-gon) (G. sweetness) Protein hormone secreted by the islets of Langerhans of the pancreas that causes the breakdown of glycogen, releasing glucose from the liver in response to low blood sugar.

Goiter (GOY-ter) (L. throat) Enlarged, hypertrophic thyroid gland producing a bulge at the base of the neck. Goiter may have several causes and may be associated with either a deficiency or an excess of thyroxine. It is often caused by inadequate iodine resulting in inability to make enough thyroxine to slow the pituitary production of thyroid stimulating hormone.

Gout (GOWT) (L. drop) Condition in which uric acid crystals form in joints (gouty arthritis) and other places in the body; a result of a disorder in metabolism of urate wastes.

Granulocyte (GRĂN-yoo-lo-sīt) (L. granule, cell) White blood cell with prominent granules in the cytoplasm. Granulocytes, or polymorphonucleated granulocytes as they are called, include basophils, eosinophils, and neutrophils.

Gustatory (GUS-tah-tor-e) (L. to taste) Relating to taste. The taste sense organs are found in the taste buds of the tongue and on the palate. They sense four tastes, sweet, sour, salt, and bitter.

Gynecomastia (gin′-e-ko-MĂS-te-ah) (G. woman + breast) Breast enlargement in males. This condition usually results from estrogens or certain drugs and sometimes indicates abnormal glandular activity.

Gyrus (JĪ-rus) (G. ring) Elevated segment of cortex of the cerebrum or cerebellum caused by folding of the gray matter. The gyri are the ridges of the brain and are surrounded by fissures or sulci.

Haversian canal (hä-VER-se-un) (Havers- English physician) Longitudinal canals in bone surrounded by concentric lamellae of bone and rings of osteocytes. Each tiny Haversian canal contains a small arteriole, venule, and nerve.

Hematocrit (he-MĂT-o-krit) (G. blood + separate) Percent volume of blood that is cellular. Blood is centrifuged with the packed cells accumulating on the bottom of the centrifuge tube and the plasma going to the top. It is a rough estimate of the red cells and normally ranges from 37% to 52%.

Hematuria (he′-mah-TUR-e-ah) (G. blood + urine) Blood in the urine. It may be visible as pink, red, or brown urine or occult, microscopic hematuria.

Hemoglobin (HE-mo-glo-bin) (G. blood) Oxygen-carrying pigment of blood consisting of a protein (globin) and four associated iron atoms (the heme groups).

Hemolytic anemia (he′-mo-LIT-ik) (G. blood + split) Condition characterized by inadequate numbers of red blood cells because of lysis (rupture or splitting) of the cells.

Hemopoietic (he′-mo-po-əT-ik) (G. blood + to make) Blood manufacturing. Used in reference to the development of the different cell lines of blood from stem cells.

Hemoptysis (he-MOP-ti-sis) (G. blood spitting) Blood in the sputum.

Hemorrhage (HƏM-o-rij) (G. blood + to break forth) Abnormally large leakage from one or more blood vessels.

Hemorrhagic anemia (hem′-o-RA-jik ah-NE-me-ah) Abnormally low number of red blood cells as a result of bleeding.

Hemothorax (he′-mo-THOR-aks) (G. blood + chest) Blood in the chest (pleural) cavity. Excessive fluid in the chest cavity may prevent the lung from expanding.

Heparin (HƏP-ur-in) (G. liver) Mucopolysaccharide produced in the liver that prevents the conversion of prothrombin to thrombin and thereby prevents formation of blood clots.

Hepatic (hə-PĂT-ik) (G. liver) Pertaining to the liver.

Herpes zoster (HƏR-pez) (L. creeping thing) (ZOS-ter) (G. girdle) Acute inflammatory disease arising along the path of sensory spinal and cranial nerves (dermatomes). It is caused by the chickenpox virus *Briareus varicellae* and produces painful skin lesions.

Hiatus hernia (hī-ĀT-us HUR-ne-ah) (L. a cleft + rupture) Prolapse of a portion of the stomach through the cleft in the diaphragm through which the esophagus passes.

Hippocampus (hi′-po-KĂM-pus) (G. sea + horse) Curved structure in the floor of a portion of the lateral ventricles of the cerebrum. It plays a part in learning, memory, and rage and is an important part of the limbic system.

Huntington's disease (American physician) Also known as Huntington's chorea or St Vitus' dance. An inherited condition involving atrophy of the cerebral cortex and caudate nucleus of the brain, resulting in uncontrolled movements.

Hyaline (HĪ-ah-lin) (G. glass) Transparent or translucent material. Hyaline cartilage is the translucent cartilage found on the articulating surfaces of bones and elsewhere in the body.

Hyaluronic acid (hī-ul-yur-ON-ik) (G. glassy) Mucopolysaccharide found in vitreous humor, synovial fluid, and other tissues.

Hyaluronidase (hī′-u-loo-RON-i-das) (G. glassy + enzyme) Enzyme that increases the permeability of tissues; hydrolyzes hyaluronic acid and is found in synovial fluid.

Hydrophilic (hī′-dro-FIL-ik) (G. water + loving) Attracted to water; absorbs or combines with water.

Hydroxyapatite (hī-drok′-se-ĂP-ah-tīt) (G. water + deceit) Hydrated crystals of calcium and phosphate ions that form the hard structural material of bones.

Hyper (HĪ-pər) (G. over) Prefix meaning above or more than normal or necessary.

Hyperglycemia (hī′-pər-glī′-SEM-e-ah) (G. over + sweet + blood) Above normal blood sugar levels.

Hyperinsulinemia (hī′-pər-in′-su-lin-E-me-ah) (G. over + L. island chemical + G. blood) Excessive insulin secretion by the pancreas resulting in hypoglycemia.

Hypermetropia (hī′-pər-meh-TRO-pe-ah) (G. excessive + measure + eye) Hyperopia or farsightedness. Inadequate refractive power of the eye, resulting in the point of focus being behind the retina.

Hypernatremia (hī′-pər-na-TREM-e-ah) (G. high + sodium + blood) Above-normal levels of sodium in the blood.

Hyperplasia (hī′-pər-PLAZ-e-ah) (G. high + formation) Abnormal increase in number of normal cells in normal arrangement in a tissue.

Hypertension (hī′-pər-TƏN-shun) (G. above + tight) Sustained abnormally high blood pressure. A diastolic pressure above 115 mmHg is usually recognized as moderate hypertension.

Hypo (HĪ-po) (G. under) Prefix meaning below or less than normal or necessary.

Hypoglycemia (hī′-po-glī-SEM-e-ah) (G. below + sweet + blood) Below-normal levels of blood glucose (blood sugar); may result from starvation, inadequate glucagon, or excessive insulin.

Hypokalemia (hī′-po-kal-EM-e-ah) (G. inadequate + L. potassium + G. blood) Inadequate potassium in the blood.

Hyponatremia (hī′-po-na-TREM-e-ah) (G. inadequate + L. sodium + G. blood) Inadequate sodium in the blood.

Hypophysis (hī-POF-i-sis) (G. under + to grow) Synonym for the pituitary gland. See pituitary.

Hypothalamus (hī′-po-THAL-mus) (G. under chamber) Portion of the floor of the third ventricle of the brain that links the pituitary and the brain as well as being a center for many functions such as the control of blood pressure, the thirst reflex, and control of reproduction.

Hypoxia (hī-POKS-e-ah) (G. under + oxygen) Inadequate oxygen levels in the lungs or bloodstream.

Idiopathic (id′-de-o-PATH-ik) (G. self-disease) Disease or condition of unknown origin or cause.

Ileocecal valve (il′-e-o-SE-kul) (L. lower small intestine, blind pouch) Constriction at the end of the ileum that prevents backward flow of the material in the cecum into the small bowel.

Ileum (IL-e-um) (L. lower small intestine) Terminal portion of the small intestine ending at the ileocecal valve. It functions mostly to absorb nutrients.

Impotence (IM-po-tens) (L. to lack + power) Inability to create or maintain a erection of the penis. Erection, or penile stiffness, is essential for penetration of the vagina and completion of coitus.

Incontinence (in-KON-tin-ens) (L. without control) Inability to prevent a natural discharge, as in urinary incontinence or fecal incontinence.

Incus (ING-kahs) (L. anvil) Middle of the three middle ear bones (ossicles); it receives vibrations from the malleus and conducts them to the stapes.

Infarction (in-FARK-shun) (L. to stuff in) Death of tissue caused by loss of blood supply. The dead tissue is often replaced with scar tissue, making old infarctions visible.

Infundibulum (in′-fun-DIB-yoo-lum) (L. funnel) Funnel-shaped entrance to a passage. The fimbriated infundibulum guides ova into the lumen of the oviduct.

Innervate (IN-ər-vat) (L. in + nerve) Supplying of nerves to a muscle or other organ. The nerves that stimulate a muscle are referred to as the muscle's innervation.

Insipidus (in-SIP-i-dis) (L. without taste) Diabetes insipidus refers to a condition which, like diabetes mellitus, involves excessive water consumption and urination, but does not involve high blood or urine sugar. It results from inadequate antidiuretic hormone or the loss of ability to respond to that hormone.

Insulin (IN-soo-lin) (L. island + chemical) Endocrine hormone secreted by the islets of Langerhans of the pancreas. Insulin increases the permeability of the body's cells to glucose, allowing the sugar to move from the blood into the tissues.

Interstitial (in′-tur-STISH-al) (L. between + to set) Area between tissues or groups of cells in tissues.

Integument (in-TəG-yu-mənt) (L. covering) Covering, usually the skin. The term integumentary system refers to the system of organs arising from ectoderm except for the nervous system. It includes all of skin including the mesodermal component and all skin derived organs.

Intercalated disk (in-TuR-kah-lat-əd) (L. inner + to call or isolate) Plate or disk-like end of a cardiac muscle cell. Each is composed of the interdigitating ends of two cardiac muscle cells.

Inulin (IN-yoo-lin) (L. a species of plant) Soluble product of fructose, not metabolized by the body. It can be injected and the rate of excretion used as a precise measure of kidney function.

Iris (Ī-ris) (G. rainbow) Circle of muscular, pigmented tissue suspended in the anterior chamber of the eye over the lens. The open center of the iris is the pupil. Eye color results from the genetically determined pigments of the iris and contraction of the muscles of the iris regulates the pupil size, limiting the amount of light that enters the eye.

Ischemia (is-KE-me-ah) (G. to suppress blood) Inadequate blood supply to a tissue or organ. Ischemia may produce pain, itching, or death of tissue.

Islets of Langerhans (LANG-ger-hans) (German pathologist) Clusters of small insulin or glucagon secreting cells scattered among the acini of the pancreas. They are endocrine tissues, lacking secretory ducts.

Isotonic (īs′-o-TAHN-ik) (G. equal, stretching) Of the same osmotic pressure. A solution is isotonic to a tissue if the osmotic pressure gradient between them is zero and water moves into the cells at the same rate as it moves out.

Jejunum (ja-JOON-um) (L. empty) Portion of the small intestine between the duodenum and the ileum where digestion is mostly completed and nutrients are absorbed.

Jaundice (JAWN-dis) (L. yellow) Yellow color seen in the whites of the eyes and skin as a result of deposition of a hemoglobin breakdown product, bilirubin. It usually indicates rapid erythrocyte destruction or liver insufficiency.

Juxtaglomerular apparatus (juk′-stah-glo-MER-yu-lar) (L. next to, small ball) Complex of specialized smooth muscle cells of the afferent and efferent arterioles in association with cells in a loop of the distal convoluted tubule, the macula densa. The juxtaglomerular cells of the kidney release renin in response to low salt concentration in the distal tubule as a result of low filtration rate.

Keratin (KƏR-ah-tin) (G. horn) Scleroprotein that contributes to waterproofing and toughening of skin; also present in fingernails, hair, and horn.

Keratohyalin (kər′-ah-to-HĪ-ah-lin) (G. horn + transparent) Precursor of keratin found as granules in the cytoplasm of the stratum granulosum of skin.

Ketosis (ke-TO-sis) Ketone, (G. excessive) Condition of having excessive ketones in the circulation, breath, or urine. The ketone often observed on the breath of diabetics is acetone.

Kyphosis (kī-FO-sis) (G. humpback) Increase in the natural thoracic curvature of the spine.

Kupffer Cells (KOOP-fer) (German anatomist) Monocytes specialized for existence in the sinusoids of the liver where they phagocytize foreign material.

Lacrimal (LAHK-rah-mahl) (L. tear) Pertaining to tears. The lacrimal gland produces tears.

Lactation (läk-TA-shun) (L. to suckle) Milk production by the mammary glands.

Lacteal (läk-TEL) (L. milk) Small lymphatic vessels that extend into the villi of the intestines and absorb fat. The droplets of fat give the lymphatic fluid a milky appearance.

Lactic acid (LÄK-tik) (L. milk) Acid produced by the souring of milk. It is toxic in tissues in large amounts and is produced in oxygen-deprives tissues from pyruvic acid. This reaction releases energy.

Lactiferous (läk-TIF-ər-us) (L. milk bearing) Producing or carrying milk as in the mammary glands or ducts.

Lactogenic (läk′-to-JEN-ik) (L. milk producing) Responsible for the production of milk. Lactogenic hormone (prolactin) produced by the anterior lobe of the pituitary stimulates milk production.

Larynx (LÄR-inks) (G. voicebox) Cartilage-covered structure that protrudes as the Adam's apple and contains the vocal cords. When one inhales, air enters through the glottis, passes the vocal cords and exits into the trachea.

Laryngitis (Lar′-in-JI-tis) (G. voicebox + inflammation) Inflammation of the larynx, producing soreness and hoarseness, sometimes with loss of voice.

Lethargy (LƏTH-ur-je) (G. forgetful) Condition of being sluggish or indifferent to events.

Leukemia (loo-KE-me-ah) (G. white + blood) Excess numbers of lymphocytes or granulocytes in the bone marrow resulting in replacement of the blood-forming tissue. Chronic lymphatic leukemia is a slowly developing form frequently found in the elderly. Myeloid leukemia evolves faster and involves excess production of granulocytes.

Leukocyte (LOO-ko-sīt) (G. white cell) White blood cells, including granulocytes, monocytes, and lymphocytes; primarily concerned with fighting disease and removing debris and toxins from the body.

Leukoplakia (loo′-ko-PLAK-e-ah) (G. white + plate) White lichen-like growth on the mucus membranes or vocal cords. It is sometimes malignant.

Levator muscle (LEV-ah-tor [in common usage]) (L. one who raises) Muscle that raises a part of the body without abducting it.

Levodopa (LEV-o-do′-pah) (G. left, + dopa) Synthetic compound, L.-deoxyphenylalanine, also called L-dopa. It is closely related to dopamine, a neurotransmitter, and is used in treatment of Parkinson's disease.

Libido (li-BE-do) (L. lust) Sexual drive; strong sexual attraction.

Limbic system (LIM-bik) (L. border) System of nerve tracts associated with the lateral and third ventricles. It includes the limbic lobe in the cerebrum, the hippocampus, the amygdaloid nucleus, thalamus, and hypothalamus, and is concerned with emotion.

Lipofuscin (l′-po-FUSH-in) (G. fat + fuscia) Colored granules that accumulate in the cytoplasm of old cells. They are considered a marker for aged cells and tissues.

Lordosis (lor-DO-sis) (G. swayback) Increase in the lumbar curvature of the spine.

Lumbar (L. loin) Part of the body from the diaphragm and rib cage (thorax) to the sacrum and ilium of the pelvis.

Lupus erythematosus (LOO-pus ər′-i-them′-ah-TO-sis) (L. wolf + G. flush of the skin) Destructive inflammatory skin condition. It has several forms but most are localized to skin. The systemic form is a generalized connective tissue disease.

Luteinizing hormone (LOOT-en-i-zing) (L. yellow + G. to cause) Protein hormone produced by the anterior pituitary. It causes the ovarian "scar", left after ovulation of an ovum, to develop into a corpus luteum and to produce progesterone and estrogens.

Lymph node (LIMF) (L. water) Gland-like structures located along lymph channels; are centers for lymphocyte activity that become enlarged when combating an infection or when they have trapped metastatic cells from a nearby cancer.

Lymphocyte (LIMF-o-sīt) (G. water + cell) White blood cells that are centers for the immune response. "B" lymphocytes become plasma cells and produce antibodies against foreign substances; "T" lymphocytes are subdivided into several forms whose stem cells originate in the thymus; NK (natural killer) lymphocytes eliminate abnormal body cells.

Lymphoma (LIMF-o-mah) (L. water + tumor) Any tumor composed of solid masses of lymphocytes or monocytes. They usually arise in the lymph nodes or the spleen. Hodgkin's disease is the most commonly cured form of lymphoma.

Macrophage (MÄK-ro-faj) (G. large + to eat) White blood cell, a monocyte, that moves through the tissues of the body consuming foreign materials and parts of dead cells.

Macula lutea (MÄK-yoo-lah LOOT-e-ah) (L. spot + yellow) Oval yellow spot in the center of the retina. It usually lacks rods and contains a high concentration of cones (color receptive cells).

Malleus (MAHL-e-us) (L. hammer) Outermost of the three middle ear bones (ossicles); receives vibrations from the tympanic membrane and conducts them to the incus.

Malignant (muh-LIG-nänt) (L. malicious) Dangerous or life-threatening. In reference to cancer, capable of invasion or metastasis. Not benign.

Mammary carcinoma (MÄM-ər-e) (G. milk gland) (kar′-si-NO-mah) (G. cancer) Malignancy arising from the lactogenic tissue of the breasts; the common form of breast cancer.

Marrow (MÄR-o) (ME the soft center of bone) Red marrow fills the center of compact bones in children and is found in some spongy bone in adults. It is the center for manufacture of blood. Yellow marrow contains a large amount of fat and is found in the centers of those bones not containing red marrow.

Mastectomy (mäst-EK-to-me) (G, breast + to cut out) Surgical removal of the breast.

Mastication (MÄS-ti-ka-shun) (L. to chew) Chewing of food.

Medulla (mə-DU-lah) (L. middle) Central portion of an organ or structure, as contrasted with the cortex or outer layer.

Medulla oblongata (awb′-long-GAH-tah) (L. middle, elongated) Base of the brain stem; it transmits impulses between the spinal cord and the rest of the brain and contains the pyramids where most of the motor fibers of the brain pass to the opposite side of the spinal cord. The medulla is the center for regulation of heart rate, breathing, and regulation of blood vessel diameter.

Megakaryocyte (məg′-ah-KAR-e-o-sīt) (G. large + nucleus + cell) Very large cell with a large, irregular nucleus, residing in the bone marrow. Megakaryocytes fragment to produce platelets (thrombocytes) that aid in the clotting process.

Meissner's corpuscle (MĪZ-nurz KOR-pus-ul) (German physiologist) Sensory nerve endings in the papillary layer of the dermis that detect light to moderate pressure and are important in detecting texture and shape of objects.

Melanocyte (məl-AHN-o-sīt) (G. black cell) Cell producing the dark brown pigment melanin that is usually found near the basal cell layer of epidermis.

Melanin (MəL-ah-nin) (G. black) Dark pigment found in skin, the choroid coat of the eye, and the substantia nigra of the brain.

Melanoma (məl′-ah-NO-mah) (G. black swelling) Tumor made up of melanocytes. The term today is synonymous with malignant melanoma, a dangerous skin cancer.

Membranous glomerulopathy (məm-BRAN-us glo-Mər′-yoo-LOP-ah-the) (L. thin transparent covering + small ball + G. disease) Common kidney disease in the elderly in which there is no inflammation but a thickening of the glomerular basement membrane. It involves deposition of an antigen antibody complex.

Meniere's disease (mən-e-ARZ) (French physician) Balance-related problem with vertigo and sometimes nausea as common symptoms; believed to result from increased endolymph pressure in the inner ear and usually results in deafness.

Meninges (me-NIN-jez) (G. membranes) Three protective membranes covering the brain: the outer, tough dura mater, the vascular arachnoid layer, and the delicate pia mater that resides adjacent to the surface of the brain.

Menopause (MəN-o-pawz) (G. menstruation + to stop) Termination of menstruation in women, usually between the ages of 46 and 50.

Merkel's disks (MER-kulz) (German anatomist) Disk-shaped sensory structures located in the epidermis; sensitive to delicate touch and concentrated in the lips and ends of the fingers.

Mesencephalon (mez-ən-SəF-ah-lahn) (G. middle + brain) Portion of the brain between the pons and the cerebrum, commonly called the midbrain, that includes the substantia nigra and red nucleus. It is a center for eye reflexes.

Mesoderm (MES-o-durm) (G. middle skin) Midmost of the three embryonic skin layers, between ectoderm and entoderm; forms muscle, bone, cartilage, fibrous connective tissue, blood, and blood vessels.

Metastasis (mə-TÄS-tah-sis) (G. beyond or after being stationary) Transfer of tumor cells (or in older usage disease such as tuberculosis) from an organ to a distant one.

Microglia (mī-KRO-gle-ah) (G. small + glue) One of the supportive cells (neuroglia) of the central nervous system. Microglia are derived from neural ectoderm and phagocytize debris and foreign material in the brain and spinal cord.

Micturition (mik-tu-RI-shun) (L. to urinate + to bend backwards) Emptying of the urinary bladder.

Midbrain Mesencephalon, the center of the three basic portions of the brain. It carries motor and sensory fibers between the pons and cerebrum and has two pigmented nuclei, the substantia nigra and the red nucleus, both important in eliminating undesired muscle contraction.

Milia (MIL-e-ah) (L. millet seed) Whitish nodules that are small retention cysts of sweat or sebaceous gland secretions.

Mitochondria (mī′-to-KAHN-dre-ah) (G. thread + cartilage) Small rod-like organelles that reside in the cytoplasm of a cell; they function as centers for cellular respiration and reproduce more or less independently of the cell nucleus although in response to the energy demands of the cell.

Mitral valve (MĪ-tral) (L. turban - Shaped like a miter, the formal head dress of a bishop) Bicuspid valve of the heart located between the left atrium and left ventricle.

Monocyte (MON-o-sīt) (G. one cell) Large phagocytic white blood cells that move through tissues as well as the bloodstream. They can develop into specialized tissue macrophages and some lines form osteoclasts.

Mononucleocyte (mon′-o-NOOK-le-o-sīt) (G one + L.nut + G. cell) White blood cell with a single, round nucleus and lacking cytoplasmic granules; includes monocytes and lymphocytes.

Mucoprotein (Myoo'-ko-PRO-ten) (L. mucus + G. first rank) Stable compound containing polysaccharides combined with protein and present in connective tissue.

Mucosa (Myoo-KOS-ah) (L. mucus) Mucus-secreting tissue layer of the digestive system lying next to the lumen; usually separated from the spongy submucosa by a thin layer of smooth muscle fibers, the mucosa muscularis.

Multinucleate (MUL-te-nu'-kle-at) (G. many + nut, kernel) Having more than one nucleus surrounded by a single cell membrane. Skeletal muscle fibers are multinucleate "cells."

Multiple myeloma (mī'-e-LO-mah) (G. marrow + tumor) Malignant tumor arising from plasma cells ("B" lymphocytes); often becomes established in bone, producing pathologic fracture and its presence can be verified by abnormal immunoglobulins in the blood or urine.

Multiple sclerosis (sklə-RO-sis) (G. many + L. hardenings) Disease characterized by the progressive destruction of the myelin sheaths of nerves in the central nervous system. The sheaths deteriorate to sclerosis or plaques in multiple regions. Remissions occur but each new attack results in the loss of additional function.

Muscular dystrophy (DIS-tro-fe) (G. negative + growth) Any of three types of genetically determined, painless diseases characterized by weakness and atrophy of the muscles without involving the nervous system.

Mutation (myoo-TA-shun) (L. to change) Any change in a gene or the location of a gene on a chromosome. Mutations can be caused by active chemical agents or by high energy radiation. Most mutated genes are recessive and are deleterious to the organism.

Myasthenia gravis (mī'-as-THEN-e-ah GRA-vis) (G. muscle weakness + L. heavy) Autoimmune condition of progressive weakness of the muscles, especially of the face or throat as a result of loss of acetylcholine receptor sites. Difficulty in swallowing often results and the disease sometimes progresses to death through respiratory failure.

Myeloid (MĪ-ə-loid) (G. marrow) Taking place in the marrow of bone. Myeloid leukemia involves replacement of the blood forming tissue by granulocytes.

Myelofibrosis (mī'-ə-lo-fi-BRO-sis) (G. marrow + L. excessive fiber) Fibrous replacement of the bone marrow resulting in failure to produce the cellular elements of blood.

Myocardial (mī'-o-KAR-de-al) (G. muscle + heart) Pertaining to the heart muscle.

Myocardium (mī'-o-KAR-de-yum) (G. muscle + heart) Tunica media of the heart, composed of cardiac muscle fibers.

Myopia (mī-O-pe-ah) (G. to shut + eye) Nearsightedness. Excessive refractive power of the eye resulting in the point of focus being in front of the retina.

Myosclerosis (mī'-o-sklə-RO-sis) (G. muscle + hardening) Also senile myosclerosis, the muscle hardening that sometimes occurs in immobile elderly patients. The limbs become flexed and straightening is accomplished only with great pain. Contractures.

Myosin (MĪ-o-sin) (G. of muscle) Densest protein in muscle; it is a globin. Myosin forms the thick filaments making up the "A" bands of striated muscle.

Myxedema (miks-ə-DE-mah) (G. mucus + swelling) Condition resulting from thyroxine deficiency and involving swelling of the face and hands.

Necrosis (ne-KRO-sis) (G. dead + state) Death of a group of cells within a living tissue. Necrotic cells in stained slide preparations typically have small, dark nuclei.

Neoplasm (NE-o-plaz-um) (G. new growth) Any benign or malignant abnormal growth of new tissue; a tumor.

Nephritis (nəf-RĪ-tis) (G. kidney + inflammation) Inflammation of the kidney.

Nephron (NəF-rahn) (G. kidney) Functional unit of the kidney. Each kidney contains approximately 1,250,000 nephrons.

Neuroglia (noo-ROG-le-ah) (G. nerve + glue) Supportive cells of the central nervous system. They originate from neural ectoderm and consist of astrocytes, oligodendrocytes, or microglia.

Neurohypophysis (NYOOR-o-hī-pof'-i-sis) (G. nerve + grow under) Posterior lobe of the pituitary; an extension of the hypothalamus. See pituitary.

Neuron (NYOOR-ahn) (G. nerve) Nerve cell. Neurons may be unipolar, having a single branch from the nucleus, bipolar having a single axon and a single dendrite or multipolar, having several processes serving as either axons or dendrites.

Neutrophil (NYOO-tro-fil) (L. neither + G. to love) Polymorphonuclear granulocyte whose granules do not take either a basic or eosin stain. They are phagocytic and make up 60-70% of the white blood cells.

Nevus (NE-vus) (L. birthmark) Usually raised, circumscribed, brown or red and brown, congenital spot on the skin. It is often called a mole.

NIDDM Abbreviation for noninsulin dependent diabetes mellitus.

Nonconjugate (KON-ju-gat) (L. not yoked together) Not working in unison; uncoupled, as in crossed eyes.

Norepinephrine (nor′-əp-ə-NəF-rin) (G. without + above kidney) Neurotransmitter found in the sympathetic nerves and adrenal glands. It differs from epinephrine (adrenalin) by the absence of the N-methyl group.

Normochromatic (norm′-o-kro-MĀT-ik) (L. normal + G. color) Normal color to the blood. This implies normal hemoglobin levels.

Normocytic (norm-o-SĪT-ik) (L. normal + G. cell) Having cells of normal size in the blood.

Nucleus (NOO-kle-us) (L. a kernel) Center or focal portion of a structure or system. The nucleus gracilis and nucleus cuneatus are two centers for routing sensory impulses in the medulla oblongata.

Nystagmus (nis-TĀG-mus) (G. to nod) Tremor of the eyeball. It may be horizontal, vertical, rotating, or mixed.

Obesity More than slightly overweight as a result of excessive fat.

Obtundation (ob-tun-DA-shun) (L. to blunt) Reduction of pain, as by the use of narcotics, with a resulting decrease in level of consciousness.

Olfactory (ol-FĀK-to-re) (L. to smell) Relating to the sense of smell. This is the most sensitive of our senses and the total number of compounds that can be detected and distinguished has not yet been determined.

Oligodendrocyte (ol′-e-go-DəN-dro-sīt) (G. few, + tree + cell) Supportive cell (neuroglia) of the central nervous system. Oligodendrocytes produce the myelin sheath surrounding axons of neurons in the brain and spinal cord.

Oogenesis (o′-o-JəN-ə-sis) (G. egg + creation) Production of ova (eggs) by the meiotic division of primary oocytes. Oogenesis begins in the ovary although the final stage of the process occurs after the ovum is released from the ovary (ovulation).

Organ of Corti (KOR-te) (Italian anatomist) Structure in the cochlear canal of the inner ear that converts vibrations to nerve impulses.

Organomegaly (or-gan′-o-MəG-ah-le) (G. instrument + large) Unusually large size of an organ; often an indication of disease.

Ossicle (AWS-si-kul) (L. little bone) Small bone, usually one of the three small bones of the middle ear, the malleus, incus, and stapes.

Osteoarthritis (aws′-te-o-arth-RI-tis) (G. bone joint + inflammation) Most common form of degenerative joint disease. Unlike rheumatoid arthritis it does not begin with inflammation of the synovial membrane. The disease progresses with degeneration of the articular cartilages, reactive bone formation, and can progress to joint fusion. This form of arthritis tends to be more severe in the weight-bearing joints.

Osteoblast (AWS-te-o-blast) (G. bone + germ) Cells derived from fibroblasts that lay down bone. When surrounded by bone they have produced, they become less active cells called osteocytes.

Osteoclast (AWS-te-o-klast) (G. bone + to destroy) Large multinucleate cell that develops from a line of monocytes and resides in bone. It responds to parathyroid hormone releasing enzymes that dissolve bone freeing calcium that diffuses into the bloodstream.

Osteocyte (AWS-te-o-sīt) (G. bone + cell) Mature osteoblast that has surrounded itself with bone. It achieves its nutrition through cytoplasmic extensions projecting through tiny canaliculi leading to Haversian canals.

Osteoma (aws-te-O-mah) (G. bone + tumor) Any tumor composed of bone. The term usually implies that the tumor is benign.

Osteomalacia (aws'-te-o-muh-LA-she-ah) (G. bone + softening) Bone degeneration associated with vitamin D deficiency in adults, sometimes called adult rickets; characterized by loss of the calcareous structure of bone; may contribute to pathologic fracture.

Osteomyelitis (aws'-te-o-mī'-ə-LĪ-tis) (G. bone + marrow + inflammation) Infection of bone and bone marrow often associated with trauma or caused by blood-borne organisms (sepsis). The most common bacterial cause is *Staphylococcus.*

Osteopenia (aws'-te-o-PEN-e-ah) (G. bone + poverty) Bone loss for any unspecified reason. It occurs after any prolonged period when osteoclastic activity exceeds osteoblastic activity.

Osteoporosis (aws'-te-o-por-O-sis) (G. bone + passage + excessive) Abnormal loss of both the mineral and organic components of bone; usually most severe in women whose estrogen levels have undergone severe decline as occurs in menopause.

Otolith (O-to-lith) (G. ear + stone) Calcareous deposit lying above the hair cells (stereocilia) in the sacculus and utriculus of the inner ear.

Oxytocin (awks-e-TO-sin) (G. quick + birth) Protein hormone produced by the hypothalamus and released by the posterior pituitary; causes uterine contractions including those that propel the fetus from the uterus.

Paget's disease of bone (PAJ-ets) (English surgeon) Also known as osteitis deformans, Paget's disease of bone involves rapid bone turnover creating areas of excessively dense but weak bone and areas of less dense bone. It is usually without symptoms and is suspected as being caused by reemergence of the measles virus.

Pancreatitis (păn'-kre-ah-TĪT-us) (G. all + flesh + inflammation) Inflammation of the pancreas. The condition is usually accompanied by abdominal pain and vomiting. Acute pancreatitis can result from digestion of the pancreas from within as a result of blockage of the pancreatic duct.

Pannus (PAN-us) (L. cloth) Abnormal vascularization of a membrane; formation may occur on the cornea of the eye producing opacity or on synovial membranes where it produces erosion of articular cartilages in rheumatoid arthritis.

Papilla (pah-PIL-ah) (L. nipple) Small nipple-like projection or elevation.

Papillary layer (PÄ-pil-ar-e) (L. nipple) Outer portion of dermis. Its outer edge forms papillae that interdigitate with projections of epidermis.

Papillary plexus (PÄ-pil-ar-e PLEK-sus) (L. nipple + braid) Network of blood vessels between the dermis and epidermis.

Parathyroid (pär'-ah-THĪ-royd) (G. beside + shield + form) Next to the thyroid. There are usually four small parathyroid glands attached to the dorsal surface of the thyroid gland; they secrete parathyroid hormone, a hormone that stimulates osteoclastic activity.

Parietal (par-Ī-ə-tul) (L. wall) Outer wall or layer. Parietal coelom refers to the mesodermal layer that lies on the outer portion of the coelomic cavities.

Parkinson's disease (English physician) Paralysis agitans. A slowly progressive disease associated with degenerative changes in the substantia nigra of the brain. The disease produces a resting tremor in the early stages and progresses to paralysis.

Parturition (par'-tu-RISH-in) (L. to desire + to bring forth) Act of giving birth.

Pathogenic (păth'-o-JEN-ik) (G. disease producing) Capable of producing disease or the symptoms of disease.

Pacinian corpuscle (pa-SIN-e-an) (Italian anatomist) Sensory nerve endings located deep within the dermis, muscles, joints, tendons and other tissues. They have an onion-like appearance when sectioned and sense vibrations and other recurring stimuli.

Pectoral (PəK-to-rul) (L. breast) Pertaining to the upper thorax or shoulder area.

Pepsinogen (pəp-SIN-o-jən) (G. digestion + creation) Inactive form of the proteolytic enzyme pepsin; secreted by the chief cells of the gastric glands and activated by hydrochloric acid.

Pericardium (pər'-e-KAR-de-yum) (G. around + heart) Tunica externa of the heart; secretes pericardial fluid into the pericardial cavity and is also known as the visceral pericardium or the epicardium. The parietal pericardium refers to the outer, secretory lining of the pericardial cavity. Together they make up the cardiac coelom.

Periodontal (pər'-e-o-DON-tul) (G. around + tooth) Situated around the teeth.

Perilymph (PəR-e-limf) (G. around + water) Fluid of the outer two canals, the tympanic canal and the vestibular canal, of the cochlea of the inner ear.

Periosteum (pər-e-AWS-te-um) (G. next to + bone) Membrane that lies adjacent to the outside of bone.

Peristalsis (pər'-i-STAL-sis) (G. pouch + contraction) Progressive wave-like muscular contractions that move material through the intestines.

Pernicious anemia (pur-NISH-us) (L. destructive) Specific form of anemia, formerly fatal. It resulted from lack of intrinsic factor and subsequent vitamin B_{12} deficiency. The red blood cells that are produced tend to be large and malformed.

Pharynx (FÄR-inks) (G. throat) Passage between the internal nasal passages and the esophagus and glottis. It is subdivided into the nasal pharynx, oral pharynx and laryngopharynx.

Pia mater (PE-ah MAH-tur) (L. gentle + mother) Innermost of the three meninges covering the central nervous system. It follows the contours of the gyri and sulci of the brain.

Pinna (PIN-ah) (L. wing) Portion of the outer ear that projects from the head. It functions to collect sound waves.

Pituitary (pi-TOO-i-tär-e) (L. the gland at the base of the brain) Endocrine gland located at the base of the brain and linked to the hypothalamus. The anterior pituitary (adenohypophysis) releases hormones that control metabolism, body repair, and reproduction. The posterior pituitary (neurohypophysis) is an extension of the hypothalamus and releases hormones that regulate blood pressure and uterine contractions.

Platelet (PLAT-lət) (G. small plate) Thrombocyte, small cellular fragments that adhere to damaged tissue and to each other. They help to prevent leaking of blood vessels and release clotting factors such as thromboplastin. Platelets are formed by fragmentation of megakaryocytes in the bone marrow.

Pleural (PLUR-al) (G. rib or side) Membrane covering the lungs and the lung cavity. The term pleural is used in reference to the cavity surrounded by this membrane or the chest cavity. Pleurisy is inflammation of the pleura.

Plexus (PLəKS-us) (L. braided) Network of intersecting nerves or blood vessels.

Pneumoconiosis (noo'-mo-ko'-ne-O-sis) (G. lung + dust + excessive + condition) Fibrotic degeneration of the lungs as a result of chronic exposure to irritating materials such as coal dust (black lung disease), silica dust (silicosis), or asbestos (asbestosis).

Pneumothorax (noo'-mo-THOR-aks) (G. lung + chest) Accumulation of air in the chest cavity outside the lung. It may escape from the lung through an abscess or through a wound created by trauma.

Pneumonia (noo-MO-ne-ah) (G. inflammation of lungs) Inflammation of the lungs involving accumulation of fluid and cellular material in the alveolae and bronchi.

Poliomyelitis (po'-le-o-mī'-e-LĪ-tis) (G. gray + marrow + inflammation) Acute viral infection also known as infantile paralysis. Motor nerve cell bodies in the ventral root gray matter of the spinal cord are destroyed, causing paralysis.

Polyarteritis (pol'-e-ar'-tər-ĪT-is) (G. many + artery + inflammation) Condition involving many inflammatory destructive lesions of arteries.

Polymyalgia rheumatica (pol'-e-mī-AL-je-ah roo-MÄT-i-ka) (G. many + muscles + pain + flux) Condition involving pain of muscles and connective tissues, especially of the neck, shoulders, upper arms, pelvis, and upper legs. It often occurs in association with giant cell arteritis.

Polymyositis (pol'-e-mī'-o'-SĪ-tis) (G. many + muscle + inflammation) Inflammation of several muscles at one time. It may produce pain, insomnia, and edema. It may involve skin inflammation as well (dermatomyositis) and is sometimes associated with undetected cancer.

Polycythemia (pol'-e-sī-THEM-e-ah) (G. many + cell + condition) Excessive blood cells, defined as a hematocrit above 53 for men and 48 for women. It may result from excessive red cells, white cells, or both or from loss of plasma volume.

Pons (PAHNZ) (L. bridge) Expanded part of the brainstem between the medulla and the midbrain. It carries nerve fibers into the cerebellum as well as between the medulla and midbrain. It also has nuclei for facial feeling, facial movements, and eye movements.

Presbycusis (prəz'-be-KOO-sis) (G. old + hearing) Decrease in hearing acuteness associated with age; the high range of sound is usually lost before the low range and ability to distinguish consonants is decreased.

Presbyopia (prez'-be-O-pe-ah) (G. old + eye) Loss of the ability to accommodate (focus) the lens on objects at varying distances. It is usually caused by loss of elasticity of the lens and associated with age. The ability to focus on close objects decreases, as does the ability to focus on distant objects but to a lesser extent.

Progesterone (pro-JəS-tər-on) (G. before + L. carry + G. tallow) Steroid hormone produced primarily by the corpus luteum of the ovary. The dominant hormone of pregnancy and of the second half of the female monthly hormonal cycle.

Prolactin (pro-LÄK-tin) (G. before + L. milk) Protein hormone secreted by the anterior pituitary; causes growth of mammary glandular tissue and production of milk (lactation).

Proteinurea (pro'-te-in-YOOR-e-ah) (G. protein + urine) Protein in the urine. It may occur from renal damage, urinary tract infection, or for no apparent reason.

Proteolytic (pro'-te-o-LIT-ik) (G. first [living matter] + to split) Action of splitting proteins, usually by the addition of water to the molecule (hydrolysis).

Proximal (PRAHK-si-mal) (L. next to) Closest to any point of reference. Closest to the point of origin of a structure. Opposite to distal.

Pruritis (PROO-RĪ-tis) (L. itch + inflammation) Sensation of itching. It is associated with skin conditions and other organic diseases as well as some neuroses.

Ptosis (TO-sis) (G. falling) Drooping of the upper eyelid, usually caused by lack of nervous innervation. Also, prolapse of an organ or other body part.

Puberty (PYOO-bər-te) (L. Maturity; the hair of maturity) Attainment of sexual maturity, marked by appearance of hair in the axillary and inguinal regions of the body.

Pupil (PYOO-pul) (L. girl) Opening in the center of the iris of the eye that transmits light to the retina.

Pyelonephritis (pī'-ə-lo-nəh-FRĪ-tis) (G. pus + kidney + inflammation) Diffuse interstitial infection of the kidney. The renal tissues are destroyed by white blood cells, producing pus. The condition may be the result of bacterial infection or drug induced.

Ramus (RA-mus) (L. branch) First branches of the spinal cord after they leave the vertebrae. Also used to refer to other branches, such as those of arteries.

Rectocele (RəK-to-sel) (G. rectum + hernia) Prolapse (hernial protrusion) of the rectum into the vagina. Usually results from atrophy or other damage to the pelvic floor, often as a result of estrogen deprivation.

Renin (RE-nin) (L. kidney) Enzyme released into the bloodstream by the juxtaglomerular cells of the kidney. It causes conversion of the blood enzyme angiotensin I to angiotensin II, resulting in constriction of blood vessels, release of aldosterone from the adrenal cortex, resultant increased absorption of salt and water from the urine, and consequent increased blood pressure.

Resuscitation (re-sus'-i-TA-shun) (L. returning to life) Returning of breathing and heartbeat to one who appeared to be dead.

Respiration (rəs'-pi-RA-shun) (L. breathing) Act of breathing; a single cycle of inhaling and exhaling is called a respiration. Chemical respiration refers to the releasing of energy by the oxidation of sugar.

Reticular (rə-TIK-yu-lar) (L. net-like) Any material arranged in a net-like pattern. The collagen fibers of the reticular dermis are arranged in a net-like pattern giving this layer its toughness.

Retina (Rə-ti-nah) (L. the light sensitive layer of the eye) Photoreceptive layer of the eye. lying between the vitreous humor and the choroid layer and connected to the optic nerve.

Retinopathy (rə'-tin-OP-ah-the) (L. the light sensitive layer of the eye + G. disease) Disease of the retina. Diabetic retinopathy involves small vessel hemorrhages in the retina.

Rheumatoid arthritis (ROO-mah-toyd) (G. flux + joint + inflammation) Degenerative joint disease that begins with inflammation of the synovial membranes and the formation of vascular pannus. Pannus formation results in degeneration of the articular cartilages and may be followed with reactive bone formation and loss of joint function. Rheumatoid arthritis is usually initially more severe in the distal joints, with the weight-bearing joints affected less uniformly.

Rhodopsin (ro-DAHP-sin) (G. rose + vision) Purple pigment of the rods of the retina in the eye. When it absorbs light it splits into the protein opsin and retinal, a derivative of vitamin A. This initiates an impulse in a neuron of the optic nerve.

Rotator (ro-TA-tor) (L. one that turns) Muscle that turns a body part, such as the head, on its axis.

Saccule (SÄK-yul) (L. a pouch) Smaller of two pouches that lie above the cochlea in the inner ear. The saccule and utricle contain otoliths and are responsible for our sense of static equilibrium.

Sacrum (SA-krum) (L. sacred) Five fused vertebrae between the last lumbar vertebra and the coccyx or tail bone. It forms the back of the pelvis.

Sarcoma (sär-KO-mah) (G. flesh + tumor) Any tumor arising from connective tissue. Many sarcomas are highly malignant. Osteogenic sarcoma is a malignant tumor arising from bone.

Sciatica (sī-ÄT-i-kah) (G. the ischium) Inflammation of the sciatic nerve. The condition usually causes pain along the course of the nerve in the thigh and leg.

Scirrhous (SKIR-us) (G. hard) Hard substance, as in scirrhous carcinoma of the breast, a hard, fibrous, invasive cancer.

Sclera (SKLə-rah) (G. hard) Hard, opaque, protective layer of the eyeball. It is continuous with the cornea and resides over the choroid layer.

Scleroderma (sklər'-o-DəRM-ah) (G. hard + skin) Skin disease producing hard, thickened, pigmented patches. It results in an increase in connective tissue in the epidermis and dermis.

Scoliosis (sko'-le-O-sis) (G. curvation) Now used primarily for abnormal lateral curvature of the spine.

Sebaceous (se-BA-shus) (L. greasy) Secreting oily or greasy material. Sebaceous glands are the lubricating glands of the skin and hair.

Seborrheic dermatitis (səb'-o-RE-ik) (G. grease + running) (der'-mah-TI-tus) (G. skin + inflammation) Inflammatory disease of skin characterized by yellowish, greasy scaling, especially on the face and scalp, and associated with overactivity of sebaceous glands.

Seborrheic keratosis (səb'-o-RE-ik) (G. grease + running) (kər-ah-TO-sis) (G. horny + condition) Benign, raised, sometimes pigmented, horny, waxy, or greasy skin lesions. They become larger and more numerous with age.

Secretin (SE-kre-tin) (L. to secrete) Hormone secreted by the mucosa of the duodenum and jejunum when the tissue comes in contact with acid chyme from the stomach. Secretin is carried in the bloodstream and causes the pancreas to release large amounts of bicarbonate, neutralizing the acidic chyme.

Sedentary (SƏD-ən-tər′-e) (L. sitting) Habitually inactive. To sit excessively or to be in a sitting posture.

Semen (SE-mən) (L. seed) Final fluid ejaculated from the penis, consisting of sperm and fluid from the prostate, seminal vesicles, and Cowper's glands.

Seminal vesicle (SE-mi-nul) (L. seed + sac) Seminal vesicles empty into the ejaculatory ducts where the vasa deferentia enter the prostate. These structures were originally believed to store sperm but it is now known that they serve only to produce fructose rich fluid that constitutes approximately 60% of the volume of semen.

Senile purpura (SE-nīl) (L. old man) (PUR-pur-ah) (L. purple) Slightly raised small vessel hemorrhages. A purple bruise-like lesion often seen in aging skin, especially on the legs.

Senility (sə-NIL-i-te) (L. old man) Characteristic of old age. Especially the noticeable physical and mental deterioration that occurs with age.

Senescence (sə-NƏS-əns) (L. growing old) Process or state of age-related deterioration.

Septicemia (səp′-ti-SEM-e-ah) (G. putrid + blood) Infection that has entered the blood stream, usually having spread from an infected organ or tissue.

Shingles Outbreak of the chickenpox virus *(Briareus varicellae)* usually in old age and always following an earlier chickenpox infection. Characterized by painful lesions along nerve pathways. Also called herpes zoster.

Sinus (SĪ-nus) (L. cavity) Cavity or hollow space, such as the frontal sinuses located in the frontal bone of the forehead. A small sinus is called a sinusoid such as the hepatic sinusoids, small areas between the lobules of the liver where the rate of blood flow greatly slows.

Sjögren's syndrome (SYE-grenz) (Swedish physician) Disease involving degeneration of the salivary glands with resulting xerostomia, conjunctivitis, degeneration of the gingiva, tooth decay, as well as other symptoms. Believed to be autoimmune in origin and may be inherited.

Solar lentigo (SO-lər) (L. sun) (LƏN-te-go) (L. skin freckle) Concentrations of pigment-laden cells in the skin. Result from age or exposure to sunlight and are also called age spots or liver spots.

Spermatogenesis (spur-mat′-o-JƏN-ə-sis) (G. seed + creation) Production of sperm (male gametes) by the meiotic division of primary spermatocytes. This process takes place in the testes.

Sphincter (SFINGK-ter) (G. binder) Ring of muscle that closes a tubular orifice.

Spirometer (spī-ROM-ah-ter) (L. to breathe + measure) Instrument for measuring lung capacity and function.

Spondylosis (spon′-de-LO-sis) (G. vertebra + condition) Fusion of a vertebral joint, usually at an abnormal angle. The condition often results in the pinching of spinal nerves.

Squamous (SKWA-mus) (L. scaly or plate like) Cells having a spherical or ovoid appearance. The squamous cell layer of skin is equivalent to the stratum spinosum and is part of the stratum germinativum of epidermis.

Stapes (STA-pez) (L. stirrup) Innermost of the three middle ear bones (ossicles). It receives vibrations from the incus and conducts them to the oval window.

Stasis dermatitis (STA-sis) (G. stationary) (dur′-muh-TĪT-tus) (L. skin inflammation) Degenerative skin condition resulting from inadequate circulation. The condition is further developed by scratching and often ulcerates.

Steroid (STƏ-royd) (G. solid) Group of compounds with a basic structure similar to and often derived from cholesterol. In addition to cholesterol, they include estrogenic and androgenic sex hormones and cortisone.

Stratum corneum (STRA-tum) (L. a layer) (KOR-ne-um) (L. horn-like) Tough, horny, outer layer of epidermis, composed of dead, keratinized cells.

Stratum germinativum (jer'-mi-nah-TE-vum) (L. sprouting) Reproductive layer of epidermis, composed of the basal and squamous or spinous cell layers.

Stratum granulosum (gran'-yoo-LO-sum) (L. granular) Layer of cells in the epidermis between the squamous cell layer and the stratum lucidum (in nonhair-bearing skin) or the stratum corneum (in hair-bearing skin).

Stratum lucidum (loo-SE-dum) (L. clear) Layer of translucent, dead cells in the epidermis between the stratum granulosum and the stratum corneum of nonhair-bearing skin.

Subclavian vein (sub-KLAV-e-un) (L. below + little key) Large vein that returns blood from the axillary vein of the arm to the superior vena cava.

Subcutaneous (sub-kyu-TA-ne-us) (L. under + skin) Occurring beneath the skin.

Substantia nigra (sub-STAN-she-ah NI-grah) (L. substance, black) Pigmented body located in the mesencephalon (midbrain) that functions to filter out unwanted impulses from motor pathways. Degeneration of the substantia nigra produces Parkinson's disease.

Sudoriferous (soo'-do-RIF-ə-rus) (L. sweat bearing) Secreting or producing sweat. The sweat glands.

Sulcus (SUL-kus) (L. furrow) Shallow groove on the surface of an organ such as the interventricular sulcus of the heart or the sulci between the gyri of the brain.

Surfactant (surf-ÄK-tənt) Surface-active agent that reduces the surface tension or cohesive force of water. A natural detergent found in the lungs that reduces the tendency of the alveoli and small bronchioles to remain filled with fluid.

Symphysis (SIM-fi-sis) (G. growing + together) Articulation in which two bones are joined by fibrocartilage. The union of the two pubic bones, the symphysis pubis, is an example.

Syncope (SIN-cop) (G. fainting) "Light headedness," fainting, the series of sensations and loss of brain function that is often associated with temporary inadequate blood supply to the brain. There are many other causes of syncope including drug effects.

Systole (SIS-to-le) (G, contraction) Contraction of the ventricles of the heart. The systolic blood pressure is the approximate maximum pressure reached in the arteries following ventricular contraction.

Tachycardia (tak'-e-KAR-de-ah) (G. fast + heart) Heart rate greater than 100 beats per minute. Atrial tachycardia refers to increased rate of the atria without being coordinated with the ventricles. Ventricular tachycardia refers to rapid rate of ventricular beat.

Tamponade (TÄM-pon-ad) (Fr. plugged) To stop the flow of blood by using pressure. Cardiac tamponade refers to the suppression of the heart by accumulation of fluid in the pericardial cavity.

Temporomandibular (təm'-pə-ro-män-DIB-yoo-lar) Pertaining to the articulation of the upper and lower jaw, as in the temperomandibular joint.

Tetany (TəH-tah-ne) (G. extreme tension) Spasmodic constriction of the muscles, usually including extreme flexion of the wrists and ankles, as may be caused by the tetanus toxin or by extreme calcium deficiency.

Thalamus (THAL-mus) (G. chamber) Part of the diencephalon that makes up a portion of the walls of the third ventricle of the brain; it serves as a distribution center for sensory and voluntary motor tracts and is important in arousal and emotional expression.

Thoracic duct (thor-AS-ik) (G. chest) Largest lymph duct; returns lymphatic fluid from most of the lymph channels to the bloodstream; located at the junction of the left subclavian and the left internal jugular veins.

Thorax (THOR-aks) (G. chest) Chest area, extending from the neck to the last (12th) rib.

Thoracocentesis (tho-rak'-o-sən-TE-sis) (G. chest + puncture) Process in which fluid is removed from the chest cavity by a needle inserted between the ribs. Organisms or cells suspended in the fluid can then be identified.

Thrombophlebitis (throm′-bo-flə-BĪT-is) (G. clot + vein + inflammation) Inflammatory condition of veins resulting in the formation of thrombi (blood clots).

Thrombus (THRAHM-bus) (G. lump) Blood clot. A thrombus is normally formed by a complex process to plug leaks in blood vessels. Under some conditions undesired thrombi may form creating a health risk.

Thymus (THĪ-mus) (G. the thymus gland) Gland located at the top of the sternum. It serves as the center for the organization of T lymphocytes and secretes the hormone thymosin. The gland begins to degenerate during the teen years.

Thyroid (THĪ-royd) (G. shield + form) Highly vascular gland at the base of the neck; secretes thyroxin, a hormone that stimulates metabolism and calcitonin, a hormone that stimulates formation and activity of osteoblasts.

Thyrotoxicosis (thī′-ro-tahx-i-KO-sis) (G. shield + form + poison + excessive) Toxic condition resulting from excessive activity of the thyroid gland, including increased metabolic rate, bulging eyes, large goiter, and nervousness.

Thyroxine (thī-RAHKS-in) (G. shield + chemical agent) Hormone derived from tyrosine and secreted by the thyroid gland; contains iodine and stimulates metabolism.

Tinnitis (TIN-ĭ-tis) (L. to tinkle) False ringing, hissing, or other perceived sound not produced by an external source. Most people experience temporary tinnitis after hearing an extremely loud sound.

Trabecula (trə-BəK-yu-lah) (L. little beam) Small projections of tissue usually forming a network around spaces as in trabecular bone or the trabecula carnae of the heart.

Trauma (traw-mah) (G. injury) Physical wound or injury.

Trypsinogen (trip-SIN-o-jen) (G. to rub + digestion) Inactive form of the proteolytic enzyme trypsin; secreted by the pancreas and is activated by enterokinase.

Tuberculosis (too-berk′-yoo-LO-sis) (L. nodule + G. excessive or disease) Bacterial disease characterized by the formation of tubercles in body organs, especially the lungs. Symptoms result primarily from the extreme inflammatory response of the body to the organism once the body is sensitized to the tuberculin organism.

Tunica externa (TOON-i-kah ex-TəRn-ah) (L. outer coat) Outer tissue of the circulatory system, composed of connective tissue and is penetrated by the blood vessels that nourish the walls of the large vessels.

Tunica intima (TOON-i-kah IN-ti-mah) (L. coat + inner) Inner tissue of the circulatory system consisting of the endothelium, myointimal cells, and elastic and other fibers.

Tunica Media (TOON-i-kah ME-de-ah) (L. coat + middle) Middle layer of arteries and large veins. It is composed mostly of smooth muscle.

Tympanum (tim-PÄN-um) (G. drum) Commonly, the ear drum. Correctly, the middle ear, including the middle ear bones.

Ulcerate (UL-sər-at) (L. to form an ulcer) Deterioration or loss of substance from an epithelial surface resulting in exposure of underlying tissues and loss of function or protection.

Urea (yu-RE-ah) (G. urine) Slightly toxic product of ammonia and carbon dioxide formed in the liver and excreted by the kidneys. Excessive amounts of urea in the blood produce uremia.

Uremia (yoo-REM-e-ah) (G. urine + blood) Excessive urea in the blood, also called azotemia. It can be fatal if not corrected.

Urosepsis (yu′-ro-SəP-sis) (G. urine + putrid) Urinary tract infection (UTI) with bacteria also present in the bloodstream.

Urticaria (ur′-ti-KAR-e-ah) (L. nettle) Slightly raised patches on the skin; hives. These usually itch and are often a response to allergies.

Utricle (YOO-tri-kul) (L. small bag) Larger of two membranous pouches above the cochlea of the inner ear. It and the saccule contain otoliths and are responsible for static equilibrium.

Varicose (VÄR-i-kos) (L. swollen vein) Veins enlarged to such an extent that the valves no longer function, resulting in tortuous, irregular enlargement of the smaller tributary veins.

Vas deferens (VĀS DəF-ər-ənz) (G. vessel + to carry away) Tube that carries sperm from the epididymis of the testis to the seminal vesicle and prostate.

Vasa vasorum (VAZ-ah vaz-OR-um) (L. vessels of vessels) Blood vessels that supply the walls of the blood vessels. The coronary circulation is sometimes called the vasa vasorum of the heart.

Vascular (VĀS-kyu-lar) (L. vessel) Pertaining to the blood vessels, arteries, veins, or capillaries.

Vasopressin (vas'-o-PRES-in) (L. vessel + press) Hormone produced by the hypothalamus and released by the neurohypophysis (posterior pituitary). Also known as antidiuretic hormone (ADH). It increases the blood volume by increasing the permeability of the collecting tubules of the kidney to water thereby increasing reabsorption of water from the urine into the blood. Trade name is pitressin.

Vena cava (VE-nah KA-vah) (I. vein + cavity) Superior vena cava brings blood from the head and shoulders. The inferior vena cava brings blood from the remainder of the body into the right atrium.

Ventricle (VəN-tri-kl) (L. the underside portion of the heart) One of two muscular lower chambers of the heart. The right ventricle receives blood from the body and pumps it to the lungs and the left receives blood from the lungs and pumps it to the body. Also the cavities of the brain containing cerebro-spinal fluid.

Vesicle (VəS-i-kl) (L. a small sack) Small blister; thin-walled, fluid filled, blister-like lesion.

Vertigo (VUR-te-go) (L. dizziness) Sensation that one is not fixed in space, either that the environment is revolving around oneself or as if one were spinning. Often associated with nausea and may be caused by inner ear, gastrointestinal, or brain problems.

Vermilion (vur-MIL-yun) (L. bright red) Vermilion border of the lip includes the red tissue extending from the junction of the facial skin to the mucosa within the mouth.

Vestigial (vəs-TIJ-e-ul) (L. animal track) Something that is left behind; a remnant of an organ that had a function in a more primitive animal or an animal of the opposite sex, e.g., the male breast or the appendix.

Villus (VIL-us) (L. tuft of hair) Small, usually vascular, projecting segment of tissue, often with the function of increasing surface area for absorption or excretion.

Visceral (VIS-ər-əl) (L. of large internal organs) Pertaining to or associated with the internal organs.

Volkman's canal (FOLK-mahn) (German biologist) Series of canals in bone carrying blood vessels to the Haversian canals.

Xerostomia (zər'-o-STOM-e-ah) (G. dry + mouth) Dry mouth, resulting when secretions of the salivary glands decline. Causes include action of some drugs, disease, and advanced age.

Xerophthalmia (zər'-op-THAL-me-ah) (G. dry + eye) Condition resulting from inadequate tear secretion. Dry eyes are more subject to infection than are normally lubricated eyes.

Index

A

Abdominal aneurysm, 179
Accommodation, 295, 306
Accumulated waste theory of senescence, 24, 25
Acetabulum, 53
Acetaminophen, 328
Acetylcholine, 77, 268, 281, 285
Acetylcholine transferase, 280-281, 285
Achlorhydria, 119, 126, 323323
Acquired immunodeficiency syndrome (AIDS), 210
  case history of, 214-215
Acral lentiginous melanoma, 44
Acromegaly, 243
ACTH; *see* Adrenocorticotropic hormone
Actin, 75
Actinic keratosis, 43, 44
Active immunity, 203
Activity and bone loss in aging skeletal system, 60
Addison's disease, 249, 259
  case history of, 262
Adenocarcinoma
  of breast, 231-233
  of colon, 130-131
  of lung, 107
  of ovary, case history of, 237-238
  of prostate, 229
  of stomach, 127-128
    case history of, 135
Adenohypophyseal hormones, 243-247
Adenohypophysis, 243, 246
ADH; *see* Antidiuretic hormone
Adrenal cortex, 249
Adrenal glands, 248-250
Adrenal medulla, 249-250
Adrenaline, 174, 249
Adrenocorticotropic hormone (ACTH), 245, 246
Adult rickets, 65
Aerobic exercise, 331, 335
Aerobic respiration, 78
Afferent arterioles, 142, 143
Age spots, 24, 35, 43

Aging, *see also* Senescence
  chemical, studies of, 14
  and death, perception of, 5-7
Aging fingernails and toenails, 37
Aging hair and hair production, 35-37
Aging sebaceous glands, 37
Aging skin; *see* Skin, aging
Aging sweat glands, 37
AIDS; *see* Acquired immunodeficiency syndrome
Airways, upper and lower, 91-93, 94
Alcohol, aging and, 329-331
Alcohol cirrhosis, 131-132, 133
Alcoholics Anonymous, 331
Aldehydes, 24
Aldosterone, 145, 146, 158, 175, 245, 249, 256, 259
All or none principle, stimulation of muscle and, 77-78
Alveoli, respiratory, 94
Alzheimer's disease, 278, 280-281, 284-285, 287, 327
Ammonia, 141
Amylase, 121
Amyotrophic lateral sclerosis, 82, 278, 281-282
Anaerobic respiration, 78
Analgesia, 305
Ancestor worship, 5
Androgens, 217-221, 249, 259
Anemia, 196-197, 204, 206-207, 325
  aplastic, 196
  autoimmune hemolytic, case history of, 211-212
  hemolytic, 196
  hemorrhagic, 196
  iron deficiency, 206, 323
    case history of, 212
  megaloblastic, 325
  nutritional, 196
  pernicious, 126, 196, 206, 323
    case history of, 134-135
  refractory, case history of, 212-213
  sickle cell, anemia and, 18
Aneurysm, 179
  berry, 283
  case history of, 186